D1269147

BEGINNING GEOLOGY

BEGINNING

London:
MACMILLAN & COMPANY LTD
GEORGE ALLEN & UNWIN LTD

New York: ST MARTINS PRESS

GEOLOGY

H. H. Read & Janet Watson

GEORGE ALLEN AND UNWIN LIMITED
Museum Street London WC 1

MACMILLAN AND COMPANY LIMITED
Little Essex Street London WC 2
also Bombay Calcutta Madras Melbourne

THE MACMILLAN COMPANY OF CANADA LIMITED
70 Bond Street Toronto 2

ST MARTIN'S PRESS INC
175 Fifth Avenue New York NY 10010

Library of Congress Catalog Card Number 66–18762

H. H. READ, F.R.S. (Royal Medal, 1963), F.R.S.E., D.Sc., A.R.C.S., F.G.S., was for many years a member of H.M. Geological Survey, and subsequently Professor of Geology in the University of Liverpool and at Imperial College in the University of London. In 1955 he became Professor Emeritus of Geology in the University of London.

JANET WATSON, Ph.D., B.Sc., F.G.S., is Research Assistant in the Geology Department at Imperial College. She is an authority on the geological history of the Pre-Cambrian rocks and specialises in the study of metamorphic rocks.

PRINTED IN GREAT BRITAIN BY JARROLD & SONS LTD, NORWICH

PREFACE

Beginning Geology is intended to provide the foundations on which the reader may build a fuller knowledge of the science of geology in general or of any of its branches. Although it is not a popular treatment aimed at the random buyer, we have tried to bring out the highlights of the subject as well as to record the necessary facts. The book should serve as a basis for a beginners' course in geology, such as that leading to the G.C.E. Advanced Level and kindred examinations. It is itself a kind of introduction to our two-volume *Introduction to Geology* (Macmillan) which is intended for the undergraduate student.

Our treatment of historical geology is concerned largely with the British Isles. This approach seems justified on several grounds. The geological record in Britain covers an extraordinarily full range of geological events and there are few topics that cannot be illustrated by reference to it. British geology has been the subject of some classic researches which have influenced the development of the whole subject and, on a lower level, Britain is the region which we ourselves know best. In any event, geological principles are the same everywhere, though the emphasis may differ in different regions. One final word of advice to the beginner in geology: reading about the subject is not enough. The outdoor world is a geological laboratory in full operation and it is here that rocks should be studied. Other things being equal, the best geologist is he who has seen most rocks.

Our thanks are due to the following, who have kindly allowed us to use original photographs: Dr Graham Evans, Dr W. G. H. Maxwell, Professor J. G. Ramsay, Dr G. P. L. Walker, the Dean and Chapter of Bristol Cathedral, and Eric Buston and Associates, Ltd. We are also indebted to Mr A. Hill, Mr J. A. Gee and Miss Margaret Doe of the Geology Department, Imperial College, for much technical help with the illustrations. Many of the drawings of fossils included in Figures 11.4, 11.6, 11.8, 11.10, 11.11, 11.12, 11.13, 11.14, 11.16, 11.17 and 11.18 are modified from the magnificent illustrations of common British species in the handbooks *British Palaeozoic Fossils*, *British Mesozoic Fossils*, and *British Cainozoic Fossils* published by the British Museum (Natural History).

H. H. READ
JANET WATSON

Imperial College, London
1966

Acknowledgements

The authors and publishers wish to thank the following who have kindly supplied the photographs on the pages quoted:

Aerofilms and Aeropictorial: pp. 3, 5, 47, 58, 71, 72 (copyright, White's Aviation Ltd), 81, 84, 92, 95, 98, 103, 158, 222, 225

Mr E. P. Bottley: p. 130

The Dean and Chapter of Bristol Cathedral: p. 57

The Trustees of the British Museum, Natural History (Crown copyright): p. 217

Dr Graham Evans: p. 2

The Director of the Geological Survey and Museum and the Controller of H.M. Stationery Office (Crown copyright): pp. 4, 20, 25, 32 (right), 34, 35 (right), 44, 45, 49, 53, 60, 61, 65, 76, 97 (both), 104, 113 (upper), 118, 145, 157, 159, 162 (both), 177 (upper), 180, 190, 211, 221

Dr W. G. H. Maxwell: p. 128 (left)

Professor J. G. Ramsay: pp. 69, 102

The Times: p. 91

Dr G. P. L. Walker: pp. 48, 93, 94, 177 (lower)

CONTENTS

Contents

Contents

1

SOME FUNDAMENTALS

The purpose of this first chapter is to introduce some of the principles on which the science of geology is based. The three classes of sedimentary, igneous and metamorphic rocks are defined and it is shown that rocks of each class can be regarded as records of particular events in the history of the earth.

1. Geology as Earth-history

The science of geology (Greek *ge*, the earth, *logos*, discourse) is concerned with the *history of the earth*. It seems certain that the earth has been in existence for thousands of millions of years and one of the geologist's main tasks is to piece together the complex history of this long period. Changes in the earth's surface, such as the crumbling away of a cliff, the filling of a lake with mud or the outpouring of lava from a volcanic cone, can be observed even within the lifetime of a single person. But even the whole span of human history is very short compared with the length of earth-history. Moreover, the fundamental happenings which control geological changes take place in the interior of the earth where they cannot be directly observed. The history of all these events is however recorded in the *rocks* which make up the earth, and the geologist aims to use these as historical documents.

2. Rocks as Records

Definition. A *rock* is a solid portion of the earth's crust which has a recognisable appearance and composition. In this geological sense, rocks are not necessarily hard; peat and mud are just as much rocks as granite and slate and, in fact, all the naturally occurring, non-living, solid matter which is found in or at the surface of the earth is rock of one sort or another. A particular kind of rock, such as clay, can be distinguished from others and *geological maps* can be made to show the distribution of rocks of different kinds.

Primary and secondary characters. In thinking of rocks as historical records, we have to disentangle the *primary characters* which were acquired when the rock was originally formed from the *secondary characters* which were impressed on it by later happenings. We can illustrate the distinction by an artificial example. Clay, in its primary state, is wet, sticky and plastic. Its character can be changed entirely by baking, as in the making of pottery, and the secondary characters so produced, in this instance man-made, are the hardness and brittleness of the pottery. It is obvious that if we can interpret both primary and secondary characters of a rock we shall obtain the record of a sequence of events, and not simply of a single incident.

3. The Sedimentary Rocks and their Interpretation

One group of rocks can be fairly readily interpreted because it is possible to observe typical members being made at the present time. These are the *sedimentary rocks* which are formed by the accumulation of materials at the earth's surface as a result of geological processes going on there. Such common rocks as clays, sandstones and limestones belong to this group. We can observe the way in which tiny particles settle out from muddy water to produce a layer of clay, or the way in which sand is spread out on a beach (Fig. 1.1), or broken shells gathered into a shell-bank. With suitable reservations, we can conclude that similar rocks have been produced by similar

Fig. 1.1
A sedimentary rock in the making: sand accumulating in the inter-tidal zone of the Wash, East Anglia

processes in the past. The present can thus be used as the key to the past. This *principle of uniformitarianism* is one of the fundamental precepts of geology.

Bedding. The material of which sedimentary rocks may be made includes all the kinds of debris, organic as well as inorganic, which are found at the earth's surface. Since this material is subject to the force of gravity, sedimentary rocks are formed in layers, each of which is spread out over a wide area but is of no great thickness. Such layers are called *beds* or *strata* (Latin *stratum*, something spread out) and because bedding is an almost universal property of sedimentary rocks these rocks are often called the *stratified rocks* (Fig. 1.2).

The law of superposition. If we consider once more how sedimentary rocks are made, we can draw a conclusion of fundamental importance for the interpretation of earth-history. It is obvious that if one bed lies, in its original position, on top of another, this upper bed must be the younger: the lower bed has to be already there before the upper one can be laid down on it. This simple argument supplies another principle of geology, dignified by being called William Smith's First Law, or the *law of superposition*.

Fossils. Many sedimentary rocks being laid down today contain the remains of present-day organisms in the form of shells, bones, plant-fragments and so on. In a similar way, more ancient sedimentary rocks often contain *fossils* which are the remains of organisms entombed in them at the time of deposition. Fossils in the oldest fossiliferous sediments do not resemble any animals or plants living today and are of primitive types. Younger rocks contain more advanced forms and the most recently deposited beds contain the fossils most like living forms. These contrasts are due to the fact that animals and plants have undergone a process of change or *evolution* during the course of geological time. It follows that each successive set of fossiliferous sedimentary rocks is characterised by a particular set of organic remains and that strata of the same age can be identified by means of these characteristic fossils wherever they are found. This is William Smith's Second Law of *strata identified by fossils*. The possible kinds of fossils which any stratum may contain are determined by the stage of evolution which had been reached when the stratum was formed: it is obvious that one of the uses of rocks considered as historical documents is as records of organic evolution.

Fig. 1.2 *Bedding or stratification in Jurassic sedimentary rocks, Kimmeridge Bay, Dorset*

Facies. Examination of a bed of sedimentary rock will supply information about the material of which it is made, the way in which this material is put together, the kinds of fossils it contains, and so on. All this information enables us to form an opinion about how and where the bed was laid down—in other words, it tells us about the *environment of formation* of the rock. The sum-total of characters which reveal the environment of formation of the rock is known as its *facies* (Latin=face). We can illustrate the word by a few examples. A certain gravel-bed is said to be of *beach-facies*: it is seen to be made of more or less rounded pebbles packed close together and mixed with broken marine shells, and has the characters of a deposit formed on an old shore-line. Another bed is seen to be made of fragments of marine shells mixed with clean sand: it was evidently formed in clear shallow waters and is of *shallow marine facies*. Still another bed is made of finely banded clay and contains no shells, but instead the remains of organisms that float in the open sea: it is of *deep-sea facies* deposited far from land.

By putting together all the information about the sedimentary facies of rocks of one particular age, we could make a *facies map* of the environments of deposition of that time. A physiographic map of the world today is a map of present-day facies on which are distinguished the environments of mountains, deserts, river-valleys, lakes, shore-lines, deep seas and so on.

4. The Igneous Rocks

Another important group of rocks which can be interpreted by the uniformitarian method is that of the *volcanic rocks*, produced by the solidification on the earth's surface of molten lava erupted from volcanoes. This hot lava is not itself a surface product. It has come from deep in the earth and it often happens that some of it

fails to reach the surface before it solidifies: consolidated portions may be seen, for example, filling fissures in the walls of old volcanic craters. Where the molten material cannot escape at the surface it is forced under pressure into weak places in the rocks through which it passes and is then said to be *intrusive*; the rocks formed by solidification of this material are the *intrusive rocks*. Volcanoes were at one time thought to be burning mountains and the name *igneous rock* (Latin *ignis*, fire) was given to their products. We now apply the name to all *the rocks produced by consolidation of molten rock material* whether this was erupted at the surface or intruded at depth. The molten material from which they are formed is known as *magma*.

Minerals of igneous rocks. So far, we have not considered in any detail the kinds of materials which make up the common rocks. Examination of a consolidated lava such as some of those erupted from Mount Vesuvius in 1944 shows that it is made up of several different minerals which can be distinguished by colour, shape and many other properties.

Minerals. Minerals are the fundamental units that make up most igneous (and many sedimentary) rocks and it is therefore necessary to know something of their nature before going any further. A mineral is an inorganic substance with a definite *chemical composition* and a definite *atomic structure*. This second statement means that the minutest particles or *atoms* within the mineral are arranged according to a fixed geometrical pattern which is the same for all minerals of a single kind. The regular internal structure of a mineral is often reflected in its outward form: the mineral then makes *crystals* bounded by regularly arranged plane surfaces. Each kind of mineral, as already mentioned, has its own particular composition. Nearly all the common minerals contain silicon and oxygen, which are the most abundant elements in the solid outer part of the earth. In addition, some of the elements, aluminium, calcium, magnesium, iron, sodium and potassium, with other rarer elements, may be present.

Fabric. As molten lava cools and solidifies, certain minerals begin to *crystallise* in it, rather as ice crystals appear when water is cooled below freezing point. The newly crystallised minerals arrange themselves in various ways according to the conditions of cooling—in some lavas, for example, all elongated crystals are arranged parallel to one another because they have been pulled into alignment by flow of the sticky lava. The arrangement of the minerals produces a pattern in the rock which is called its *fabric* and, naturally, this fabric is related to the conditions under which it was formed. It is found that lavas piled up around modern volcanoes show characteristic kinds of fabrics and by applying the uniformitarian method we can identify older

Fig. 1.3 *A dyke of igneous rock intruded into a sedimentary series, Isle of Arran*

Fig. 1.4
Folded bedding, Iran. The cliff in the foreground is about two miles in width

rocks showing similar compositions and fabrics as ancient lavas even when the volcanoes from which they were erupted have long since been worn away.

Forms of igneous rocks. Lava which has been erupted from a volcano spreads out and consolidates in a fairly thin sheet. The primary form of a *lava flow* is therefore not unlike that of a bed of sedimentary rock, and it is easy to see that when successive eruptions take place, younger flows must be piled up on top of older ones. Lava flows therefore obey William Smith's Law of Superposition. *Intrusive igneous rocks* have different primary forms because they fill spaces opened up in pre-existing rocks. If the magma is squeezed into a vertical crack, it will consolidate to make a thin vertical sheet of igneous rock, called a *dyke*, along the line of the crack (Fig. 1.3): if it is intruded in other ways it will form masses of different shapes. Thus, the primary forms and fabrics of igneous rocks can be used to interpret the way in which they were produced.

5. Secondary Characters

Because the primary characters of sedimentary and igneous rocks which are being formed at the present day can be established by direct observation, it is possible to show that older rocks of similar kinds must have been modified by happenings which took place after the time of their formation.

Disturbances of strata. Some of the most important evidence of this sort is provided by changes in the attitude of strata. We have seen that each bed of a sedimentary series is laid down as a flat, almost horizontal layer. But in many series of ancient sedimentary rocks it is found that the beds slope at a steep angle or are vertical or even bent (Fig. 1.4). If we stick to the uniformitarian argument that these ancient beds must originally have been flat, we arrive at the conclusion that they have been tilted or bent at some later stage in their history.

Many other lines of evidence show that natural forces at work in the earth are capable of moving enormous masses of rocks. This process is demonstrated most dramatically by *earthquakes*, which are the shocks produced by sudden sharp earth-movements along fractures. Less dramatic but even more impressive evidence of earth-movement is provided by the occurrence of ancient sedimentary rocks of marine facies containing characteristic marine fossils far above sea-level in the interiors of the continents. These rocks have been forced upwards, sometimes for many thousands of feet, above the level at which they were originally deposited.

In the light of this sort of evidence it is easy to see the significance of these secondary modifications of the rocks. They provide *records of earth-movements*. Furthermore, the kind of modification gives a clue to the conditions under which movement took place. Under pressure, rocks either *bend* or *break*. They may bend when the pressure is applied slowly and regularly, and especially when they are plastic or are softened by heat. Bending results in tilting of strata and often in the production of *folds*. Rocks break under sudden stresses and when they are brittle: the result is the development of fractures or *faults* along which one mass of rock has moved relative to another.

The effects of folding and faulting are assessed by comparing the secondary arrangement of rocks with what we believe to have been their primary arrangement. We must use as guides structures whose original forms are known. The most important of such structures are the *beds* of stratified rocks and the *bedding-planes* which separate adjacent beds, but other structures, including even the fossils in a rock, may give information of the same kind.

Mobile belts. When the effects of earth-movements are studied, it is found that strata of a particular age may show very little sign of disturbance in one part of the world and yet may be violently folded in another. At any given time, some parts of the earth tend to remain *stable* while others are much more *mobile* and prone to disturbance. At the present day it is possible to recognise *mobile belts* in which the rocks have been strongly folded in geologically recent times and which are often still subject to frequent earthquakes: the mountain chains of the Alps and the Himalayas are parts of these recent mobile belts. From the principle of uniformitarianism, we might expect to find records of more ancient mobile belts preserved in old rocks.

6. The Metamorphic Rocks

The final group of rocks which we shall consider is one which originates beneath the surface. We have seen that rocks laid down in the sea may be uplifted to a great height. When movements of this kind take place, the uplifted rocks begin to be worn away by the scouring action of mountain torrents and glaciers and by all the other forces which cause *erosion* or removal of material from the land-surface. As erosion removes more and more debris, rocks from lower and lower levels become exposed to view. The combined effects of uplift and erosion therefore make it possible for us to examine rocks which were once buried at depths of many thousands of feet below the earth's surface. Among these rocks from deep levels are found many which, though recognisable as originally sedimentary or igneous, have been so transformed that they have acquired very conspicuous secondary characters. These transformed rocks constitute our last great group, the *metamorphic rocks* (Greek *meta*, signifying change, *morphe*, form).

Since the metamorphic rocks were formed below the surface they cannot be interpreted directly by uniformitarian methods—nobody has ever seen a metamorphic rock being made. Sometimes, however, it is possible to trace rocks in which primary igneous or sedimentary characters are still well preserved into more completely transformed rocks and to show that the process of change or *metamorphism* has involved the growth of *new minerals* and the development of *new fabrics* of special types. Moreover, the situations in which metamorphic rocks occur tell us something about their origin. Certain types of metamorphic fabric are only seen in rocks which lie close to large igneous intrusions, and it is concluded that the changes that produced them were due to the heating-up caused by the intrusion of the molten magma: metamorphism in these conditions is really a process of cooking. Other types of metamorphic fabrics are confined to rocks which make parts of ancient mobile belts and it is therefore thought that earth-movements played a part in producing them. In general, metamorphism takes place when rocks are subjected to high temperatures and pressures deep in the earth's crust. Under such conditions, the primary constituents and fabrics are no longer stable and are replaced by new minerals and fabrics in harmony with the new environment.

7. Reaction to Environmental Change

These metamorphic changes illustrate the working of another fundamental principle of geology. This principle of *reaction to changes in environment* states that when the physical or chemical environment of a rock is altered, then reactions will go on which tend to establish a new harmony or *equilibrium* between the rock and its new environment. The long geological history of the earth is very largely a record of the responses to changes of environment.

8. The Three Classes of Rocks

We can now summarise the three classes of rocks as follows:

1. *The sedimentary rocks* formed by the deposition of material at the earth's surface through the agency of geological processes working there.
2. *The igneous rocks* formed from molten rock-substance or magma consolidating either at the earth's surface or within the crust of the earth.
3. *The metamorphic rocks* produced deep in the earth by the transformation of previously existing rocks through the action of heat and pressure.

9. The Scope of Geology

The geologist, from the fundamental principles discussed in this chapter, has several operations to perform. First, he must *describe his material*, the rocks, the minerals of which they are made, the fabrics they exhibit, and the fossils that they contain. The study of the rocks is *petrology* (Greek *petra*, a rock), of minerals *mineralogy* and of fossils *palaeontology* (Greek *palaeo*, ancient, *onto*, past participle of 'to be'.)

Second, the geologist has to determine the size and shape of rock-bodies and the way they fit together. This is the province of *structural geology* and leads to a consideration of *tectonics*, the study of the forces controlling the structure. Finally, he must combine all these lines of inquiry and bring in the dimension of *time* to interpret rocks and structures in terms of their history. *Historical geology* aims to provide a record of all kinds of happenings in the earth from the time when geological processes began to operate. Side by side with these inquiries, the geologist must explore special problems important to human beings: it may for instance be necessary to establish the structure of a bed of coal which is being mined, or to search for fertilisers or for rocks containing valuable metals, or to assess the stability of the foundations for a skyscraper. In this field of *applied geology* the geological principles and methods are used in the solution of problems of immediate practical importance.

Geology is largely an observational science: most of what is known of earth-history has been learnt by looking at rocks as they appear at the earth's surface, by making maps of their distribution and by studying specimens in more detail in the laboratory. In addition, complex instruments are being used more and more to supplement the geologists' observations, for example in studying the effects of earthquakes, and experimental techniques are beginning to make possible the production of man-made minerals and rocks for comparison with the real thing. But experiments and instruments cannot by themselves tell us anything of the earth's past history, though they may help in its interpretation. The first step must be to become familiar in the field and the laboratory with the rocks which are the historical documents. Success in geology is, literally, in the student's own hands.

2

THE EARTH: ITS CONSTITUTION, ORIGIN AND AGE

The purpose of this chapter is to give an account of the physical and chemical properties of the earth as a whole before we get down to dealing with the accessible parts of it. Its size, shape and density are considered and its internal structure deduced from the study of earthquakes. The chemical composition of the earth is discussed and its age incidentally mentioned.

1. Size and Shape of the Earth

Measuring the earth. From classical times it has been known that the earth was spheroidal in shape, but the accurate determination of this shape is a modern achievement. The radius of the earth is calculated separately in all latitudes, and from the results the shapes of all radial sections of the earth can be found. The results show that the earth is roughly spheroidal and slightly flattened at the poles. The actual measurements now accepted are:

Equatorial axis 6,378,388 metres (3,963·5 miles).
Polar axis 6,356,912 metres (3,950·2 miles).

The difference in length of the two axes is thus just over 21 kilometres (13·3 miles).

Surface relief. The measurements given above are calculated on the assumption that the earth has a perfectly smooth surface. Since we live on that surface, we know that this assumption is only an approximation: in some places the surface reaches up to form high mountains, in others it falls below the depths of the sea. In human terms this surface relief seems very considerable, but in relation to the size of the earth it is in fact of little importance. The difference in level between the summit of Mount Everest and the floor of the deepest ocean is about 20 kilometres, or less than 1/300 of the diameter of the earth.

Continents and oceans. One aspect of the surface relief is, however, of importance to geologists because it is an expression of the internal structure of the earth. Any map of the world shows that most of the dry land is massed in seven great *continents*—Europe, Asia, Africa, North and South America, Australia and Antarctica—usually fringed by shallow seas. Separating the continents are enormous *oceans*: the Pacific, the Atlantic and the Indian oceans, within which only small islands project above the water. This distinction between the continental areas, with the shallow seas that fringe them, and the great oceans will be dealt with in more detail later.

2. Density of the Earth

Weighing the earth. The weight of a given mass varies with its distance from the centre of the earth—a conclusion derived from Newton's Law of Gravitation which states that the force of attraction between two bodies is proportional to the product of their masses divided by the square of their distance apart. The earth can be weighed like any other body. In one way of doing this a heavy mass is weighed on a very sensitive balance; a second heavy mass of known weight is then placed a known distance below it, and the weight of the first mass determined again. Since the radius of the earth is known, from these measurements the earth's weight can be calculated. When this weight is divided by the earth's volume, we obtain the *density* of the earth, its weight per unit volume. The accepted value is 5·516 grams per c.c.; the density of pure water is 1 gram per c.c.

Let us consider for a moment this value of 5·5. The density of most of the rocks that can be found at the earth's surface does not exceed 3: for example, the density of sandstone is about 1·9–2·4, of limestone 1·9–2·7, of the very common rock granite 2·6–2·7, of the equally common rock basalt 2·8–3·0. This fact must mean that there must be material of greater density at deeper levels in the earth. The earth may therefore be made of different materials at different levels. It is thought, in fact, that a number of concentric *shells* of fairly uniform properties enclose a central core. The evidence which has led to this idea comes largely from the study of earthquakes, which we now examine.

3. Earthquakes

Stresses developed in the earth may become great enough to break the rocks and move them along the resulting fractures or *faults*. The sudden snapping causes a jar to travel through the adjacent portion of the earth and an *earthquake* results. The study of earthquakes is *seismology* (Greek *seismos*, a shaking) and has its own value and interest.

Since the masses of rock involved in these movements are very large, the damage caused by an earthquake may be colossal—in the Japanese earthquake of 1923 many more than 100,000 people perished. The damage is greatest near the centre or *focus* of the earthquake and becomes less farther away. Lines of equal damage or *isoseismal lines* rather like contour lines can be recorded on a map, as illustrated for the Inverness earthquake of 1890 (Fig. 2.1). From this figure it will be seen that the earthquake is obviously connected with movement on a fault running north-east/south-west. This fault, called the Great Glen Fault, has been active off and on for hundreds of millions of years: repeated movements have weakened the rocks along it and, as a result, erosion has dug along the line of the fault the valley containing Loch Ness. In Fig. 2.1 the *epicentre* is the point vertically above the focus. In the famous 1906 earthquake of San Francisco, movement on the San Andreas fault was shown by the displacement of fences and other boundaries and amounted to some 21 feet, 6·4 metres.

When we plot the localities of modern earthquake centres we find that they fall within belts that coincide with known zones of geological disturbance such as mobile belts.

Earthquake waves and records. The shock of an earthquake is transmitted by three kinds of waves, one that travels around the surface of the earth and the other two, the *body waves*, that travel through the interior of the earth. These body waves are of most interest to us. One type is transmitted by vibrations oscillating to-and-fro in the direction of transmission; this is called the *primary*, *compressional*, *P or push* wave. The second body wave is due to vibrations at right angles to the direction of propagation and is the *secondary*, *shear*, *distortional*, *S or shake wave*. These many names tell a great deal about these two waves (Fig. 2.2). The S waves cannot be transmitted by a liquid—a liquid cannot be

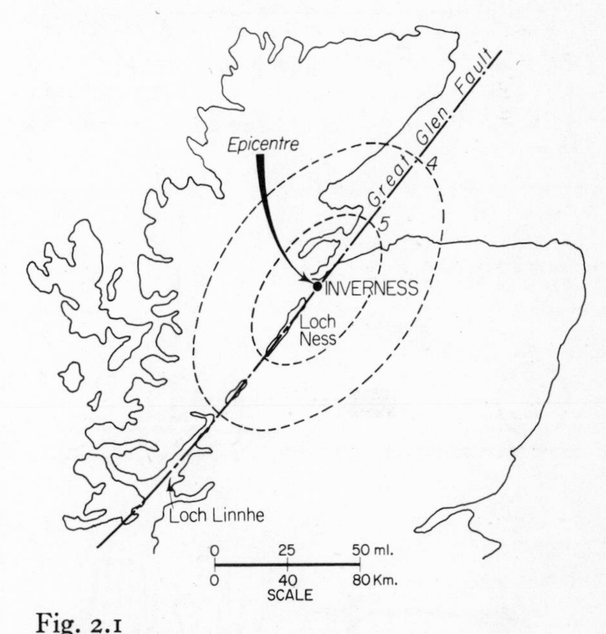

Fig. 2.1

Isoseismal lines (based on Davidson) for the Inverness earthquake of 1890: a standard scale of intensity is used which is indicated by figures from 1 (weakest) to 10 (strongest)

Chapter two

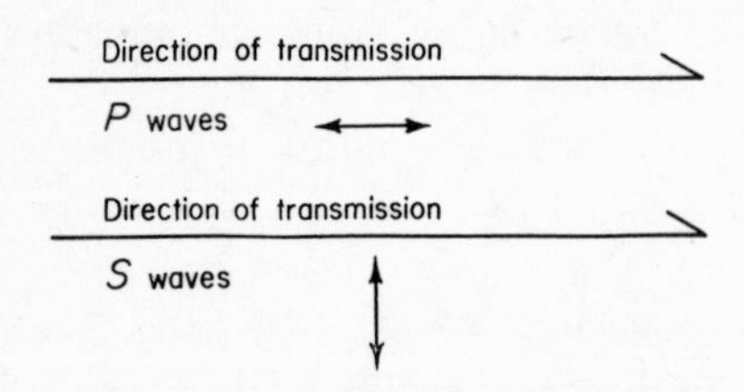

Fig. 2.2 *Earthquake vibrations*

sheared or distorted. If therefore we find that a part of the earth does not transmit S waves, then we have reason for suggesting that that part has the physical properties of a liquid.

Earthquake-waves are recorded by instruments known as *seismographs* which are really pendulums of various kinds (Fig. 2.3A). The vibration of the wave moves a beam carrying a heavy weight and this slight jerk is magnified and recorded photographically on a moving sheet of paper. The succession of movements caused by the arrival of the different waves is shown on this *seismogram* (Fig. 2.3B).

Seismographs set up at different stations receive the P and S waves from a single earthquake at different times, according to their distance from the focus. When the times of emission and arrival

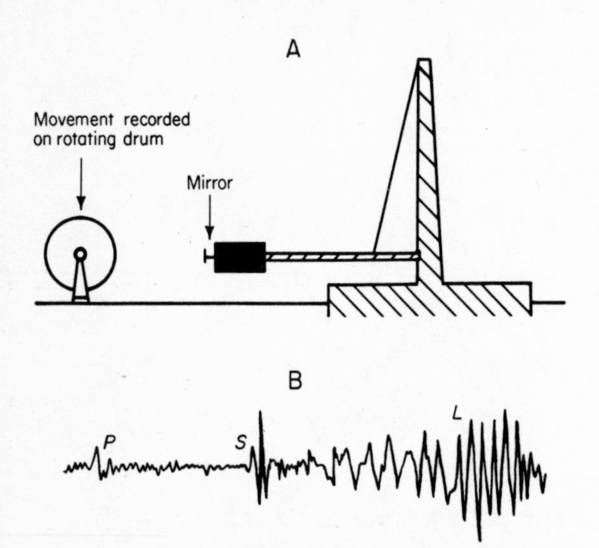

Fig. 2.3 *Recording earthquakes*

A. The construction of a seismograph

B. A seismogram, showing the P and S waves

are known, the speeds with which the waves have travelled can be calculated. These *velocities* depend on the physical properties of the rock material through which the waves have passed and, since it is possible to measure the velocities of artificially produced waves in rocks of known density, rigidity and compressibility, we can make some suggestions about the properties of the materials through which natural waves travel with similar velocities.

4. The Constitution of the Earth from Earthquake Records

The core

Earthquake records have provided information about the central parts of the earth because it has been found that the S waves are lost when they enter these parts. Study of innumerable records shows that when the paths which the waves must follow to reach the recording station take them deeper than about half the radius of the earth, the S waves do not arrive at the station. Between 105° and 142° of arc from the epicentre, there is a *shadow zone* so far as S waves are concerned. The P waves come through the central region, but are suddenly slowed down as they enter it. The shadow zone and its interpretation are shown in Fig. 2.4. It will be remembered that S waves cannot pass through a liquid. It is therefore thought that the earth has a *core*, with radius about 3,500 kilometres, which has the physical properties of a liquid. The P waves have a low velocity in this core.

As we have seen earlier, the density of the earth as a whole is greater than that of the common rocks making its outer parts. We may conclude that the material of the core is heavy and it is calculated that its density must be about 12. It is thought that this material is a nickel–iron alloy and from the chemical symbols (Ni and Fe) of these two elements the core is sometimes called the *nife*. Alloys of the kind suggested make some sorts of meteorites and are thus known to occur naturally.

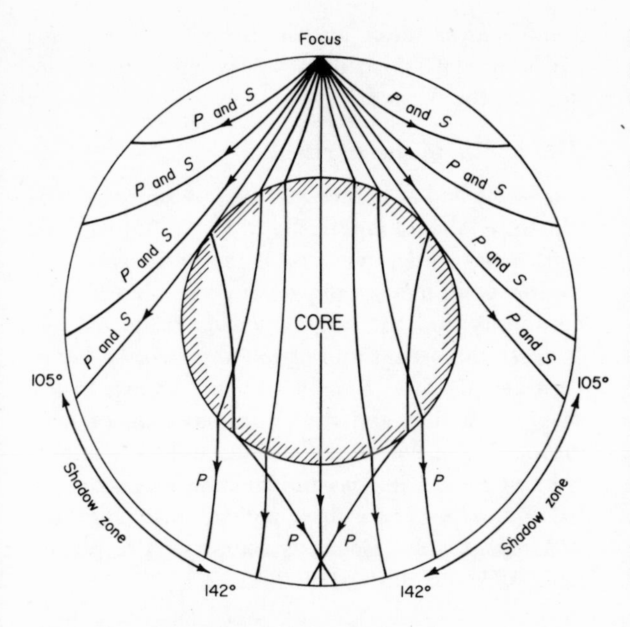

Fig. 2.4 *The shadow zone*

The crust and mantle

Information concerning the composition of the *outer parts of the earth* comes from records made at stations not far from the focus of an earthquake. To reach these stations the body waves must travel in the outer layers of the earth. From such *near earthquakes*, both P and S waves are received, and it is therefore concluded that these layers have the properties of solids. Both P and S waves are found to arrive as three pulses which have travelled with slightly different velocities. This multiplication of the number of pulses is due to the presence of earth-shells of different composition at whose margins the earthquake-waves are refracted and change direction as shown in Fig. 2.5.

Of special interest is the layer in which the pulses labelled P and S travel. This lies at some depth and earthquake-waves entering it from above suddenly speed up to travel with increased velocity. The sudden change indicates that higher and lower layers are rather sharply separated. The boundary surface or *discontinuity* is very important to geologists because it marks the lower limit of the kinds of rocks which make the outermost earth-shell on which we live. It is known as the *Mohorovičić or M discontinuity*—or in conversation 'the Moho'. Above it lies the *crust* made of all the rocks with which geologists are familiar. Below lies the *mantle* which has never yet been examined by man (at the time of writing, American scientists plan to drill down into the mantle: this is the 'Mohole' project). The composition of this mantle is unknown, but the velocities of earthquake-waves within it and other evidence suggest that it may be made of the rock *peridotite*, whose chief mineral is the iron–magnesium silicate olivine (the gem variety of this mineral is peridot). Meteorites of this composition are known to exist. The density of the mantle increases downward from about 3 near the M discontinuity to about 5·7 near the core.

The crust

The *crust* of the earth, as the word is used by geologists, is the outermost solid shell which lies above the M discontinuity. The two remaining pairs of earthquake-wave pulses P*, S* and P^g, S^g travel within it (Fig. 2.5) and their behaviour tells us something of the structure of this outer shell. Without going into details, we may say that the part of the crust which lies beneath the continents itself appears to consist of two layers, perhaps not sharply separated, which transmit earthquake-waves with different velocities. The parts of the crust which make the floors of the great oceans, on the other hand, usually consist of only one important layer identical in properties with the lower crustal layer beneath the

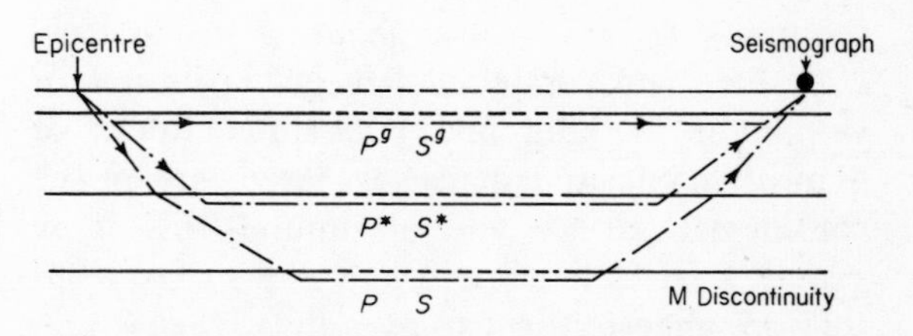

Fig. 2.5 *The paths of waves from near earthquakes (based on Jeffreys)*

continents. The thicknesses of these layers and of the crust as a whole can be measured by various methods—for example, by making artificially produced shock-waves 'bounce' back from the boundary surfaces and measuring the time it takes them to go there and back. By these means it has been found that the *continental crust* is much thicker than the *oceanic crust* and, moreover, that the continental crust is even thicker than usual beneath the great mountain ranges. The thickness of the continental crust usually lies between 25 and 40 kilometres, that of the oceanic crust averages 5 kilometres.

Sial and sima. We know more about the composition of the crust than about that of the mantle and core, because suggestions based on the velocities with which earthquake-waves are transmitted can be checked by observations on the parts of the crust accessible to man. The velocities of waves travelling in the oceanic crust and in the lower parts of the continental crust would be appropriate for a layer made of the rock *basalt*. Basalt is a dark, compact, rather heavy igneous rock (density about 3·0) made largely of silicates of magnesium, aluminium, calcium and iron. It is by far the commonest substance erupted from volcanoes in oceanic regions and is also very widespread in continental regions. It is therefore suggested that most of the oceanic crust and the lower part, some 15 to 20 kilometres thick, of the continental crust form a *basaltic layer* which is given the name *sima* from two of the principal components (*si*lica and *ma*gnesia) of basalts.

The velocity of waves in the upper part, 10 to 20 kilometres thick, of the continental crust is appropriate for the rock *granite* which is pale-coloured, light (density about 2·7), coarsely crystalline and composed largely of silicates of aluminium, sodium and potassium. Rocks of granitic composition make very large parts of the continental surface and are found still more widely a few kilometres below the surface: but they are almost absent from islands lying within the great oceans. It is thought therefore that the upper part of the continental crust constitutes a *granitic layer* often termed the *sial* (from *si*lica and *al*umina) which does not extend everywhere beneath the oceans.

The shells of the earth

The shelled structure revealed by seismology is illustrated diagrammatically in Fig. 2.6. From the centre outwards, we have the liquid *core*, the *mantle* bounded on the outside by the M discontinuity and the *crust* which is made of *sima* beneath the oceans and of *sima overlain by sial* in continental areas. Finally, and not shown in the diagram, a thin and discontinuous skin of *sedimentary rocks* lies on the surface: a discontinuous layer of water (the seas and lakes) which is known as the *hydrosphere* lies above this: and the *atmosphere*, the gaseous envelope of the planet, surrounds the entire structure.

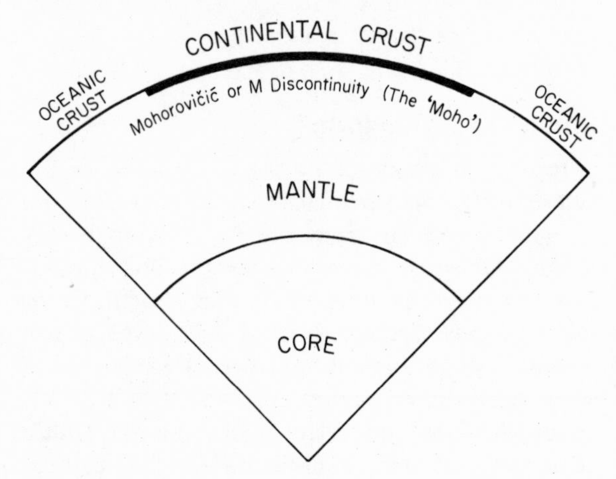

Fig. 2.6
The shells of the earth

5. Balance in the Crust: Isostasy

We must now look at the variations in the earth's crust from another point of view. We know that the surfaces of the continents stand at high levels relative to the floors of the oceans and that the continental crust is thicker than the oceanic crust: furthermore, where the surface rises to great heights in mountain ranges, the base of the crust dips down to unusually great depths. We arrive at the conclusion that there is a connection between the *thickness* of the crust and the *height* of its surface. What this connection means can be understood by considering the distribution of densities. In oceanic regions where the crust is thin and its surface low, it is made of dense basaltic material. On the continents, the much thicker crust is made in part of less dense granitic material. For this reason, the *mass* of a vertical column of material measured from some fixed level in the mantle to the surface of the continent is no greater than the mass of a similar column taken from the same fixed level to the surface of the ocean: the continental column would be longer, but it would also be less dense and we can think of its material as being less concentrated than that of the short dense oceanic column. By this distribution of densities, the different parts of the crust achieve a state of balance or of *isostatic equilibrium* (Fig. 2.7).

The reality of this balance has been demonstrated largely by the findings of seismology, but the first idea of isostasy arose from much simpler observations made some two hundred years ago. It was found by early surveyors that when a plumb-bob was suspended near a mountain, the attraction exerted by the mountain was very much less than was expected; indeed, in some experiments the plumb-bob was actually repulsed. It was realised that this might mean that the mountain was underlain by a 'root' of light material so that the total mass beneath the mountain was no greater than that beneath the adjacent low-lying land.

Measurements taken during the last century show that the great features of the crust such as continents, mountain plateaux and ocean basins are in isostatic equilibrium. But geological happenings from time to time upset this equilibrium—for example, if great volumes of erosion debris are washed down from a mountain range and deposited in the sea, the crustal column under the mountains is made lighter and that under the sea becomes heavier. To compensate for these changes, slow vertical movements begin to raise or lower the parts of the crust affected in such a way as to restore their isostatic balance and slow lateral movements of dense material beneath the sial adjust the deeper layers. We can see, therefore, that the workings of isostasy provide one source of earth-movements and may be responsible for such things as the uplifting of marine sediments to high levels.

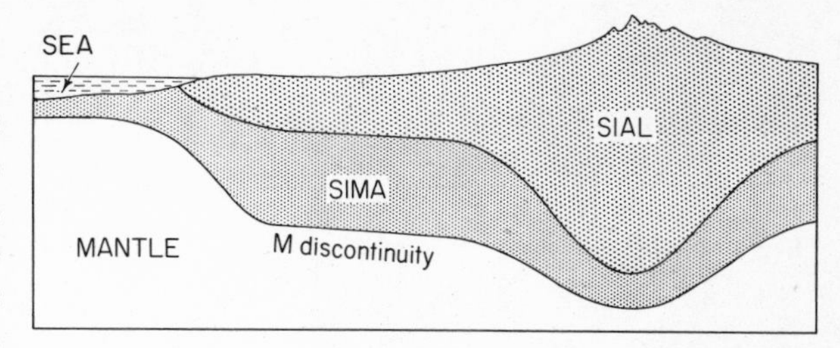

Fig. 2.7 *Isostatic equilibrium: the thick sialic rafts making the continental masses float isostatically in the sima and mantle*

6. Surface Features of the Earth

Having learned something about the internal structure of the earth, we can now go back to a consideration of its surface features. The meaning of the distinction between continental and oceanic areas mentioned on page 8 is now clearer. The *continents* are great raft-like masses of sial floating, as it were, on the underlying sima which is pressed down deeper into the mantle beneath them (Fig. 2.7). The true edges of the continents are not the shores of the land areas but the edges of the sialic rafts which extend out for up to a few hundred kilometres under the fringing seas. These submarine extensions of the continental crust form the *continental shelf*, which is covered by seas less than 200 metres deep and slopes gently oceanward with a general gradient of about 1 in 1,000. The outer margin of the continental shelf—the real limit of the sialic rafts—is marked at the *shelf break* by a sudden steepening of the gradient to more than 1 in 40. This pronounced *continental slope* makes a great step in the surface of the solid earth at which the sea-floor falls in a few hundred kilometres to depths of 3,000 metres

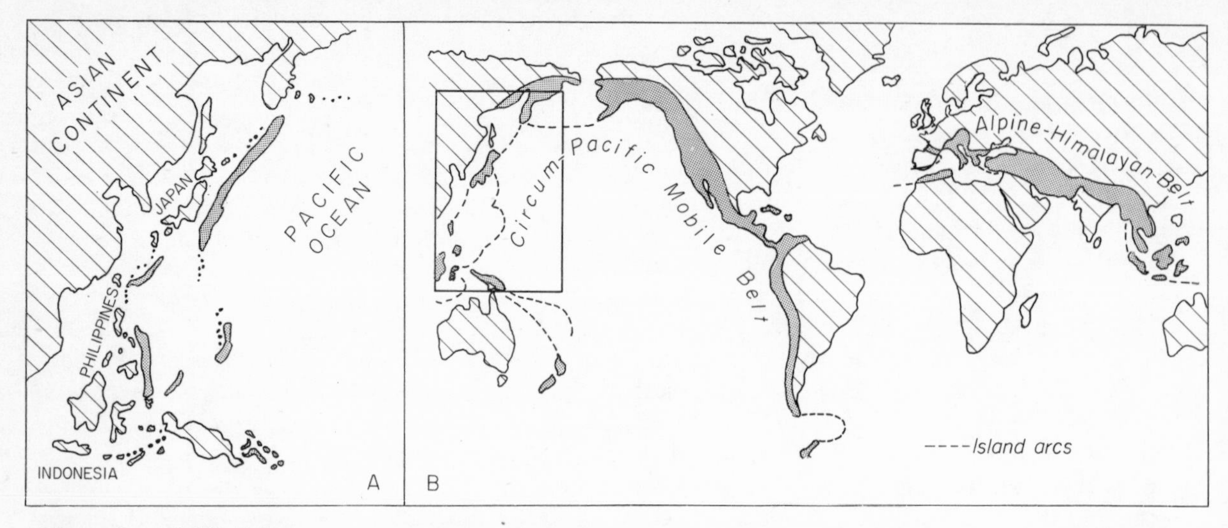

Fig. 2.8 *Mobile belts*

A. Island arcs and oceanic trenches (stippled) off eastern Asia

B. The circum-Pacific and Alpine-Himalayan mobile belts. The frame shows the area of Fig. 2.8A

or more. Beyond the continental slope the gradient flattens off again and the surface continues gently downward into the floor of the true *ocean basins*. These basins, floored almost directly by sima, are usually covered by more than 4,000 metres of water.

A relief map of the world shows that the greater part of the surface of continental regions lies not far from sea-level—just above on land and just below on the continental shelf. Measurement of the percentage of the surface lying at different levels shows that the most common level is about 100 metres above sea-level. The most common depth of the floors of the ocean basins is about 4,700 metres. Surfaces at very high or very low levels are by comparison extremely uncommon. The highest mountain, Everest, is 8,840 metres, 29,140 feet (nearly 6 miles) above O.D. and the deepest oceanic hollow, the Mariana Trench off New Guinea, is 10,790 metres, 35,400 feet (approaching 7 miles), below O.D.

Surface features of mobile belts

When we trace out the highest parts of continents and the deepest parts of oceans on a map, we find that they both have linear forms—that is, mountains are arranged in chains or *belts* much longer than they are broad and ocean deeps form *trenches* of similar shape. The mountain belts, when examined in detail, are found to be made up of folded and disturbed rocks which have clearly undergone large-scale earth-movements: in fact, they mark the site of *mobile belts* (p. 6). The oceanic trenches also mark mobile belts: the M discontinuity beneath them is depressed below the usual oceanic level, isostatic equilibrium is disturbed and seismic activity is high. *Island arcs* in which active volcanoes are common are often associated with the trenches.

We can trace the extent of present-day mobile belts by following out the lines of mountains, island arcs and oceanic deeps where volcanic activity and seismic disturbances are recorded (Fig. 2.8). One such belt encircles the Pacific Ocean, following the great mountains of the Rockies and Andes and the island arcs and oceanic trenches fringing the coast of Asia. A second runs eastward through Europe and Asia making the Alps, Caucasus and Himalayas and joining the circum-Pacific belt in eastern Asia. Outside these two mobile belts, the continental areas are much more stable. Their relief is generally less pronounced and they are usually free from severe earthquakes and volcanic activity. The structure of the older rocks in these *stable areas*, however, provides evidence that they have been mobile in past times, and we conclude that the positions of mobile belts have changed during geological history, old belts becoming stabilised and hitherto stable areas becoming mobile. The pattern of mobile belts as we see it today is only a temporary one.

7. Origin and History of the Continental Masses

Early history of the earth

We have seen in Section 6 that the sialic material which makes the continents is not spread in an even layer over the earth's surface, but is gathered in relatively small thick rafts covering only about a third of the surface. The reason for this remains one of the unsolved problems of geology, and to understand the possibilities we must consider very briefly the early history of the earth.

Whatever its ultimate origin, it is likely that at a very early stage the earth was a spinning globe of hot molten material, of the same composition throughout. As it began to cool the heavier, less easily oxidised elements such as iron and nickel became concentrated near the centre to produce the core or nife, while the remaining material formed a shell of molten silicates. With further cooling, minerals began to crystallise in this shell and the first to form were heavy silicates such as olivine, rich in magnesium. These crystals sank through the liquid shell and so built up the mantle and sima. The material which was left on the outside of the globe was a residue rich in lighter silicates of aluminium which finally solidified as a thin sialic skin.

Most geologists consider that this sialic skin was originally continuous, and several suggestions have been put forward as to how it became concentrated in the discontinuous continents. It may have been thickened, and therefore reduced in area, by being crumpled as a result of folding in mobile belts, or it may have been massed together like scum on boiling jam by the drag of slow convection currents in the mantle. Another suggestion is that a part of the sialic skin was torn away by the escape of the moon from the earth and still another that the sial was discontinuous from the first.

A problem of more immediate interest to geologists concerns the *permanence* of the continents and oceans: have the main ocean basins and continental rafts remained in the same places throughout geological history? Until comparatively recently, many geologists believed that great land-masses could sink beneath the ocean floor and disappear. We now know (Section 6) that the floors of the true ocean basins contain almost no sialic continental material and it therefore seems impossible that they can harbour sunken continents. Again, the present continents do not commonly exhibit ancient sedimentary rocks whose facies is that of the abyssal ocean basins, and we may therefore conclude that the floors of these basins have not at any time been uplifted to make new continents. The evidence thus suggests that the *material* making the continents has always remained the same.

Continental drift

It is by no means certain, however, that this continental material has remained fixed in *position* throughout geological time. We have likened the sialic masses to *rafts* floating on the substratum of the sima and mantle and there is strong evidence that these rafts have been able to glide slowly over the substratum, thus changing both their relative situations and their positions with relation to the north and south poles. According to one widely accepted suggestion, at a time about 300 million years ago Africa, South America, Australia, Antarctica and India were massed together in a huge sheet known as *Gondwanaland* in the southern hemisphere, while North America, Europe and Asia formed another great cluster. Gradually, these sheets began to split and the fragments so produced drifted apart to their present positions (Fig. 2.9).

In support of this idea of *continental drift* many geological and geophysical arguments can be put forward. There is, for example, a great similarity in the geological structure and history of the lands

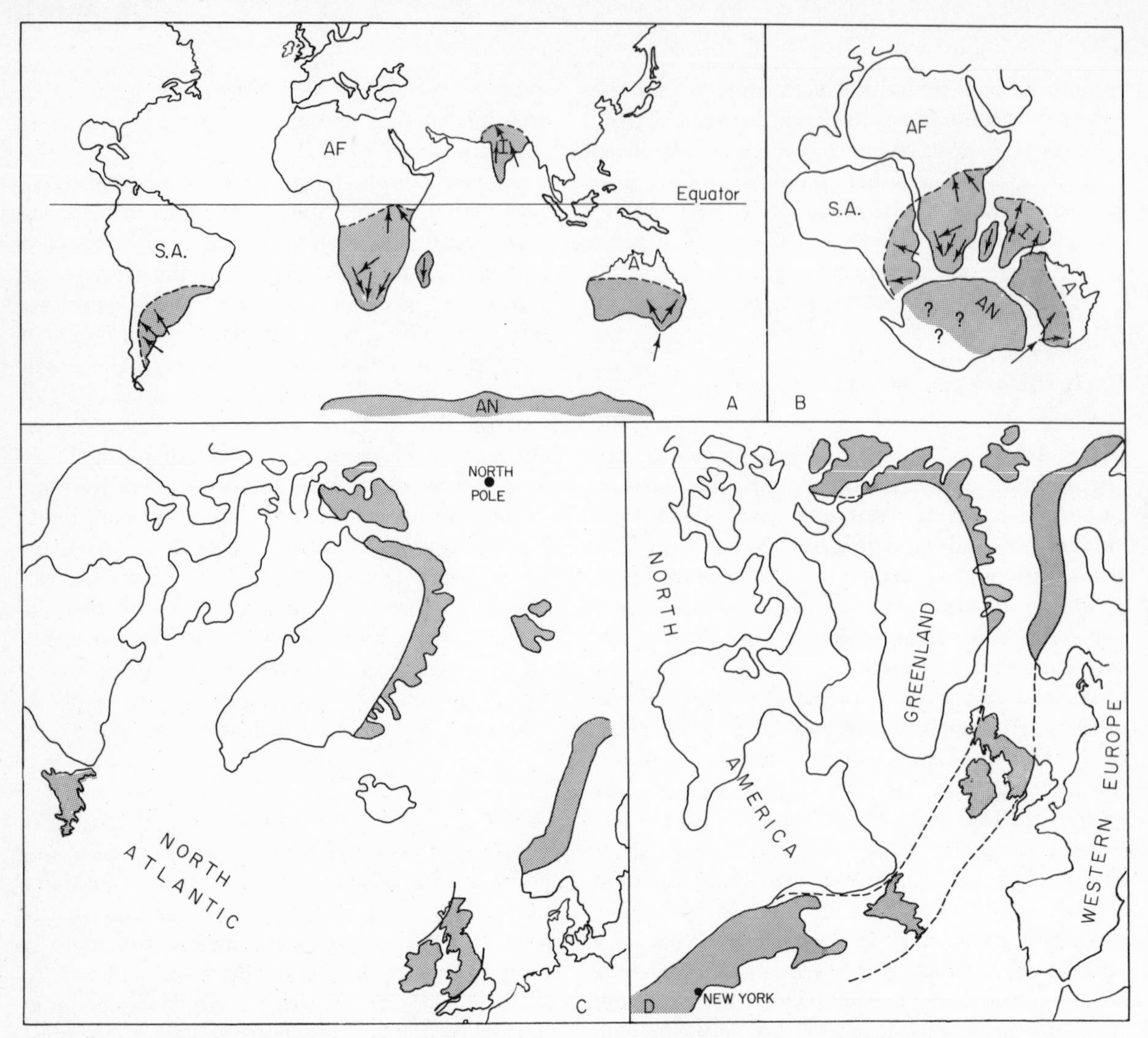

Fig. 2.9 *Continental drift:* A *and* B, *Gondwanaland;* C *and* D, *the North Atlantic region*

A. The areas containing evidence of Permo-Carboniferous glaciation (arrows show direction of ice-flow)

B. The continents reassembled to show their possible relationships at the time of the glaciation (after J. T. Wilson)

C. The distribution of ancient mobile belts (the Lower Palaeozoic Caledonian belts) in the lands bordering the North Atlantic: note that the belts seem to be cut off at the margins of the ocean

D. The North Atlantic lands reassembled to show their possible arrangement before drifting: note that the mobile belts fit together to produce a simpler pattern

west and east of the Atlantic Ocean and this similarity would be easier to explain if the Americas were once much nearer to Europe and Africa. One can say, with a little exaggeration, that the patterns of old mobile belts on either side of the North Atlantic could be fitted together like the torn halves of a bank-note. Again, sediments of glacial origin (that is, formed by the action of ice) are known to have been deposited about 300 million years ago in parts of South America,

South Africa, Australia, Antarctica and India. As the continents are now arranged, these sediments are scattered over an enormous part of the earth's surface and many of them lie in tropical latitudes. Their origin would be much easier to understand if, at the time of deposition, the continents had been clustered together near the south pole (Fig. 2.9). Thirdly, coal-seams, which are sedimentary rocks of tropical swamp facies, are now found in polar regions, as in Spitzbergen, where it is very unlikely that tropical forests ever flourished.

Evidence of other kinds comes from recent work on *palaeomagnetism*. The earth can be regarded as a permanent magnet whose magnetic poles are near the geographical poles. When certain kinds of rocks are formed, magnetic particles in them are lined up in the earth's magnetic field, their exact orientation depending on their position relative to the magnetic poles. Ancient rocks can thus have locked up in them a record of their position with relation to the poles as these were situated when they were made. By measuring the orientation of the magnetic particles it has been found that the direction of magnetisation in many ancient lavas and sediments is not that appropriate to their present positions. Furthermore, the magnetic orientations of rocks of particular ages in North America are consistently different from those of rocks of the same age in Europe. It seems reasonable to suppose that the continents have drifted relative to the poles and to each other since these rocks were formed. Additional support for drifting comes from the consideration of the wind-systems of past ages as explained on p. 105 and Fig. 9.6.

8. Chemical Composition of the Earth and its Shells

From what is known of the constitution and dimensions of the various shells or zones proposed for the earth, its chemical composition can be estimated. The core and mantle form over 99% of the earth's mass and, using the appropriate compositions for these, Brian Mason has estimated that the *composition of the whole earth*, in percentages of elements, is roughly: Fe 35, O 28, Mg 17, Si 13, Ni 2·7, S 2·7, Ca 0·61, Al 0·44, Co 0·2, Na 0·14, Mn 0·09, K 0·07, Ti 0·04, P 0·08, Cr 0·01. Mason has also estimated the relative abundance of the fourteen most common elements of the *crust* as: oxygen O, silicon Si, aluminium Al, iron Fe, calcium Ca, sodium Na, potassium K, magnesium Mg, titanium Ti, hydrogen H, phosphorus P, manganese Mn, sulphur S, carbon C.

Chemical analyses of rocks are usually given in oxide percentages and, in the Table below, the composition of the *crust* as calculated by Arie Poldervaart is so presented.

Composition of the Crust

SiO_2	55·2
Al_2O_3	15·3
Fe_2O_3	2·8
FeO	5·8
MgO	5·2
CaO	8·8
Na_2O	2·9
K_2O	1·9
TiO_2	1·6
P_2O_5	0·3
MnO	0·2

We may form an opinion on the composition of the material that the geologist has to deal with in the crust by using the chemical composition of the *average igneous rock*, since the material of sedimentary and metamorphic rocks is derived ultimately from igneous rocks. This is given below, expressed in various ways:

Average Igneous Rock			*Weight Percentage*	*Volume Percentage*
SiO_2	59·12	O	46·60	93·77
Al_2O_3	15·34	Si	27·72	0·86
Fe_2O_3	3·08	Al	8·13	0·47
FeO	3·80	Fe	5·00	0·43
MgO	3·49	Ca	3·63	1·03
CaO	5·08	Na	2·83	1·32
Na_2O	3·84	K	2·59	1·83
K_2O	3·13	Mg	2·09	0·29
H_2O	1·15			
TiO_2	1·05			
P_2O_5	0·30			
MnO	0·12			
	99·50		98·59	100·00

Chapter two

The dominance of oxygen in this average means that the crust is largely made up of oxygen atoms between which are arranged the atoms of the seven other common elements. The most abundant minerals must be oxygen compounds, either *silicates* (containing silicon and oxygen linked with other elements) or *oxides*. Since the eight elements listed make over 99% of the average igneous rock, it is obvious that all other elements must be either absent or present only in minute quantities. These scarcer components are known as *trace elements*.

It will be noticed that the eight common elements of the average igneous rock include very few of the metals and other elements that are useful in modern industry. These rare and valuable metals can only be extracted for use by man when they have been concentrated into *ore-deposits* by natural geological processes, and it is one of the tasks of geologists to seek out and exploit such deposits.

9. Age of the Earth

The age of the *universe* is calculated on various astronomical and chemical grounds to be between 3,000 and 5,000 million years. Some meteorites which have reached the earth are thought to be about 4,500 million years old. The early stages in the development of the earth are not of great importance to geologists because the conditions during those stages were so different from those of later times that they cannot be studied by geological methods. It has been suggested that the *earth's crust* was formed about 4,000 million years ago and at some time after this the first rains fell and geological history proper began. The oldest known rocks are dated at about 3,500 million years and are types which can be interpreted on uniformitarian lines: so far as is known, nothing is left of the rocks formed during the original consolidation of the molten planet. The length of time with which the geologist is concerned is therefore 3,500 to 4,000 million years.

The *method of dating* rocks and minerals depends on the presence of minerals containing *radioactive elements* such as uranium, thorium and certain isotopes of rubidium, potassium and carbon. Radioactive elements disintegrate at a steady rate, leaving behind new elements formed as *breakdown products*: uranium and thorium give lead and helium, potassium gives the gas argon, rubidium gives strontium and radiocarbon ordinary carbon. If we know the rate of disintegration and can measure the proportions of the radioactive element and its breakdown product contained in a certain mineral, we can determine how long the radioactive element has been locked up in the mineral. This method has been used to establish the ages of thousands of minerals and, by inference, of the rocks containing them.

For geological purposes, the *radiometric age* of the mineral or rock must be related to its *geological age*. We may be able to date radiometrically a certain lava flow at, say, 200 million years. From the law of superposition, it follows that lavas or sedimentary rocks resting on this flow will be younger, and similar rocks beneath it will be older than 200 million years; an intrusive igneous rock squeezed into fractures in the lava will also be younger than 200 million years. The sequences of volcanic activity, igneous intrusion and deposition recorded by the geological relationships of the rocks can thus be tied to the date given by the sample. In this sort of way it is possible to build up geological history in terms, firstly, of the *relative ages* of rocks revealed by their arrangement in the crust and, secondly, of the *real ages* expressed in millions of years. We shall come back to this history in later chapters.

3

MINERALOGY

This chapter provides a background of information about the composition and properties of the minerals which make up the common rocks. It begins with a definition of a mineral and an illustration of the connection between atomic structure and the properties of minerals. Crystals are then dealt with and the subjects of crystal symmetry, crystallographic notation and the crystal systems are examined. Some of the physical properties which serve to identify minerals are mentioned—these include cleavage, fracture, hardness, specific gravity, colour and other optical properties. In the remainder of the chapter an account is given of the atomic structure of the rock-forming silicate minerals, together with a description of the six main groups—quartz, feldspars, micas, amphiboles, pyroxenes and olivines—and notes on other common silicate and non-silicate minerals.

1. Minerals

Most rocks are made of associations of minerals arranged in various ways, and before we can begin to consider the rocks we must therefore know something about the *rock-forming minerals*. This chapter deals with the chemical composition and internal structure of the minerals and with the physical properties, dependent on this composition and structure, by which the geologist recognises them. The student can only understand the details given here by examining collections of minerals for himself. Such collections can be seen in museums, can be purchased from mineral dealers or, better still, can be made in the field. It may be useful to treat much of the material in this chapter as a background to the practical study of natural minerals and of models and to refer back to it as each new group of minerals is encountered.

Definition. A *mineral* is a naturally formed inorganic substance which possesses a definite chemical composition and a definite atomic structure. This last specification means that the atoms of which it is made are arranged in a regular three-dimensional pattern which is characteristic for each family of minerals. Atoms of several elements which resemble each other in size and properties may be able to replace one another in the pattern, so that a little latitude is allowed in the requirement of a definite chemical composition.

This definition may be illustrated by the example of two minerals which have the same chemical composition but different atomic structure. The minerals *diamond* and *graphite* are both made of the pure element carbon: their chemical compositions are therefore identical. The arrangement of the carbon atoms is however very different in the two substances as is shown in Fig. 3.1, and accordingly we are dealing with distinct minerals. The carbon atoms—and indeed the atoms of all elements—are too minute to be seen even with the aid of powerful microscopes and our knowledge of their arrangement in crystals comes from studies of the way in which they modify beams of X-rays. In the figure, the atoms are represented diagrammatically as small spheres: they are held in position in the pattern by electrical forces whose effects are represented in the figure by lines linking neighbouring atoms.

Atomic structure and physical properties. The atomic structure of a mineral controls its physical characters. We can hardly imagine two

Chapter three

minerals differing more than diamond and graphite; diamond is hard, transparent and dense, graphite soft, opaque and flaky. This difference is related to the difference in atomic arrangement (Fig. 3.1): the structure of diamond is compact and tightly bound, that of graphite is in open sheets so that the mineral readily separates into flakes parallel to the sheets. Another characteristic of minerals that is of the greatest importance for the geologist is connected with the regular atomic structure. This is their tendency to form *crystals*, which we now consider in some detail.

2. Crystals

Crystals are solid forms bounded by plane surfaces arranged in a regular and symmetrical manner (Fig. 3.2). These surfaces or crystal *faces* are parallel to planes in the atomic structure in which the atoms are closely packed. The arrangement of the faces of a crystal is therefore an expression of the internal atomic structure and its regularity gives *symmetry* to the crystal. This symmetry will clearly differ according to the atomic structure, but of course will always be the same in any one particular mineral.

Crystal symmetry. We need a way of describing the symmetry of any crystal and this is done by means of three *criteria of symmetry*. First is the *plane of symmetry*, an imaginary plane dividing

DIAMOND | GRAPHITE

Arrangement of Carbon Atoms

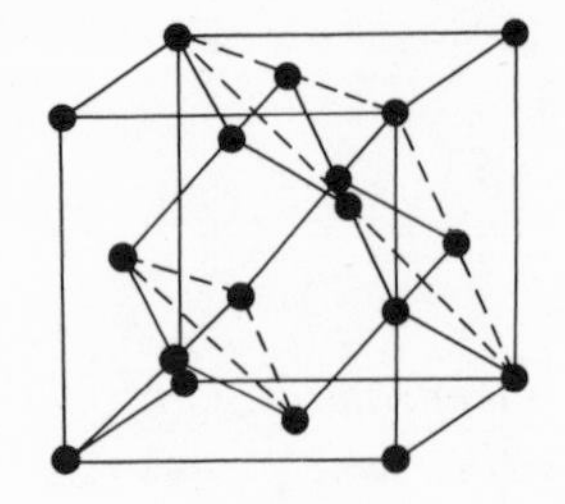

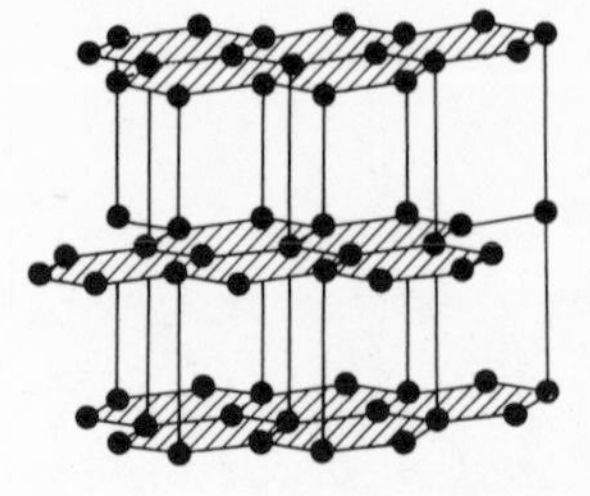

Crystal Shape

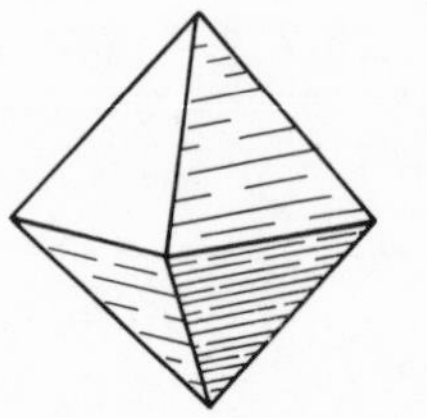

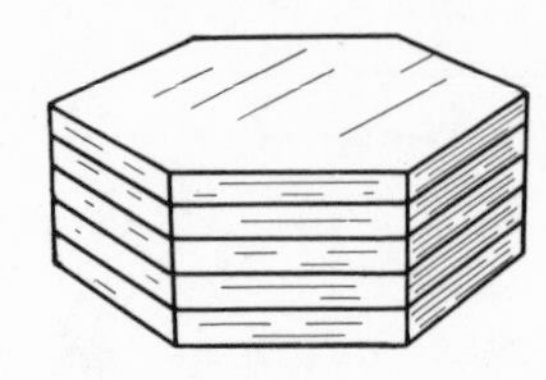

SOME PHYSICAL PROPERTIES

Diamond	Graphite
Cubic crystallisation	Hexagonal crystallisation
Octahedral cleavage	Flaky cleavage
Brittle	Flexible
Hard	Soft
Spec. grav. 3·5	Spec. grav. 2·2
Colourless	Black
Translucent	Opaque
Brilliant lustre	Metallic lustre or dull

Fig. 3.1 *Diamond and graphite: two minerals made of the element carbon, but with different lattice structure*

Fig. 3.2 *Calcite crystals from Cumberland*

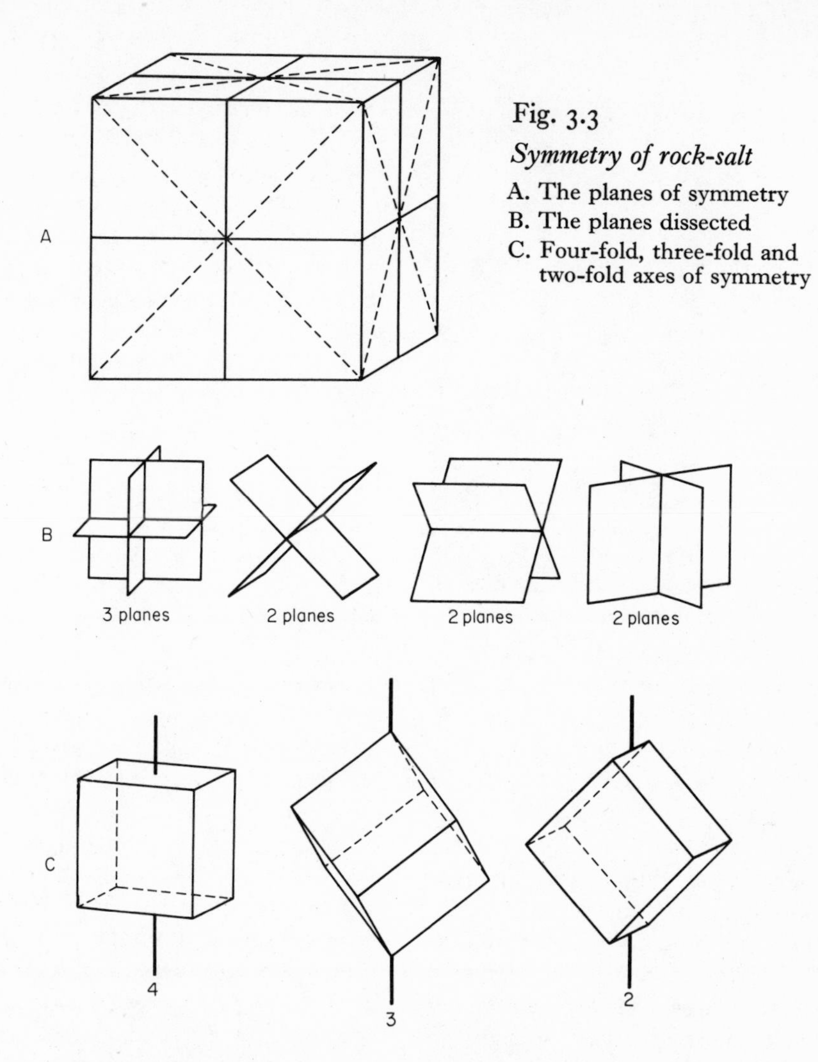

Fig. 3.3
Symmetry of rock-salt
A. The planes of symmetry
B. The planes dissected
C. Four-fold, three-fold and two-fold axes of symmetry

the crystal into two halves so that one half is the mirror image of the other. Second is the *axis of symmetry*, a line about which the crystal can be rotated to present the same appearance more than once in a complete rotation. Depending on the degree of symmetry, a crystal may come to occupy the same position 2, 3, 4 or 6 times in a complete rotation: the axis is then said to be of two-fold, three-fold, four-fold or six-fold symmetry. The third criterion of symmetry is the *centre*. A crystal has a centre of symmetry when corresponding faces, edges or points are arranged in pairs in corresponding positions on opposite sides of a centre point.

The criteria of symmetry will be better understood by working out the symmetry of a couple of minerals, rock-salt or sodium chloride, NaCl, and gypsum or hydrous calcium sulphate, $CaSO_4 . 2H_2O$. Rock-salt occurs in nature in crystals of cube shape; inspection of the cube (Fig. 3.3) shows that it has:

(1) 9 planes which divide it into two halves so that one is the reflection of the other;
(2) 13 axes of symmetry, 3 of four-fold, 4 of three-fold and 6 of two-fold symmetry, and
(3) a centre of symmetry.

This symmetry is written down as:

Planes 9 (3 axial, 6 diagonal), Axes 13 (3^{iv} 4^{iii} 6^{ii}), Centre.

Chapter three

Fig. 3.4
Symmetry of gypsum

The student should manipulate a cube and satisfy himself that this symmetry of rock-salt is really present. The planes of symmetry are shown in Fig. 3.3A and dissected in Fig. 3.3B, and the three types of axes in Fig. 3.3C. A typical crystal of our second example, gypsum, is shown in Fig. 3.4. Only one plane of symmetry is present. At right angles to this plane is the one axis of symmetry; rotation about this axis causes the gypsum crystal to take up the same position twice in a complete rotation—it is therefore an axis of two-fold symmetry. Lastly, the faces, edges and corners of the crystal are obviously arranged in pairs on either side of a central point and the crystal therefore has a centre of symmetry. The symmetry of gypsum can then be written as:

Planes 1, Axes 1^{ii}, Centre.

Its symmetry is obviously much lower than that of rock-salt, that is, it has fewer planes and axes of symmetry.

It is essential that the student becomes familiar with natural crystals and with wooden, cardboard or glass models of crystals. By handling these he can acquire a practical knowledge of crystallography that reading alone cannot give. 'Plasticine' models that can be sliced to show planes of symmetry or stuck through with knitting needles to show axes can be utilised. Some useful exercises can be performed with crystals or models of rock-salt and gypsum like those we have shown in Figs. 3.3 and 3.4.

Inspection of the cube of rock-salt will show that all the crystal faces are the same and all are interchangeable—they are said to be *like faces*; on the other hand, the gypsum crystal is seen to be made up of several sets of faces with different shapes and relations to one another: one set includes the two front vertical faces and the corresponding two faces at the back, another set is made by the two side vertical faces and a third set by the two top faces and the two corresponding bottom faces. Each set made up of like faces is called a *simple form*; the rock-salt cube is a simple form. A crystal such as that of gypsum made up of several forms is a *combination*. Two faces meet in an *edge*.

Geometrical and crystallographic symmetry. Before we go any farther, it will be as well to realise that, so far, we have dealt with perfect crystals or models, in which each like face is developed to the same extent—we have been dealing with obvious geometrical symmetry. But as crystal symmetry really depends on the internal atomic structure, it is not affected by the fact that many crystals are imperfect. Accidental differences in the rate of growth may cause some of a set of like faces to grow larger than others, but this does not change their fundamental symmetry.

As a further consequence of the fixed atomic structure of a mineral, the faces of a crystal must bear fixed geometrical relations to one another. This gives the *Law of Constancy of Interfacial Angles*. The *interfacial angle* is the angle between the perpendiculars to two intersecting faces, as shown in Fig. 3.5. It is measured by *goniometers* (angle measurers). A simple goniometer can be made with a protractor to which is pinned a moving arm of celluloid. The student should satisfy himself that all the interfacial angles of the rock-salt cube are right angles, and that the interfacial angles between corresponding faces are constant in a variety of gypsum crystals.

Lastly, the student should learn to make reasonably accurate drawings of crystals or

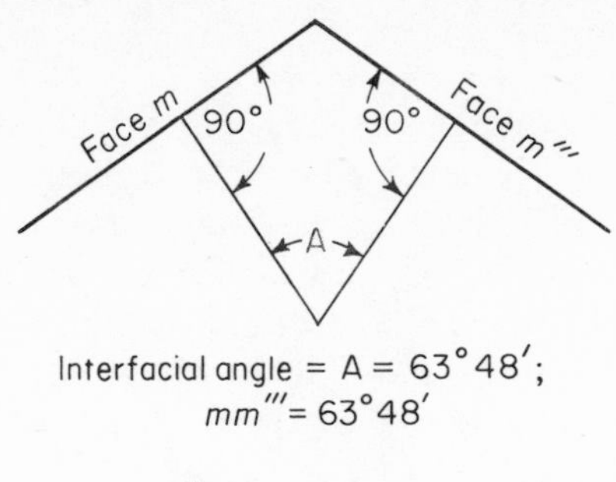

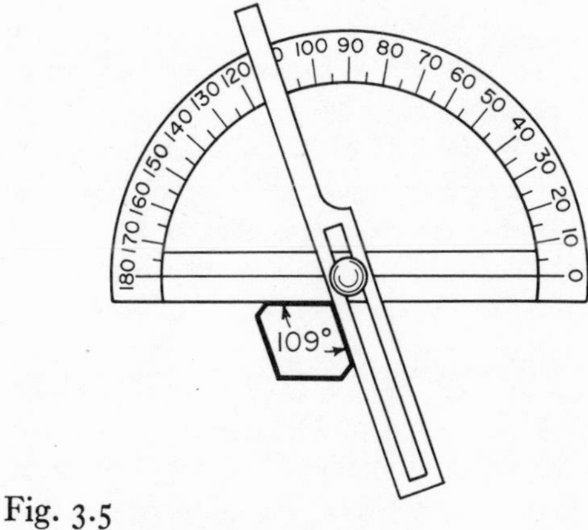

Fig. 3.5
The interfacial angle and a simple goniometer

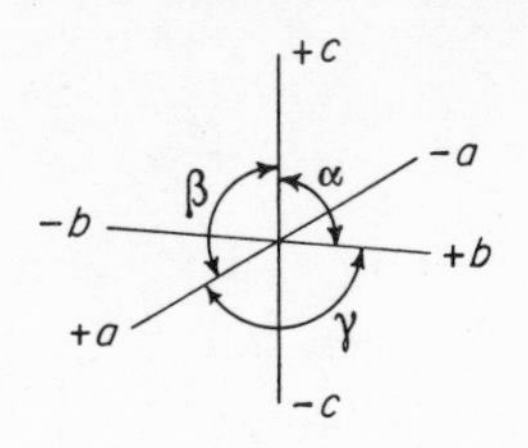

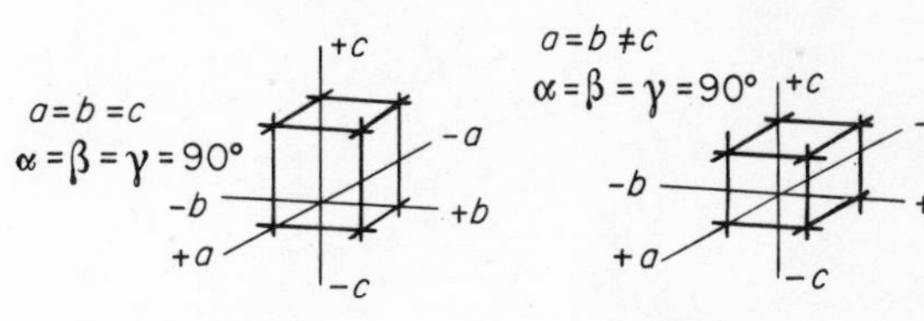

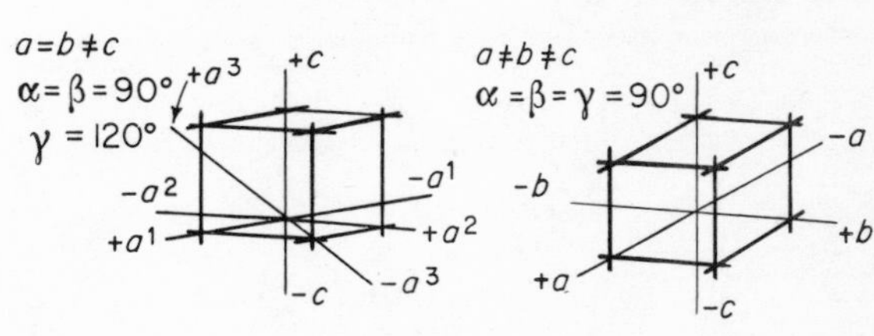

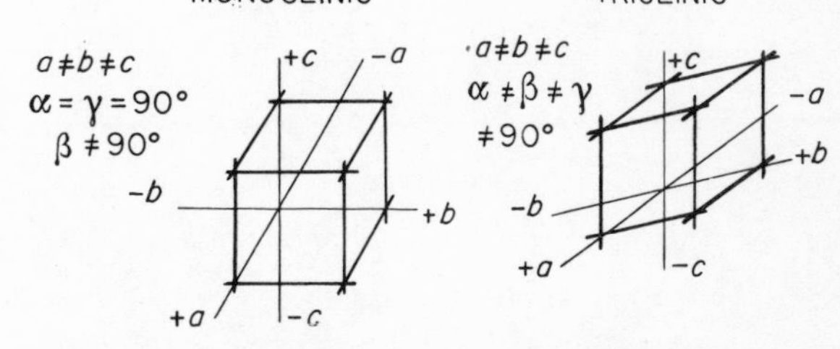

Fig. 3.6 *The unit cell types and the corresponding crystal systems, showing the arrangement and nomenclature of the crystallographic axes*

models. Two methods are noted here (there are others more complicated). In the *orthographic projection*, a plan or elevation of the crystal is made. This does not give a clear picture of the crystal and the *clinographic projection* is more often used, as in the drawings in this book; in this projection a standard rotation and tilt of the crystal are made and a parallel perspective drawing produced.

3. Unit Cells and Crystal Systems

Although an enormous number of atoms go to make up a single crystal, in order to understand the atomic structure it is only necessary to think about the arrangement of the small group of atoms which makes the basic unit of the pattern. The pattern of this *unit cell* is repeated endlessly in the complete crystal, just as a small design may be repeated many times in a strip of wallpaper. Examination of minerals by means of X-rays makes it possible to determine the shapes and sizes of their unit cells. These can be represented by simple geometrical shapes which share many elements of symmetry with the crystals built up from them.

Unit cell types. In this book we have to deal with six different types of unit cell, whose shapes are shown in Fig. 3.6. Each of these types is

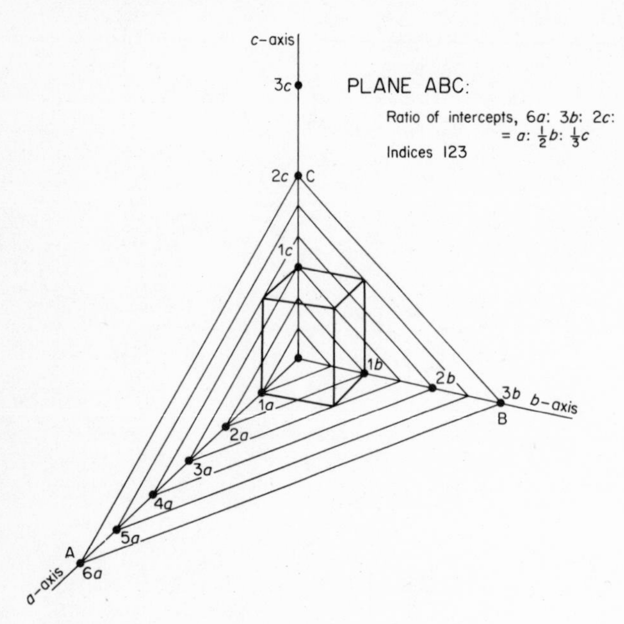

Fig. 3.7 *The derivation of Miller indices*

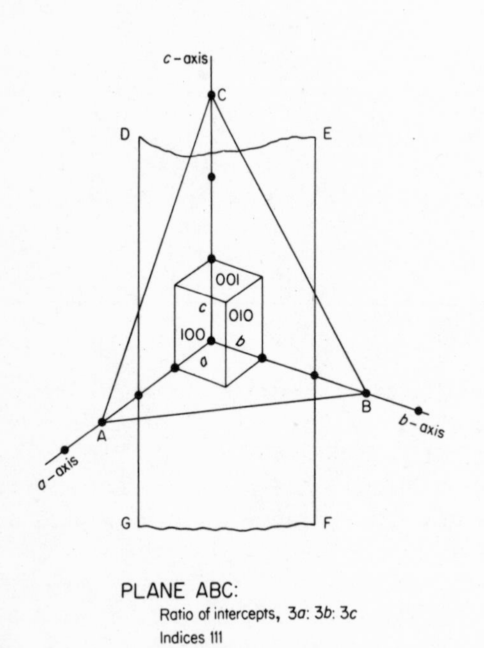

Fig. 3.8 *Miller indices of various crystal faces*

capable of building up crystals of several forms and can give rise to several kinds of symmetry according to the way in which it is repeated: but all crystals built from one unit cell type have certain symmetry elements in common and they are all said to belong to the same *crystal system*. From the six unit cell types, therefore, are derived six crystal systems named the *cubic* or isometric, *tetragonal*, *hexagonal*, *orthorhombic*, *monoclinic* and *triclinic* systems.

The edges of the unit cell, when extended, can be used as axes of reference to describe the position of any crystal face. There are three *crystallographic axes*, *a*, *b* and *c*, defined in this way for all crystal systems except the hexagonal, where it is convenient to use four. The angles between the axes depend on the shapes of the unit cells as shown in Fig. 3.6. The relative lengths of the different axes are taken to be proportional to the lengths of the corresponding edges of the unit cell, and the length of each edge is used as a unit of measurement along the appropriate axis.

Crystallographic notation. In describing the arrangement of faces in a crystal, it is customary to make use of a standard series of symbols which, once their significance has been understood, can convey a lot of information very concisely. This *crystallographic notation* depends on the geometrical relationships between the crystal faces and the crystallographic axes. The position of a crystal face can be defined by the ratio of its *intercepts* made on the crystallographic axes: the units of measurement along each axis are the lengths of the corresponding sides of the unit cell. A face which is not parallel to any crystallographic axis must, if extended, cut across all three axes; and the point at which it meets each axis gives the intercept on that axis. Fig. 3.7 shows a plane *ABC*, which is part of a crystal face, in relation to the crystallographic axes *a*, *b* and *c* of the same crystal. The unit cell is shown and we can count up along each axis the number of units of measurement cut off by the plane *ABC*. The ratio of the intercepts is seen to be $1a, \frac{1}{2}b, \frac{1}{3}c$. This formula is rather clumsy and is converted into a simpler form by using the *reciprocals* of the intercepts: these are called *Miller indices* and for the plane *ABC* will clearly be 123 (one, two, three). The Miller indices can be derived directly from Fig. 3.7 by noting the number of planes of *ABC* character which are crossed in the length of the cell side along the *a*, *b* and *c* axes in turn—one along *a*, two along *b*, and three along *c*.

By using the shorthand of the Miller indices, we can define the essential geometrical features of any crystal face. There is no need to go into details here, but it may be useful to note three types of indices with special and simple meanings (Fig. 3.8). Firstly, any face, like that illustrated in Fig. 3.7, which intercepts all three axes will have a symbol consisting of three numbers. If it intercepts each axis at unit distance its symbol must be 111. Secondly, any face which is parallel to one crystallographic axis will have an intercept on that axis of infinity (that is, it will not cut the axis). The reciprocal of infinity being zero, the figure 0 (nought) will appear in the symbol: thus the face 110 is parallel to the *c*-axis. Thirdly, a face parallel to two axes will have two noughts in its symbol: 100 cuts *a*, 010 cuts *b*, 001 cuts *c*. Finally, like faces on opposite sides of the crystal cut opposite ends of the crystallographic axis: they are distinguished by adopting the convention of calling one end of each axis positive and the other end negative and by inserting minus signs above appropriate figures in the index. Thus, the face $00\bar{1}$ cuts the lower end of the *c*-axis.

With this information, the student should be able to understand the general significance of the symbols for crystal faces which are used in this book. These symbols should be studied and their correctness confirmed. It is often helpful to work with models of crystallographic axes and to use sheets of cardboard or celluloid placed in different

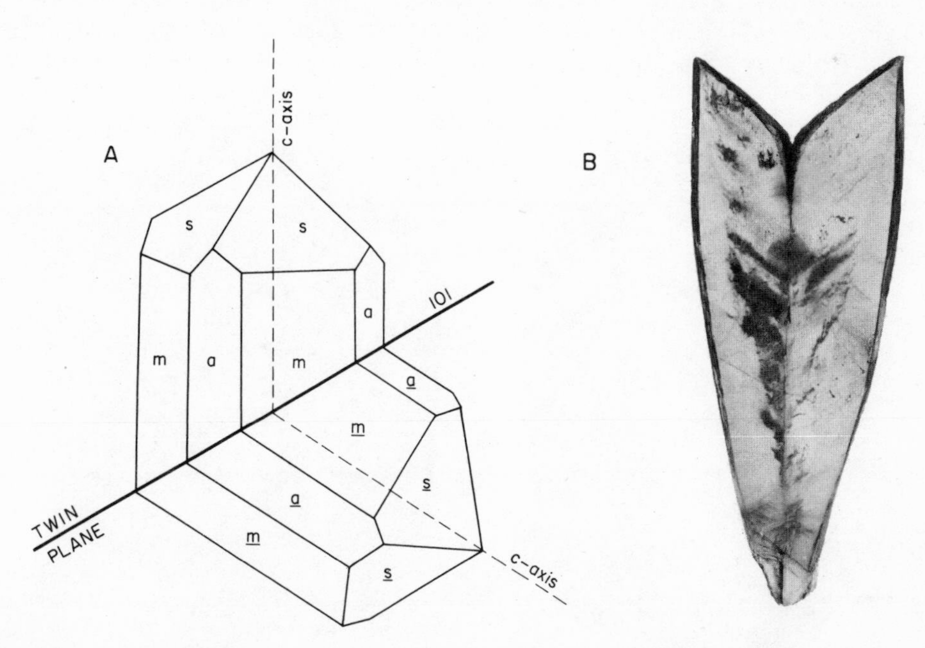

Fig. 3.9 *Twinning*

A. In rutile

B. In gypsum: a contact swallowtail twin revealed in a cleavage flake. The cleavage surface (010) faces the observer. Note that the traces of a second cleavage and dark zones of impurities show different orientations in the two halves of the twin. The rounding of the crystal outline is due to solution

positions to represent crystal faces. The kind of index numbers to be expected for different arrangements can be established, and if units of measurement are marked off on the axes the actual indices can be calculated. The minerals which are described in the later part of this chapter include examples of all the crystal systems and provide illustrations of the symmetry and forms of crystals in each system.

Twinned crystals. When a crystal grows or *crystallises* from a melt or solution, it may happen that the starting-point of its growth is a sheet of atoms, and that growth on the two sides of the sheet, while of course keeping to the appropriate structural pattern, proceeds with different orientations. Thus, in the tetragonal crystal of rutile, TiO_2, shown in Fig. 3.9, the atomic structure is the same in the two parts but the orientation of the *c*-axis is different. The result is a *twinned crystal* or *twin*. One half of the twin is a reflection of the other half about a certain *twin-plane* which is a plane of symmetry for the twin. A twin-plane cannot be a plane of symmetry of the untwinned crystal (work out why not). There are various types of twins, some of which we meet later in the descriptions of the minerals fluorspar, calcite, aragonite, gypsum (Fig. 3.9), hornblende, augite, orthoclase (Fig. 3.14) and plagioclase.

4. Additional Properties of Minerals

Before describing the common minerals we have to mention some additional physical properties that may help in their identification. These properties are: cleavage and fracture; hardness; specific gravity; colour and certain other optical characters.

Cleavage. Some crystals break along certain definite planes, the *cleavage-planes*, that are closely related to the atomic structure. The cleavage-plane is one in which the atoms are closely packed and which is widely separated from its parallel neighbours. The cleavage of graphite, illustrated early on in Fig. 3.1, is an example of this control. Cleavage is described by stating the crystal face parallel to which the cleavage takes place and adding whether it is perfect or good or poor.

Fig. 3.10 *Cleavage*

Left In galena (cubic), cleavage parallel to (100)

Right In mica (monoclinic), cleavage parallel to basal pinacoid (001). The observer is looking directly at the cleavage surfaces

Examples of cleavage in minerals are given and described in Fig. 3.10, and other examples are mentioned later on in this chapter. *Cleavage fragments* must be examined by the student.

Fracture. Some minerals break along surfaces that are not cleavage-planes and may do so in a consistent and diagnostic manner. Such a *fracture* is that shown by quartz, SiO_2, and flint, which break with a *conchoidal fracture* like that of bottle glass.

Hardness. The hardness and cohesiveness of a mineral depend again on its atomic structure. In a general way, hardness increases with the density of atomic packing. A rough *scale of hardness* is that of *Mohs* in which ten minerals are arranged in order of increasing hardness as follows: 1 talc, 2 gypsum, 3 calcite, 4 fluorspar, 5 apatite, 6 orthoclase, 7 quartz, 8 topaz, 9 corundum and 10 diamond. Hardness is tested by trying to scratch these minerals in turn with the mineral under examination. Thus, if a mineral scratched orthoclase but was scratched by quartz its hardness is between 6 and 7. A good penknife blade scratches up to about 6½, a finger-nail up to about 2½. The student should test the hardness of such minerals as he has available. He must make sure that a scratch has actually been produced on the mineral by licking the place and examining it with a pocket lens.

Specific gravity. Specific gravity is the ratio of the weight of a substance to the weight of an equal volume of water. In minerals, it depends on the kind of atoms present and the way they are packed in the unit cell, as illustrated in the table below:

1. *Diamond*, dense packing, 3·52; *graphite*, looser packing, 2·3 (both C).
2. *Calcite*, hexagonal, 2·71; *aragonite*, orthorhombic, 2·94 (both $CaCO_3$).
3. *Anhydrite*, $CaSO_4$, 2·93; *barytes*, $BaSO_4$, 4·5; *anglesite*, $PbSO_4$, 6·3. (All have similar structure but barium and lead are heavy elements.)

As a generalisation, we may say that non-metallic-looking minerals—the 'sparry' minerals—have a specific gravity of 2·6–2.8 whereas the metallic, ore, minerals have one of about 5. The student must make himself familiar with the relative weights of comparable pieces of the common minerals and rocks. Remember that lead and barium minerals, though often 'sparry', are heavy.

Specific gravity is determined by weighing the specimen in air and in water: $s.g. = W^a/(W^a - W^w)$. The ordinary balance, specific gravity bottle, Walker's steelyard or Jolly spring-balance can be employed.

Colour and streak. A pure mineral possesses its own inherent colour; minerals containing aluminium, the alkalies, or alkaline earths are light coloured, those containing iron, chromium, manganese, nickel or cobalt deeply coloured. But the presence of impurities affects the colour of minerals; the ruby and sapphire are both varieties of one mineral, corundum, Al_2O_3, and their colour is due to the presence of different kinds of impurities.

The colour of a finely powdered mineral is called its *streak* and is an important diagnostic character. Thus magnetite, Fe_3O_4, and hematite, Fe_2O_3, may both be black in mass but the streak of magnetite is black and of hematite cherry-red.

Other optical properties: thin slices. The optical properties of minerals, in addition to their colour, are very important in the recognition of minerals—but for this advanced work one needs a petrographical microscope fitted with a polar (a device for transmitting only certain kinds of light vibrations) above and below the stage. The methods are beyond the scope of this book but, nevertheless, it is worth mentioning the *thin slice* which is used in this kind of work.

A thin slice is a parallel-sided plate of a mineral or rock, of a thickness of about 1/30 millimetre, which is mounted on a glass slip and covered by a thin glass coverslip. Such a slice is translucent and, when it is held up to the light and viewed by the naked eye or a pocket-lens, it is possible to distinguish the constituent minerals of coarse rocks and, especially, to see how they are fitted together in the rock, i.e. its *texture* or *fabric*

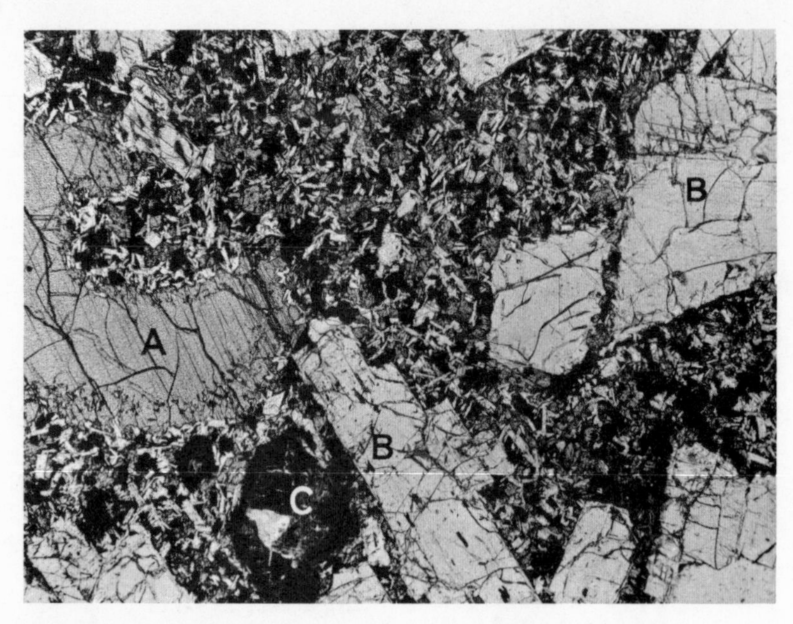

Fig. 3.11 *The microscopic examination of a lava (magnification* × 15)

The thin slice shows:

A. Large crystals or phenocrysts of pyroxene (irregular outline, strong cleavage, high relief)

B. Plagioclase (good crystal form, low relief)

C. Olivine (good crystal form)

(see Fig. 3.11). If a petrographical microscope is available, many optical properties of minerals, such as relative refractive index, colour, cleavage and twinning, can be examined without much bother. If a mineral has a refractive index markedly higher than those of the adjacent minerals, it appears to stand out in relief and its surface seems pitted. Some of these optical properties are illustrated above.

5. The Rock-forming Minerals

We have seen that the average igneous rock has 47% oxygen, 28% silicon and 24% Al, Fe, Ca, Mg, Na and K, so that the common minerals making these rocks will be *silicates* of Al, Fe, Ca, Mg, Na or K and *silica*. It is found that seven families of these minerals make up 99% of the igneous rocks. In addition to *quartz*, SiO_2, they are the silicate minerals *feldspars* and *feldspathoids*, *micas*, *amphiboles* (hornblende), *pyroxenes* (augite) and *olivines*. We shall not deal with the feldspathoids in this book.

The common minerals of the sedimentary rocks include some, such as quartz and feldspar, which persist from the parent igneous rock, and a number of new minerals formed at the earth's surface. Chief among the new minerals are the *clay-minerals*, hydrous aluminium silicates, and the *carbonates*, *calcite* and *dolomite*. Under special conditions, *rock-salt* and *gypsum* may be deposited from saline waters.

The common minerals developed in the metamorphic rocks include members of the same families as those of the igneous rocks, and, in addition, the silicates *serpentine*, *chlorite*, *talc* and *garnet* and the aluminium silicates, *andalusite*, *sillimanite* and *kyanite*. The carbonates are the main constituents of metamorphosed limestones.

In addition to the minerals listed above, we deal here with a few more that the student may come across. The ore-minerals are mentioned later in connection with ore-deposits. The complete list of rock-forming and other minerals now to be described is given on the facing page.

Silicates

Quartz, orthoclase feldspar, plagioclase feldspar, micas, amphiboles e.g. hornblende, pyroxenes e.g. augite, olivine.

Chlorite, talc, serpentine, clay-minerals, tourmaline, garnet, zircon, aluminium silicates (andalusite, sillimanite, kyanite).

Non-silicates

phosphate —apatite
carbonates—calcite, dolomite, aragonite
sulphates —gypsum, barytes
halides —rock-salt, fluorspar
oxides —corundum, magnetite, hematite, limonite
element —graphite

6. The Silicate Minerals

The classification of silicate minerals is most easily understood by considering their atomic structure. In all the families, the fundamental unit is the *SiO_4-tetrahedron*, a cluster of charged atoms or *ions* in which a single small silicon cation (with positive charge) lies at the centre and four large oxygen anions (with negative charges) lie at the corners (Fig. 3.12). This cluster is not stable by itself because the electric charges on the component ions do not balance one another. Stable structures are built by the linking together of the SiO_4-tetrahedra to produce larger groups in or between which cations of other elements can be accommodated. The electric charges of all the ions must balance and this necessity controls the number of ions of different kinds which can be fitted into the silicate lattice. We note five types of silicate-structure.

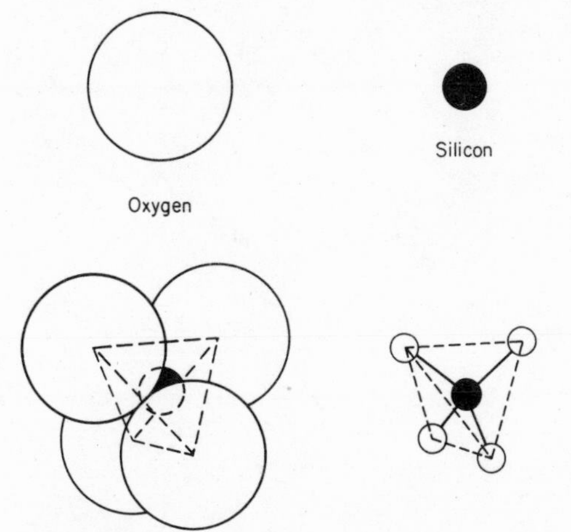

Fig. 3.12
The SiO_4-tetrahedron
Above The oxygen and silicon ions represented on the same scale
Below The tetrahedron to scale (*left*); its structure displayed diagrammatically (*right*)

(1) In the simplest type, the SiO_4-tetrahedra remain separated from one another and cations of iron, magnesium, calcium or other elements are fitted in between them. These *island* silicates are exemplified by olivine, garnet and zircon.
(2) In a second type, SiO_4-tetrahedra are linked in lines by sharing oxygen ions with their neighbours, and the charges are again neutralised by the introduction of other cations. This *chain* structure is characteristic of the pyroxenes.
(3) A *ribbon* structure, produced by further sharing of oxygen ions to tie two chains together side by side, is seen in the amphiboles. This structural type often gives rise to elongated crystals, the long axis of the crystal being parallel to the ribbons.
(4) The next structural type is one in which three oxygen ions of each SiO_4-tetrahedron are shared with adjacent tetrahedra to give a *sheet* structure, the additional cations being packed between the sheets. This structure gives rise to flaky minerals including mica, chlorite, talc and the clay-minerals. A perfect cleavage is usually developed parallel to the sheets of tetrahedra.
(5) In the last group, all the corners of the tetrahedra are linked to other tetrahedra to form a three-dimensional *framework*: quartz (which can be thought of as a silicate of silicon) and the feldspars belong here.

Chapter three

The rather concentrated information on diagnostic characters now to be given will be made more intelligible if specimens of the minerals are available when it is being read. It may be helpful too to turn back to this section when the igneous and metamorphic rocks are being studied to check up on the minerals of which they are made.

7. The Common Silicates

Quartz

This very common mineral has the composition silicon dioxide, SiO_2, and as we have just said has a framework structure. The commonest crystals appear as six-sided prisms closed in above and below by six-sided pyramids as shown in Fig. 3.13. At first sight it often looks as if these crystals possessed an axis of six-fold symmetry parallel to their length. The crystals do indeed belong to the hexagonal system but their symmetry is not the highest possible for this system. Close inspection may often reveal small faces belonging to forms with much lower symmetry; and the six apparently similar faces enclosing the prismatic part of the crystal are shown to belong to more than one crystal form by their behaviour towards etching reagents—their *etch-marks* are different. These facts are summarised in Fig. 3.13 where the names and indices of some of the forms present are given.

Pure quartz is colourless, glassy and almost transparent, as in the variety *rock-crystal*, but it may be coloured by impurities to produce gem varieties: *amethyst* is purple, *rose quartz* pale pink, and *cairngorm* smoky yellow. Quartz has no cleavage and breaks with a characteristic conchoidal fracture (p. 27). Its specific gravity is 2·65 and its hardness 7—too hard to be scratched by a knife. In thin slice, quartz is clear and colourless and looks inconspicuous because its refractive index is similar to that of the mounting medium.

Well-formed crystals of quartz are found in veins or projecting into cavities in certain rocks. In igneous and metamorphic rocks, quartz is often abundant and usually occurs in a *granular* form showing no crystal faces: its three-dimensional framework type of atomic structure favours the growth of rounded or irregular grains.

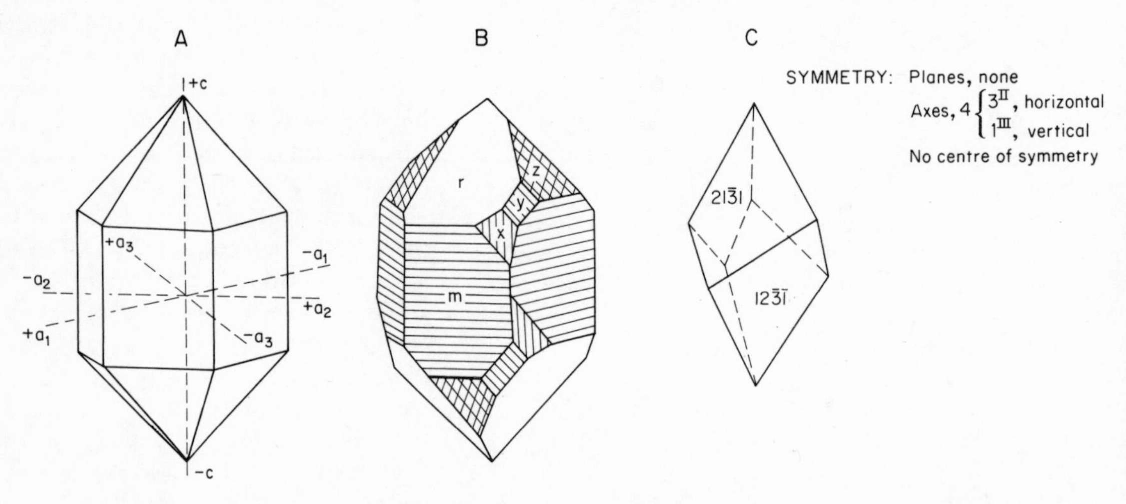

Fig. 3.13 *Quartz*

A. Crystal of a common type apparently showing highest hexagonal symmetry

B. More complex crystal revealing true symmetry. The forms represented are the prism, m($10\bar{1}0$); rhombohedron, r($10\bar{1}1$); rhombohedron, z($01\bar{1}1$); trigonal trapezohedron, x($51\bar{6}1$); and trigonal pyramid, y($11\bar{2}1$)

C. Trigonal trapezohedron ($21\bar{3}1$) of quartz type showing the true symmetry

Quartz is chemically very stable, and when igneous or metamorphic rocks crumble and decay at the earth's surface, the quartz grains therefore survive unchanged to provide the chief components of sand and of the sedimentary rock sandstone derived from it. Quartz is used industrially as an abrasive (because of its hardness), in the manufacture of pottery and silica-bricks and as a flux and filler.

Chalcedony, *flint* and *jasper* are varieties of silica consisting of mixtures of crystalline quartz and amorphous (Greek, 'without form'), non-crystalline hydrated silica or *opal*. These varieties are minutely crystalline or *cryptocrystalline*, and occur as nodules in sedimentary rocks. Their colour is varied: white, grey, brown or black. *Agate* is a banded form of chalcedony filling steam-holes in lavas. *Flint* is black or grey in colour and occurs as nodules in the Chalk of England; it breaks with a good conchoidal fracture to give sharp cutting edges and for this reason was widely used by prehistoric man for making implements and weapons. *Chert* is a variety of flint that breaks with a flat fracture. *Jasper* is usually red in colour. Lastly, *opal* is the amorphous hydrated form of silica, $SiO_2.nH_2O$. It is compact, often banded by colour, and displays opalescence in the gem variety *precious opal*. It is softer than quartz and has a lower specific gravity, 2·2.

The feldspars

The feldspars are the most important group of rock-forming minerals, occurring in abundance in most igneous and many metamorphic rocks and in certain sedimentary rocks derived from these. They are silicates with the framework structure in which some of the silica in the SiO_4-tetrahedra has been replaced by aluminium. This replacement upsets the balance of charges and requires the introduction of cations of potassium, sodium or calcium to maintain equilibrium in the structure.

The three fundamental feldspar types resulting from this are as follows.

Orthoclase, potassium feldspar, $KAlSi_3O_8$ — Alkali feldspars (with albite)
Albite, sodium feldspar, $NaAlSi_3O_8$ — Plagioclase feldspars (with anorthite)
Anorthite, calcium feldspar, $CaAl_2Si_2O_8$

In the alkali feldspars, note that one atom of silicon has been replaced by aluminium and balanced by potassium or sodium, in anorthite two atoms are replaced by aluminium and balanced by calcium. A continuous series between albite and anorthite is produced by the introduction of both Na and Ca into the silicate framework. The members of this series are the *plagioclase feldspars*.

The colour of the feldspars is whitish, greyish or a pale shade of red. In thin slice they appear colourless, often a little cloudy, and are inconspicuous because their refractive index is near that of the mount. Their hardness is about 6, specific gravity 2·5–2·76. Their crystals, whether monoclinic (orthoclase) or triclinic (plagioclase) are squarish or flattish, the prism (110), side pinacoid (010) and basal pinacoid (001) being dominant (Fig. 3.14). Two principal cleavages are parallel to (010) and (001) and are at right angles or nearly so. Twinning is common and takes place on several different patterns. *Simple twinning* produces crystals made of only two parts; *repeated twinning*, seen especially in plagioclase, builds a compact individual made of a large number of narrow twin-lamellae. We add a few notes on the main feldspar species:

Orthoclase, potassium aluminium silicate, $KAlSi_3O_8$, crystallises in the monoclinic system, typical crystals being shown in Fig. 3·14, where some of the forms present are indicated. Orthoclase twins in three main ways, a simple twin being shown in Fig. 3.14. There is another kind of potassium feldspar called *microcline* which crystallises in the same pattern as orthoclase but is triclinic. It is distinguished by possessing two sets of repeated twins at right angles, giving two sets of fine striations on the basal pinacoid (001). Orthoclase is formed at rather higher temperatures

Fig. 3.14 *Feldspars*

Left A group of potash feldspar crystals showing the prism (110), and pinacoids (010), (001)
Right Interpenetration Carlsberg type twin of orthoclase: the twin axis is the crystallographic *c*-axis, the composition plane is the clinopinacoid (010)

than microcline and has a less perfectly arranged atomic structure; microcline is said to be well *ordered*, orthoclase less so.

Plagioclase feldspars are triclinic. They show a gradation in properties between the two end-members, albite, $NaAlSi_3O_8$ and anorthite, $CaAl_2Si_2O_8$. Thus, the specific gravity of albite is 2·605, of anorthite 2·765. The plagioclases show simple twinning as in orthoclase and are often repeatedly twinned as well. This repeated twinning may appear on the appropriate crystal face as a series of fine striations.

Micas

The micas are the chief representatives of the *sheet* silicates. They crystallise in the monoclinic system to form flat crystals usually almost hexagonal in outline. The large dimensions of the crystals are parallel to the sheet structure of the lattice. They possess a perfect basal cleavage (Fig. 3.10) along which thin elastic flakes can be peeled off. We may mention two varieties:

Muscovite, white mica, potassium aluminium silicate with hydroxyl, $KAl_2[(AlSi_3)O_{10}].(OH)_2$. *Sericite* is a fine flaky variety.

Biotite, black mica, magnesium iron potassium aluminium silicate with hydroxyl, $K(Mg,Fe)_3[(AlSi_3)O_{10}].(OH)_2$.

Biotite is rather harder and heavier than muscovite, the average hardness being 2·5–3 as against 2–2·5 and specific gravity 2·7–3·1 as against 2·76–3.

The micas occur abundantly as glistening flakes in igneous rocks such as granite and in many metamorphic rocks. Muscovite is a constituent of the micaceous sandstones and it coats the splitting planes of flagstones. It is obtained in large sheets from veins of pegmatite, which is the residual portion of granitic magma. It is used extensively as an insulating material in the electrical industry and in ground-up form in a number of industries.

Hornblende and the amphiboles

The amphibole family of silicates with *ribbon* structure comprises a number of complex replacement series in which cations of several elements enter the structure in varying proportions. We select the common mineral *hornblende* as representative. It is a silicate of aluminium, calcium, magnesium, iron and sodium and its complexity can be judged from the general formula $(Ca,Na,Mg,Fe,Al)_{7-8}[(Al,Si)_4O_{11}]_2\cdot(OH)_2$. Hornblende is monoclinic, a typical crystal being shown and annotated in Fig. 3.15; the angle between the prism faces is about 56°. The prismatic crystals are elongated parallel to the ribbons of the structure—this elongation is shown in perfection by *asbestos* which forms silky fibres and is used to make fire-resisting materials. Twinning is simple on 100, the cleavage is perfect prismatic giving two sets of cleavage planes meeting at about 120° (Fig. 3.15). Hornblende is black in colour, green or yellow in thin section; its hardness is 5–6, specific gravity 3–3·47. Hornblende occurs in many igneous rocks and in metamorphic rocks derived from them.

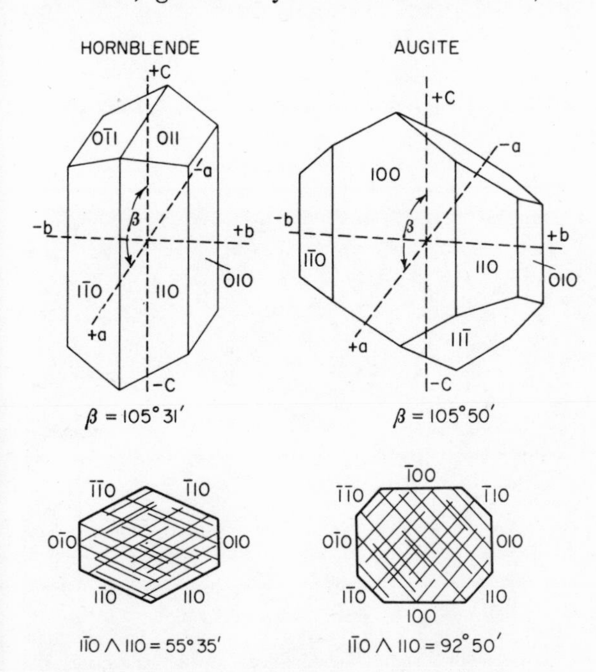

Fig. 3.15
Above Typical crystal forms
Below Sections perpendicular to C showing prismatic cleavage. The forms are the prism (110), orthopinacoid (100), clinopinacoid (010) and clinodome (011)

Augite and the pyroxenes

The pyroxenes are silicates with a single *chain* structure. They form a number of series, analogous to those of the amphiboles, and *augite* is a common representative.

Augite is a complex anhydrous silicate of calcium, magnesium, iron and aluminium, represented as $(Ca,Mg,Fe,Al)_2(Al,Si)_2O_6$. It crystallises in the monoclinic system, a common type of crystal being shown in Fig. 3.15. The prism angle is nearly 90° and there is a perfect cleavage parallel to the prism faces. Augite is often simply twinned on 100. Its colour is black, hardness 5–6, specific gravity 3·2–3·5. In thin slice it is pale greenish brown and sometimes shows a conspicuous rectangular pattern of cleavage cracks. Augite occurs in many igneous rocks and a few metamorphic rocks derived from them. Other pyroxenes are prominent in metamorphosed limestones.

Olivine

Our last group of common rock-forming silicates is the olivine family, in which the SiO_4-tetrahedra occur as separate *islands* in the structure. Common olivine is an iron magnesium silicate, $(Mg,Fe)_2SiO_4$. It belongs to the orthorhombic system and forms short prismatic crystals. Its colour is greenish, hardness 6–7, specific gravity 3·2–4·3, increasing with the iron content. Members of the olivine family occur in igneous rocks and in metamorphosed impure limestones; they easily alter along cracks to *serpentine* (p. 34).

8. The Less Common Silicates

First among the minor group of silicates come a number of *sheet-silicates*, such as chlorite, talc, serpentine and the clay-minerals.

Chlorite, a hydrous silicate of aluminium, iron and magnesium, is related to the micas in composition but has no alkalies. Chlorite forms greenish flakes with a perfect basal (001) cleavage that gives flexible but not elastic plates (remember mica plates are elastic). Chlorite is soft, 2 on Mohs' scale, and has a specific gravity of 2·6–2·9. It occurs as a common constituent of metamorphic rocks and as an alteration product of biotite and hornblende.

Talc is a hydrous magnesium silicate occurring in white or greenish compact masses of flakes. It is greasy and soft, being 1 on Mohs' scale, is easily scratched by the finger-nail and has a perfect basal cleavage; its specific gravity is 2·7–2·8. Talc is a metamorphic mineral of magnesium-rich rocks. *Steatite or soapstone* is massive talc. It is used as a filler, lubricant and absorbent and for making acid-resistant slabs.

Serpentine is another hydrous magnesium silicate occurring in massive, platy or fibrous form of a green or variegated colour. Its hardness is 3–4 and it can be cut with a knife; its specific gravity is 2·5–2·6. It is polished for use as a building or ornamental stone. *Chrysotile* is a fibrous form of serpentine that provides a great deal of commercial asbestos. Serpentine results from the reaction of watery vapours with rocks rich in magnesium, i.e. rocks containing olivine, pyroxenes or amphiboles.

Clay-minerals, Kaolinite. The clay-minerals are hydrous aluminium sheet-silicates that constitute the clays. The chief member is *kaolinite*, $Al_4Si_4O_{10}.(OH)_8$. They occur in exceedingly small flakes and are produced by the modification of other aluminium silicates. Their identification is best done by X-ray analysis.

Tourmaline. Our next minor silicate is *tourmaline*, which has a type of atomic structure not so far mentioned: lines of linked SiO_4-tetrahedra are curled up to make closed rings and these rings are piled one above the other to give elongated crystals. Tourmaline, though not particularly abundant in rocks, is represented in even the smallest mineral collections because it habitually makes good crystals and, besides, illustrates the crystallographic phenomenon known as *hemimorphism*. In composition, it is a complex borosilicate of aluminium, sodium, calcium, iron and magnesium. It crystallises in the hexagonal system, in a symmetry class in which there are no horizontal planes of symmetry and no centre of

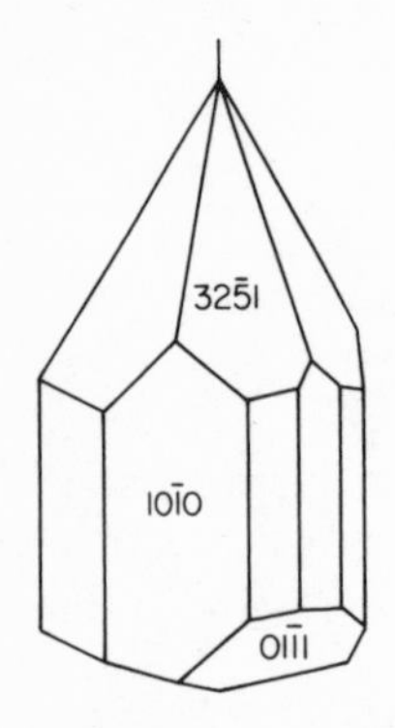

PLANES, 3 diagonal
AXIS 1 3-fold: NO CENTRE
(Forms differ at ends of unique c-axis = HEMIMORPHISM)

Fig. 3.16 *Tourmaline*
Left A hemimorphic crystal showing symmetry
Right Striated prismatic crystal associated with quartz

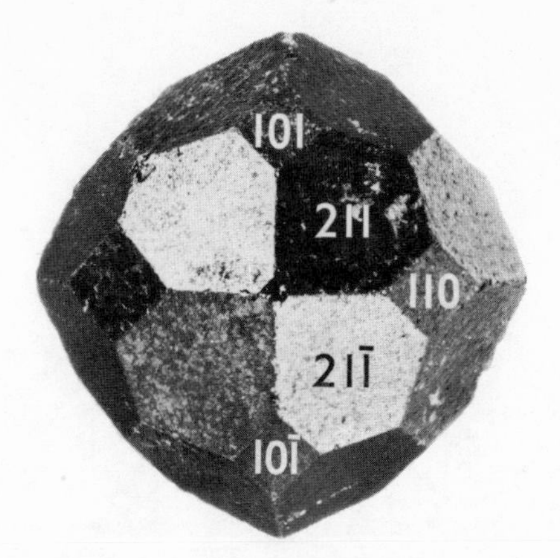

Fig. 3.17
Garnet
Left Rhombdodecahedron (110)
Right Combination of rhombdodecahedron (110) and trapezohedron (211)

symmetry; in consequence the crystals are hemimorphic, with different crystal forms at opposite ends of the vertical axis (Fig. 3.16). Tourmaline occurs as prismatic crystals, often striated along their length and giving triangular cross-sections, or as needles or grains. It is usually black in colour, its hardness is 7–7·5, specific gravity 2·9–3·2. Tourmaline is found especially where gas action has taken place at the contacts of igneous rocks and clay-rocks. *Schorl* is the black columnar variety found in the Cornish granites.

We next deal with a couple of silicates with 'island' structure, separate SiO_4-groups: they are *garnet* and *zircon*.

Garnet. The island structure of this important family is indicated in their general formula, $X_3Y_2(SiO_4)_2$, where X can be calcium, magnesium or iron and Y can be iron, aluminium or chromium. The garnets (Fig. 3.17) crystallise in the cubic system, their common forms being the rhombdodecahedron (110) and the trapezohedron (211). The colour of a garnet depends on its composition; calcium-bearing garnets are pale green but the common garnet is red or brown. They show no cleavage. Their hardness is 6·5–7·5, specific gravity 3·5–4·3. In thin slices they have a high relief—since their refractive index is much higher than that of the mount they show well-marked boundaries and pitted surfaces. Garnet is common in metamorphic rocks of many kinds. It is used, because of its hardness, as an abrasive and also as a gemstone.

Zircon, another silicate with island structure, is described here as a representative of the tetragonal crystal system. Its composition is zirconium silicate, $ZrSiO_4$. A typical crystal is illustrated in Fig. 3.18 from which an idea of the normal tetragonal symmetry can be obtained. Zircon occurs in small colourless prismatic crystals as an accessory component of many igneous and metamorphic rocks. Its hardness is 7·5, specific gravity 4·7.

Aluminium silicates. The last group of silicates to be mentioned have independent SiO_4-groups in their structure but also oxygen atoms outside these. They are the aluminium silicates, Al_2SiO_5, represented by three minerals of different atomic structure, *andalusite, sillimanite* and *kyanite*. Andalusite and sillimanite are orthorhombic, kyanite is triclinic.

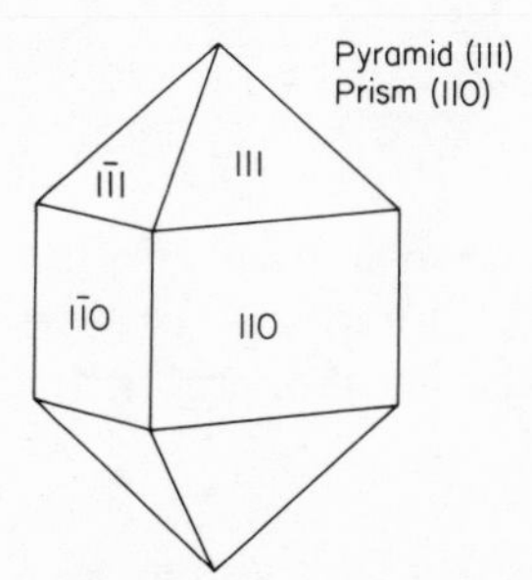

Fig. 3.18
Zircon crystal illustrating symmetry of a tetragonal mineral

PLANES 5, 3 axial, 2 vertical diagonal
AXES 5, 1 four-fold, 4 two-fold
CENTRE

Fig. 3.19
Chiastolite: cross-sections showing crystal form and inclusions in thin slice (×15)

Andalusite forms prismatic crystals, grey, red or purple in colour. The variety *chiastolite* (Fig. 3.19) found in most mineral collections, is characterised by black inclusions arrayed in a cross-pattern. The hardness of andalusite is 7·5, specific gravity 3·1–3·3.

Sillimanite occurs in grey prisms, needles or fibres, with a perfect (010) cleavage visible in cross-sections of the prisms. Its hardness is 6–7, specific gravity 3·2.

Kyanite forms bladed crystals showing several cleavages. It is grey or pale blue in colour. Its hardness varies on different faces, from 4 on some to 7 on others; an old name for the mineral is *disthene* (Greek, 'two strengths'). Its specific gravity is 3·6–3·7.

The aluminium silicates occur in metamorphic rocks of clay composition, the species present depending on the conditions of metamorphism.

9. The Non-Silicate Minerals

The non-silicate minerals that we mention here are grouped under their chemical compositions—phosphate, carbonates, sulphates, halides, oxides and one element.

Apatite is a *phosphate* of calcium with some fluorine, chlorine or hydroxyl as represented by its formula, $Ca_5(F,Cl,OH)(PO_4)_3$. The PO_4-groups play a part in the structure similar to the SiO_4-groups in the silicates. Apatite crystallises in the hexagonal system, commonly making little hexagonal prisms, and is green or yellow in colour. Its hardness is 5, specific gravity 3·17–3·20. It occurs as an accessory constituent of igneous rocks.

Calcite is calcium carbonate, $CaCO_3$. It crystallises in the rhombohedral class of the hexagonal system and forms good crystals of many habits, three special types having received the popular names of *nail-head spar*, *dog-tooth spar* and *prismatic crystals*; these are illustrated in Fig. 3.20; in the latter figure, other crystallographic details of calcite are given. Lamellar twinning is common. Cleavage of calcite gives perfect rhombohedral fragments, bounded by the faces of the $(10\bar{1}1)$ form; powdered calcite consists of minute cleavage rhombohedra. The colour of calcite is white or grey; a transparent variety in large crystals is called *Iceland spar*. Calcite can be scratched with a knife, its hardness being 3; its

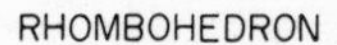

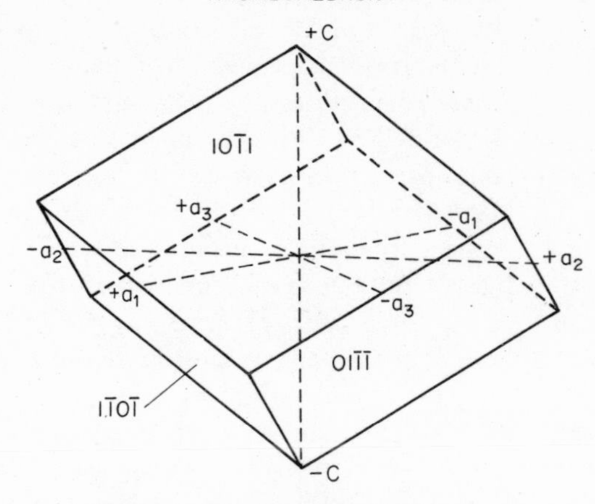

SYMMETRY OF CALCITE TYPE:

Planes 3 vertical diagonal

Axes, 4 { 3^{II}, horizontal crystallographic axes; 1^{III}, vertical crystallographic axis

A Centre of Symmetry

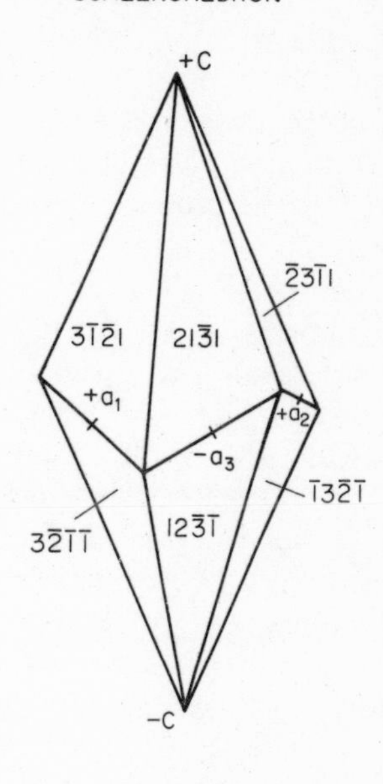

NAIL-HEAD

DOG-TOOTH

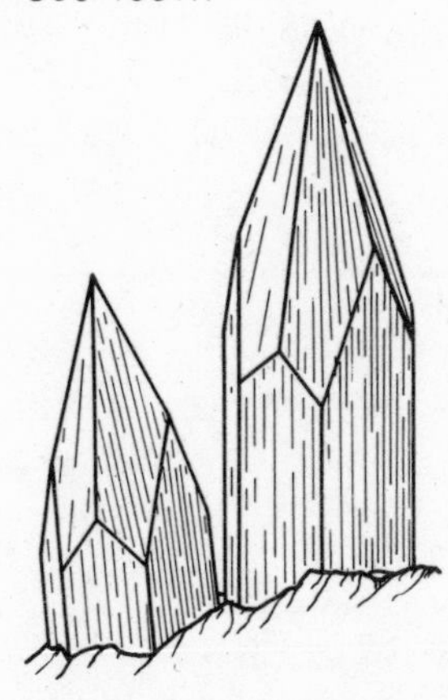

PRISMATIC

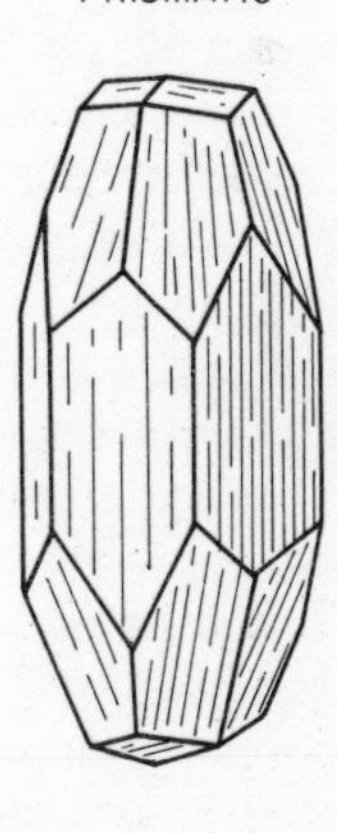

Fig. 3.20 *Calcite*

Above The characteristic forms, rhombohedron and scalenohedron, illustrating the symmetry of the calcite type. The rhombohedron shown is the unit rhombohedron in which the faces intersect the crystallographic axes at unit distances. The axial ratio $a : c$ is 1 : 0·8550; distances along the a-axes are therefore measured in multiples of 1, along the c-axis in multiples of 0·8550. In the scalenohedron illustrated, each face intersects the a-axes at distances of 1, 2 and 3 units and the c-axis at a distance of $3 \times 0{\cdot}8550$ units

Below Three common types of calcite crystal. The *nail-head spar* is a combination of a flat rhombohedron (10$\bar{1}$2) and hexagonal prism (10$\bar{1}$0); the *dog-tooth spar* is a combination of scalenohedron and hexagonal prism; and the prismatic type a combination of all three forms

specific gravity is 2·71. Calcite effervesces with cold dilute acid with the evolution of carbon dioxide. It is the main constituent of the important sedimentary rock *limestone* and of its metamorphic derivative *marble*. The uses of calcium carbonate are listed when the limestones are described (p. 112).

A different form of calcium carbonate is provided by the mineral *aragonite* which we mention because, with calcite, it gives an additional example of *polymorphism*, the occurrence of two substances with the same composition but different atomic structures, crystal forms and physical properties (p. 19). Aragonite is orthorhombic and occurs in prismatic crystals often repeatedly twinned to give a pseudohexagonal crystal. Its hardness is 3·5–4, specific gravity 2·94.

Dolomite is carbonate of calcium and magnesium, $CaMg(CO_3)_2$. It crystallises in the hexagonal system but in a different symmetry class from calcite. It occurs commonly in granular masses, or in rhombohedral crystals of $(10\bar{1}1)$ form—calcite rarely occurs in this form. Dolomite has a good rhombohedral cleavage. It is white, yellow or brown in colour, has a hardness of 3·5–4, specific gravity 2·8–2·9. It does not effervesce with cold dilute acid (cf. calcite). Dolomite forms extensive sedimentary beds resulting from the *dolomitisation* of original calcite limestones through the activity of magnesium carbonate solutions derived from sea-water. It is important in industry as a building material and refractory and as a source of carbon dioxide.

Gypsum is hydrated calcium sulphate, $CaSO_4.2H_2O$. It crystallises in the monoclinic system, the common type of crystal being shown in Fig. 3.4. Gypsum twins on two chief laws, giving *swallow-tail* twins with twin-plane (100) as shown in Fig. 3.9 and *butterfly* twins on (101). It also occurs in compact masses or in fibrous forms when it is called *satin spar*. The cleavage is perfect parallel to (010) giving thin flexible non-elastic flakes. It is colourless or white when pure. Its hardness is 2, and it can be scratched by the finger-nail; specific gravity 2·3. Gypsum is among the *saline residues* or *evaporates* which are formed when a salt lake or a cut-off portion of the sea dries up. In the London Clay of south-east England good crystals of gypsum are common: these are produced by the action of sulphuric acid, generated by the decay of pyrites (iron sulphide), on fossil shells made of calcium carbonate. Gypsum is a very important industrial mineral, especially in the building trade where it is used for the manufacture of cements and plasterboards of all kinds. *Plaster of Paris* is gypsum from which much of the water has been driven off.

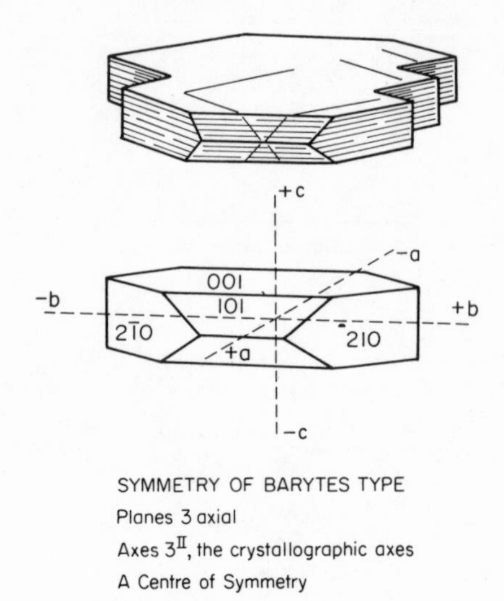

Fig. 3.21

Barytes, illustrating the orthorhombic system

Above Crystals in parallel growth

Below Combination of prism (210), dome (101) and basal pinacoid (001)

Barytes is barium sulphate, $BaSO_4$, and provides us with an illustration of the orthorhombic system in its highest symmetry class (Fig. 3.21). Barytes possesses perfect cleavages parallel to (001) and parallel to the prism (210). It is colourless, white or yellow, its hardness is 3–3·5, specific gravity 4·5, hence the name (Greek *barys*, heavy) and its alternative *heavy spar*. It occurs in veins and is a common mineral in lead and zinc veins. It is used in the manufacture of white paint and paper, and as a drilling mud in the oil industry.

Rock-salt, Halite is sodium chloride, NaCl. It crystallises in the cubic system (Fig. 3.3), the common form being the cube (100). Its cleavage is perfect cubic. It is colourless or white when pure, has a salty taste, hardness 2–2·5, specific gravity 2·2. Rock-salt is formed as an *evaporate* resulting from the evaporation of bodies of salt water. It is used very extensively in the chemical industry and for cookery and preserving.

Fluorspar has the composition calcium fluoride, CaF_2. It is a cubic mineral, the common form being the cube; it also occurs massive. Its cleavage is perfect octahedral, its hardness 4, specific gravity 3–3·25. Fluorspar occurs as a vein-stone associated with lead and zinc minerals. It is used in glass manufacture, as a flux and for lenses.

Corundum, Al_2O_3, is a hexagonal mineral, which commonly forms barrel-shaped or pyramidal crystals, without cleavage. Its colour is greyish but the gem variety *ruby* is red and *sapphire* blue as a result of the presence of impurities. Its hardness is 9, next to diamond, specific gravity 3·9–4·1. *Emery* is corundum mixed with magnetite and hematite. Corundum is an original constituent of certain rocks and is found also in metamorphosed limestones, as in the Burma gem-deposits. Its hardness makes it a valuable abrasive used in grinding wheels and the like.

Magnetite has the composition iron oxide, Fe_3O_4. It crystallises in the cubic system, the octahedron (111) being the commonest form. It also occurs in massive and granular condition. Its colour and streak are both black, hardness 5·5–6, specific gravity 5·18. Magnetite, like hematite and limonite, remains opaque even in thin slice. It is strongly magnetic. Magnetite occurs as a primary constituent of most igneous rocks, and great segregations of magnetite in such rocks provide valuable deposits of this iron-ore.

Hematite is an iron oxide with the composition Fe_2O_3. It crystallises in the hexagonal system in rhombohedra and often in thin tabular form; it also occurs as granules, or in kidney-shaped masses (*kidney iron-ore*). Its colour is steel-grey or black but its streak is cherry-red. Its hardness is 5·5–6·5, specific gravity 4·9–5·3. Hematite occurs in pockets or hollows replacing limestones as in North Lancashire or as beds of sedimentary origin. It is an important iron-ore.

Limonite is a hydrous iron oxide, perhaps represented by $2Fe_2O_3 . 3H_2O$, resulting from the alteration of other iron minerals. It has no crystal structure and is amorphous, earthy or concretionary in form. Its colour is brownish or yellowish, with a streak of the same colours, hardness 5–5·5, specific gravity 3·6–4. Limonite is another important iron-ore; *bog iron-ore* is a limonite deposited in swamps and lakes.

Graphite, carbon, C, crystallises in lamellae with a perfect basal cleavage, expressing the manner in which the carbon atoms are arranged (see Fig. 3.1). It is black and opaque, with a black streak. It occurs in veins or scattered through sedimentary and metamorphic rocks. It is used for crucibles, electrodes, paint and as a lubricant.

A *Reference Table for Silicate Minerals* is given on the next two pages.

REFERENCE TABLE FOR

Mineral	Atomic Structure and Composition	Crystal System	Habit and Crystal Form
	3-dimensional framework		
QUARTZ	SiO_2	hexagonal	usually granular; crystals prismatic with pyramidal terminations
	3-dimensional framework		
FELDSPARS: ORTHOCLASE (ALKALI FELDSPARS)	$KAlSi_3O_8$	monoclinic	often granular; crystals showing prism and pinacoid faces
MICROCLINE (ALKALI FELDSPARS)	$KAlSi_3O_8$	triclinic	
ALBITE (ALKALI FELDSPARS; PLAGIOCLASE)	$NaAlSi_3O_8$ plagioclase series contains Na and Ca	triclinic	
ANORTHITE (PLAGIOCLASE)	$CaAl_2Si_2O_8$ plagioclase series contains Na and Ca	triclinic	
	sheet structure		
MICAS: MUSCOVITE	$KAl_2(AlSi_3O_{10}).(OH)_2$	monoclinic	flaky; crystals platy, pseudohexagonal
BIOTITE	$K(MgFe)_3(AlSi_3O_{10}).(OH)_2$	monoclinic	
	ribbon structure		
HORNBLENDE (example of AMPHIBOLES)	hydroxyl-bearing silicate of Ca,Al,Fe,Mg	monoclinic	elongated or acicular or fibrous; crystals prismatic
	chain structure		
AUGITE (example of PYROXENES)	silicate of Ca,Al,Fe,Mg	monoclinic	granular; crystals stumpy prismatic
	island structure		
OLIVINE	$(MgFe)_2SiO_4$	orthorhombic	granular; crystals stumpy prismatic
	sheet structure		
CHLORITE	hydroxyl-bearing silicates of Al,Fe,Mg (chlorite) or Mg (serpentine, talc)	monoclinic or orthorhombic	flaky
SERPENTINE			massive, flaky or fibrous
TALC			massive, flaky or fibrous
	sheet structure		
CLAY-MINERALS	e.g. kaolinite $Al_4Si_4O_{10}(OH)_8$	monoclinic	flaky
	ring-structure		
TOURMALINE	borosilicate of Al,Na,Ca,Mg,Fe, OH,F	hexagonal	crystals prismatic, often acicular, hemimorphic
	island structure		
GARNET	silicates of Ca,Mg,Fe,Al,Cr	cubic	granular; rhomb-dodecahedron, trapezohedron
	island structure		
ZIRCON	$ZrSiO_4$	tetragonal	small prisms
	island structure		
ANDALUSITE	Al_2SiO_5	orthorhombic	squarish
KYANITE	Al_2SiO_5	triclinic	bladed
SILLIMANITE	Al_2SiO_5	orthorhombic	acicular

SILICATE MINERALS

Cleavage	*Twinning*	*Colour*	*Hardness*	*Specific Gravity*	*Relief in thin slice*	*Other Properties*
none	none	transparent, usually colourless	7	2·65	low	
two almost at right angles	simple repeated repeated lamellar	white, pink or buff opaque in hand specimen	about 6	2·56–2·58 2·6 2·76	low moderate	
one basal, perfect	not common	white brown	2–2·5 2·5–3	2·76–3·0 2·7–3·1	moderate	flakes elastic
two prismatic at 120°	sometimes simple or lamellar	dark green	5–6	3–3·5	moderate	
two prismatic at 90°	sometimes simple or lamellar	dark green to black	5–6	3·2–3·5	moderate	
two poor	none	green	6–7	3·2–4·3	high	
one basal, perfect	not visible	green green and red whitish	2 3–4 1	2·6–2·9 2·5–2·6 2·7–2·8	low	flakes non-elastic soapy feel
one basal, perfect	not visible	whitish			low	
poor	none	black, dark brown, green or red	7–7·5	2·9–3·2	moderate or high	
none	none	red, brown or green	6·5–7·5	3·5–4·3	very high	
none	simple	yellow or grey	7·5	4·7	very high	
parallel to length of crystals	none lamellar none	red blue white	7·5 4–7 6–7	3·1–3·3 3·6–3·7 3·2	moderate high high	

4

IGNEOUS ACTIVITY AND THE IGNEOUS ROCKS

This chapter deals with the various kinds of igneous activity which result from the production of molten magma. The conditions which lead to the formation and migration of magma in the crust are discussed and the primary shapes of lavas and intrusions are described. Volcanoes and their mode of eruption are then dealt with and their products, the lavas, pyroclastic rocks and volcanic gases, are mentioned. The igneous rocks as a whole are classified according to composition, mineral constituents and textures, and in the last part of the chapter descriptions are given of the two main groups of igneous rocks—the granites and their relatives and the basic rocks and theirs.

1. Igneous Activity

We have already defined the *igneous rocks* as those produced by the consolidation of molten rock-material or *magma*. We deal with these rocks at an early stage in the book because they are the ultimate source of the material which makes the sedimentary rocks: igneous rocks represent new material brought up from deep in the earth, whereas sedimentary rocks can only be made from material which is already present at the earth's surface (Fig. 4.1).

The properties of igneous rocks—their composition, fabric and structure—reflect the way in which they were made and to understand them we must know something about the whole process of *igneous activity*. As has been pointed out already (p. 3), the igneous activity which is revealed by volcanic action can be studied by uniformitarian methods because it takes place at the earth's surface. The deep-seated igneous processes which operate far below the surface must be investigated by other methods. Geophysical measurements, such as the earthquake studies discussed in Chapter 2, tell us something about the temperature and pressure in the crust and mantle: the structures and textures of the deep-seated rocks themselves may indicate how the rocks were formed, or they may be found to resemble those of volcanic rocks produced under known conditions at the surface: finally, the experimental melting and consolidation of rock-material in controlled conditions in the laboratory may throw light on the behaviour of natural magmas.

2. Production and Movement of Magma

From the general composition of igneous rocks, and from observation of active volcanoes and of the effects produced by intrusions of magma on their surroundings, it can be deduced that magma is a *silicate melt* which carries in solution considerable quantities of water-vapour and other volatile compounds. The igneous rocks formed on consolidation are composed mainly of rock-forming silicates and quartz. They do not as a rule contain all the water-vapour and the volatile compounds present in the magma—these *volatiles*, being very mobile, often escape from the cooling magma into the surrounding rocks or the atmosphere.

The active and recently extinct volcanoes of the present day are not haphazardly distributed over the earth's surface—they are restricted to a number of broad zones. Study of the distribution of ancient igneous rocks shows that at all times in geological history igneous activity has been similarly restricted. From earthquake records we know that the crust and upper mantle are usually composed of solid material. Both these lines of evidence indicate that fluid magma is developed only in exceptional conditions and we have next to inquire how such conditions arise.

The temperature within the earth is difficult to estimate, but from the *thermal gradient*—the rate of increase of temperature with depth—in the outer skin of the earth, it is suggested that at say 35 kilometres depth (that is, near the base of the continental crust) the temperature might be 500–600° C. This temperature is not high enough to melt crustal rocks under the pressures operative at that depth. For magma to form there must therefore be a local increase of temperature by an influx of heat or a local decrease of pressure, or both. Many proposals for a surge of heat into a particular part of the crust have been put forward—from radioactive concentrations, from deep-seated convection currents and so forth. For our purposes, we can agree that the conditions necessary for magma production could often be produced with the help of local but intensive earth-movements. We find, in fact, that a relationship exists between various kinds of crustal movements and the appearance of igneous rocks.

When magma has been formed, it tends to move upwards along fractures or weak zones, propelled either by the weight of the rocks above it or by earth-movements. It thus comes into colder regions of the crust and is subjected to diminishing pressures. The magma loses heat to the surrounding rocks and as it cools it becomes more viscous, that is, thicker and less fluid. The lowering of pressure allows volatiles in the magma to escape and their loss brings about a further increase in viscosity. These changes may cause the magma to solidify within the crust with the production of an *intrusive* body of igneous rock completely shut in by the surrounding *country-*

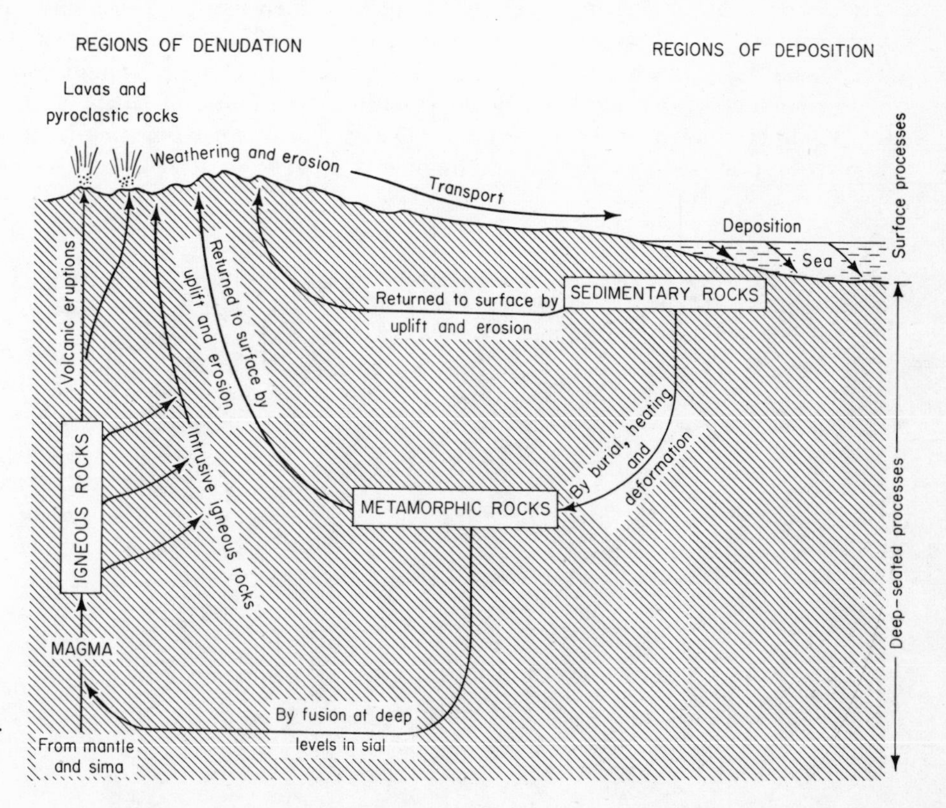

Fig. 4.1
The cycle of the rocks

rocks. On the other hand, solidification may be delayed until the magma reaches the earth's surface and flows out to produce *extrusive* igneous rocks or *lavas*. It is obvious that a single magma can give rise to both intrusive and extrusive rocks.

3. Extrusive and Intrusive Rocks

Lavas. The primary shapes and relationships of bodies of extrusive and intrusive igneous rocks are fundamentally different. The *extrusive* magma flows out from the volcanic vent to form an extensive sheet if the magma is fluid or a short thick lens or dome if it is viscous. This *lava flow* must rest on older rocks and may in time be covered by younger lavas or sediments: extrusive rocks therefore obey William Smith's Law of Superposition. The rocks underlying the lava may be baked by contact with the hot magma but, obviously, rocks above the lava cannot be affected because they were not in existence at the time of the eruption. These overlying rocks indeed may be made of weathered or crumbled lava or may fill in cracks and hollows in the lava surface.

Some products of volcanic action do not emerge quietly, but are blown out of the volcano during explosive eruptions as showers of solid fragments broken from earlier lavas and from the walls of the volcano, or as clouds of magma droplets. All this material falls back to earth to make the *pyroclastic rocks* ('fire-broken' rocks).

Igneous intrusions. The primary forms and geological relations of the intrusive igneous rocks are very different. To begin with, these rocks must be younger than the country-rocks which they intrude. Their intrusive relations are shown in several ways which are illustrated in Fig. 4.2. The contact of the igneous body may *cut across* the bedding of the country-rocks; the igneous rocks may contain fragments or *xenoliths* (Greek *xenos*, stranger) broken off from the country-rock walls; off-shoots or *veins* of igneous material may penetrate the country-rocks; the igneous rock may show a *chilled edge* of finer grain where it has cooled rapidly against the country-rock; and the country-rock may be *baked* or *contact-metamorphosed* adjacent to the igneous contact.

The primary forms of intrusions depend on their size and position in the crust, the forces at work and the nature of the magma involved. A distinction is sometimes drawn between *minor intrusions*, of small size filling restricted spaces, and *major intrusions* that have volumes of many cubic miles.

Fig. 4.2
Evidence of intrusion: the pale intrusive granite carries xenoliths of darker country-rocks and its contact cuts the bedding in these xenoliths

Fig. 4.3 *A minor intrusion: the Whin Sill of dolerite in northern England*

Minor intrusions have two main forms. *Dykes* (Fig. 1.3) are steeply inclined or vertical wall-like bodies formed by magma filling fractures opened by tangential tensions in the crust. Since such tension usually produces many fractures, dykes generally occur as large numbers of parallel or radiating intrusions which together constitute a *dyke-swarm*. *Sills* (Fig. 4.3) are sheet-like intrusions which are roughly horizontal; the magma making these intrusions must exert enough pressure to lift the overlying rocks, and as a consequence sills cannot be emplaced at great depth. Many sills are intruded along bedding-planes and are then said to be *concordant* with the bedded rock. *Veins* are narrow intrusions following irregular fractures; this form of intrusion is favoured by granitic and pegmatitic material (pp. 52–53) and by ore-bearing fluids. Finally, *plugs* are small cylindrical bodies often occupying the passages through which volcanic vents were fed.

Major intrusions are formed by the accumulation of great volumes of magma. Because of their size, they cool slowly and have time to develop coarse textures which distinguish them from the rapidly cooled lavas and minor intrusions. The heat they give out may also soften and transform the country-rocks with the production of a *metamorphic aureole* around the intrusion. The coarse-grained rocks of major intrusions are termed *plutonic rocks* in contrast to the fine-grained *volcanic rocks* making lavas and minor intrusions.

Major intrusions can be considered in two groups, *forceful intrusions* and *permitted intrusions*. The magma of **forceful intrusions** (Fig. 4.4) makes way for itself by pushing the country-rocks upward or aside. When it follows bedding-planes, it forms thick *concordant* sheets or lenses which may bulge up their roof to give *laccoliths* and *domes*. More intense pressure of magma may rupture the roof and allow the magma to force its way up as a *discordant* cylindrical mass; such masses are called *bosses*, *stocks*, *diapirs* and *plutons*.

Permitted intrusions are allowed to come into place by using space made available by the subsidence of crustal masses on ring-fractures, the mechanism being shown in Fig. 4.5. This *cauldron subsidence* gives rise to thick *ring-dykes*

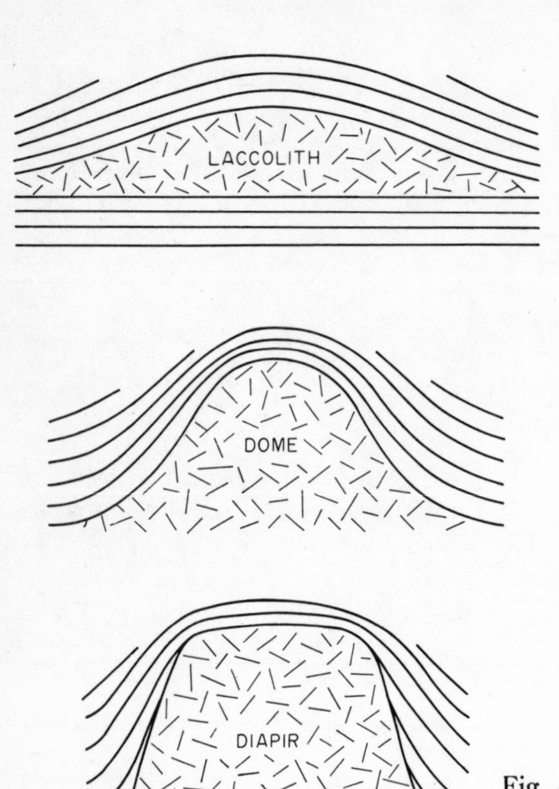

Fig. 4.4 *Three types of forceful intrusion: the laccolith and dome are concordant, the diapir partly discordant*

Fig. 4.5 *Permitted intrusions*

A. Formation of a caldera by cauldron subsidence
B. Formation of a ring-dyke not connected with the surface
C. A ring-complex made of several ring-dykes
D. Cone-sheets

or *ring-intrusions* and is often accompanied by the formation of radial dyke-swarms centred on the intrusion. *Cone-sheets*, minor intrusions filling conical fractures in the country-rocks of the roof, are also commonly associated with ring-complexes. Bodies of magma emplaced in any way may be enlarged by *stoping*, the foundering of blocks of country-rock from the roof and walls as the magma eats its way upwards.

4. Volcanoes

Fissure and central eruptions. A *volcano* is an opening at the earth's surface from which magma escapes. With repeated *volcanic eruptions*, lavas and pyroclastic rocks become heaped up around the opening to produce a *volcanic cone* pierced by a central depression, the *crater*.

Two main types of volcano can be recognised, those erupting from long cracks in the earth and those erupting from roughly circular openings. These *fissure* and *central* volcanoes are probably both related to deep fractures in the crust. Repeated eruptions from fissure volcanoes build up great lava piles, hundreds of square miles in area and thousands of feet in thickness: such *plateau lavas* are usually made of basalt (p. 54) and are illustrated by the Deccan Traps of India. Volcanoes of central type form cones of varying shape depending on the style of the eruption—this in turn depending, as we see in a moment, on the viscosity of the magma involved.

Volcanic forms. Very fluid lavas spread far from their source and build up great *shield volcanoes*, exemplified by the broad low cones of the Hawaiian Islands. Viscous lava piles up near the orifice to produce steep-sided *volcanic domes* as seen in the *Puys* of central France. Very viscous lava may solidify in the neck of the volcano and then be forced up to the surface by the pressure of volcanic gases trapped underneath it. This happened in the 1902 eruption of Mont Pelée in Martinique in the West Indies and a *lava-spine* was squeezed up out of the orifice, reaching a height of more than 800 feet before it finally broke up and began to crumble.

Explosive eruptions take place where trapped volcanic gases escape violently to the surface (Fig. 4.6). These eruptions are accompanied by the throwing out of pyroclastic material which piles up to form *ash-cones* or *cinder-cones*. Explosions of catastrophic violence are produced when dissolved gases come out of solution simultaneously through a great body of magma. The magma is converted into a kind of froth with an enormous increase in volume and is blown out of the volcano as a mass of intensely hot droplets, a *nuée ardente* or glowing cloud. The eruption of this material may empty the magma chamber and be followed by collapse of the roof, with the production at the surface of a circular hollow or *caldera* several miles in diameter.

Volcanic eruptions. The style of the *volcanic eruption* is also dependent on the viscosity of the magma, and a series of kinds of eruption related to increasing viscosity can be recognised. The *Hawaiian type* is characterised by quiet outpourings of very fluid lava. Stages of increasing viscosity, with explosions increasing in violence, are exemplified by the types of eruptions shown by *Stromboli*, *Vulcano* and *Vesuvius* and culminate in disastrous eruptions of *Peléan type*. This last type, exhibited by Mont Pelée in its eruption of 1902 mentioned above, is associated with the emission of *nuées ardentes* which burst out again and again from Mont Pelée over several months; a single nuée, taking a different course from the others, obliterated the town of St. Pierre with its 30,000 inhabitants. The 1883 eruption of Krakatao in the East Indies was also of Peléan type. Its explosions were heard 3,000 miles away and 5 cubic miles of magmatic material were blown 50 miles into the air; eruption culminated in the formation of a caldera and the collapse of portions of several islands beneath the sea.

Fig. 4.6 *A volcanic eruption: Vesuvius*

During the dying phases of volcanic activity little or no lava reaches the surface, but volcanic gases are emitted from gas-vents or *fumaroles* around which small cones or encrustations may build up.

The behaviour of volcanoes is a matter of great practical importance to the human populations living near them—and since the new igneous material gives very fertile soils, their neighbourhood is often thickly populated. As a rule, volcanoes go on behaving in much the same way over periods of at least some hundreds of years, but if the kind of magma supplying them changes, the style of eruption may be expected to change also. Volcanoes emitting lavas of low viscosity tend to erupt frequently but fairly quietly, and do not cause much damage except in the paths of the lava-flows themselves. Volcanoes fed by viscous magmas become choked between eruptions

Fig. 4.7
A flow of basaltic lava with a ropy surface

by congealed magma. A volcano of this kind exhibits a periodicity in its eruptions, a phase of activity being followed by a much longer phase of quiescence which may last for hundreds of years and lead people to conclude that the volcano is extinct. The longer the quiet period, the more violent is likely to be the ultimate explosion. The eruption of Vesuvius which destroyed the Roman town of Pompeii in A.D. 79 seems to have followed centuries of quiet, and Krakatao had been dormant for two centuries before its eruption of 1883. Warning of volcanic eruptions may be given first by relatively small explosions, by frequent local earthquakes and sometimes by measurable tilting or sagging of ground near the volcano caused by movements of the magma below.

Active or recently extinct volcanoes are found mostly in the parts of the world prone to earthquakes—that is, they are concentrated *in belts of crustal disturbance* which provide the right conditions for the generation of magma at depth. Volcanic activity is conspicuous in the 'Girdle of Fire' surrounding the Pacific Ocean and in the Caribbean and Mediterranean regions: all these lie within *mobile belts*. It occurs also along great systems of fractures which mark the African *rift-valleys* (p. 173), and rifts in the ocean floors.

5. Products of Vulcanicity

The products of superficial volcanic activity are the *lavas*, *pyroclastic rocks* and *volcanic gases*. We have seen that lavas form thin sheets or thicker tongues, lenses or domes. Their surfaces may be *ropy*, preserving the twisted shapes produced by flow of sticky lava (Fig. 4.7); *blocky* or *slaggy* surfaces are produced where the flowing lava carried on its back the broken-up fragments of its solidified crust. A special type of structure is shown by *pillow lava* made of a pile of sack-shaped masses each a foot or two long; this type is formed when magma bubbles out into water or wet mud. Lava often becomes filled, while still fluid, with bubbles of gas coming out of solution from the magma and these may be frozen into the solid lava as *vesicles*; they may be filled in later by silica, calcite or other minerals with the production of small pale nodules known as *amygdales*. Very frothy lava formed by rapid exsolution of gases solidifies as light, spongy *pumice*.

Pyroclastic rocks are made of the materials blown apart by exploding volcanic gases. They usually consist of jumbled angular pieces of the country-rocks pierced by the volcanic vents and of previously solidified lavas. Very coarse *vent-agglomerate* may choke the mouth of the vent and similar coarse fragmental material, known as *agglomerate* or *volcanic breccia*, may be heaped up around the aperture. Finer material tends to travel further from the vent and settles out to give bedded deposits of *tuff* or *volcanic ash*. As we have seen above, violent explosions may throw up to the surface portions of the liquid magma beneath the volcano. Large clots solidify in flight and fall as streamlined *volcanic bombs*. The vast clouds of magma droplets which make *nuées ardentes* give on solidification a mass of minute bubbles of *volcanic glass* (p. 51) which shatter as they fall to make a deposit of splintery glass shards: the residual heat from the magma often welds them into a compact bed of *welded tuff*. Pyroclastic rocks in general (Fig. 4.8) can be identified by the high proportion of glass and igneous minerals which they contain, and by the angularity of the fragments and mixing of pieces of different sizes which result from their mode of accumulation.

The chief *volcanic gas* is water vapour with, in addition, such gases as carbon dioxide, carbon monoxide, sulphur dioxide and sulphuretted hydrogen. These gases either escape by gas-vents or fumaroles or become dissolved in water to give *mineral springs* or *hot springs*. *Geysers* are hot springs in volcanic regions which erupt periodically as a result of the spontaneous transformation of their water into steam; the superheated water reaches a temperature at which the pressure of expanding steam is sufficient to throw out the column of water in the channel leading to the surface—thus pressure is reduced and the operation begins to repeat itself.

6. The Igneous Rocks

We now give a formal account of the igneous rocks. It is hoped that the student will have available collections or museum samples of the common igneous rocks and, if possible, corresponding thin slices—these are useful for illustrating the fabrics of the rocks. The important characters of igneous rocks are the *chemical composition*, the *mineral composition* and the *texture*—we examine these in turn.

Chemical composition. The *chemical composition* depends of course on that of the magma from which the igneous rocks are derived.

Fig. 4.8 *A pyroclastic deposit: volcanic ash with volcanic bombs*

Chapter four

Reference to the composition of the average igneous rock given on page 17 will remind us that 99% of the total bulk is made up of only eight elements. Of these, oxygen is dominant, next comes silicon and then follow aluminium, iron, calcium, sodium, potassium and magnesium. In terms of oxides, silica SiO_2 is by far the most abundant, ranging from 40% to 75% of the total. The *silica-percentage* is the basis of a four-fold chemical classification of the igneous rocks, the limits being given below:

Classification of Igneous Rocks based on silica-percentage

Over 66% SiO_2—ACID
52–66% SiO_2—INTERMEDIATE
45–52% SiO_2—BASIC
under 45% SiO_2—ULTRABASIC

Mineral composition. The *mineral composition* depends largely upon the chemical composition. The chief minerals present will naturally be silicates of the six common elements together with quartz, when silica is present in excess. Which minerals actually form will be controlled by the silica-percentage and the relative abundance of the six major cations. For instance silica-poor silicates such as olivine will be most abundant in the ultrabasic and basic rocks and absent from the silica-rich acid rocks. The chief minerals in fact are our old friends, quartz, orthoclase and plagioclase feldspars, micas, amphiboles, pyroxenes and olivines. Their distribution in the four groups established by silica-percentage is shown diagrammatically in Fig. 4.9. Many of the names given to igneous rocks are defined according to the presence of two or three particular minerals which are called the *essential minerals* for that rock-type. Other *accessory minerals* may also be present in small quantities. To give an example, the essential minerals of granite are quartz, feldspar and mica; common accessories are zircon and iron oxide.

The predominant minerals of an igneous rock often determine its general appearance and one

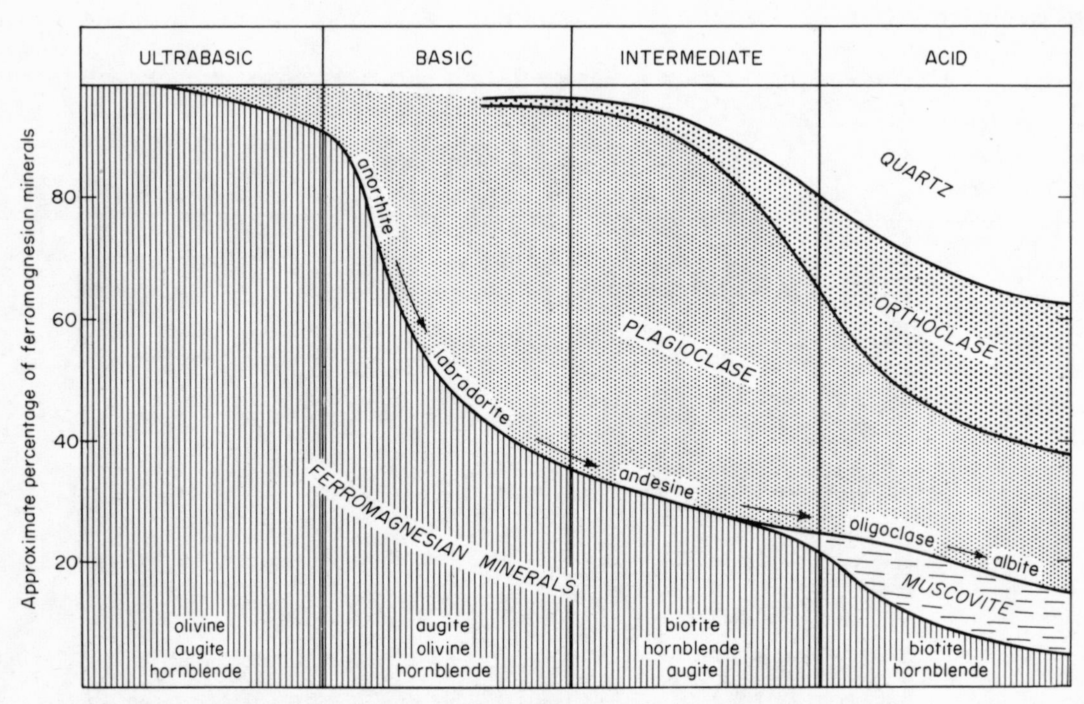

Fig. 4.9 *The chief minerals of igneous rocks*

may therefore be able to get some idea of its composition from its colour and density. Quartz, as we have seen (p. 31), is usually colourless and transparent, feldspars opaque but pale-coloured. Rocks made mostly of these minerals (that is, acid and intermediate rocks) are therefore usually pale in colour and relatively light in weight. The coloured *ferromagnesian* silicates, olivines, pyroxenes and amphiboles, are abundant in basic and ultrabasic rocks which are usually dark and relatively heavy. In using these criteria one must allow for two facts: firstly, very fine-grained or glassy rocks tend to look dark whatever their composition: secondly, weathering changes the colours of minerals and it is therefore necessary to look at freshly broken surfaces.

Texture. The *texture* of a rock is shown by the arrangement of the constituent minerals and the relation of each mineral to its neighbours. The first textural character to be observed is the *grain-size*; in a general way this depends on the rate of cooling of the magma. *Coarse-grained* rocks are the result of slow cooling which allowed time for the growth of large crystals; *fine-grained* rocks are produced by rapid cooling, as is exemplified by the chilled edges mentioned on page 44. By extremely rapid cooling, no time at all is given for crystallisation and *glasses* are formed. *Holocrystalline* rocks are entirely crystalline, *hypocrystalline* are partly crystals partly glass. The following definitions are proposed: fine-grained <1 millimetre, medium-grained 1–5 millimetres, coarse-grained >5 millimetres in grain size.

A distinctive texture is the *porphyritic* texture (Fig. 4.12) in which crystals of two different sizes occur: large *phenocrysts* are scattered through a finer-grained or glassy *groundmass*. This texture suggests that there were two stages in the cooling-history of the rock, as illustrated by the outpouring and sudden cooling of partly crystallised lava. The enclosure of many crystals of one mineral in a single crystal of another gives a *poikilitic* texture, a variety of this being the *ophitic* texture of dolerites in which large grains of pyroxene enclose a number of small tabular plagioclase crystals.

Classification. By assembling the criteria set out in the foregoing paragraphs, we can erect a classification of the common igneous rocks on the basis of grain-size and silica-percentage which is given in the table below. The characteristic minerals of rocks of different compositions are shown in Fig. 4.9 which should be studied with the Table.

CLASSIFICATION OF IGNEOUS ROCKS

Decreasing grain-size →	*Ultra-basic*	*Basic*	*Intermediate and Alkaline*	*Acid*
Fine-grained or glassy		BASALT	ANDESITE (*intermediate*) TRACHYTE (*alkaline*)	RHYOLITE
Fine- or medium-grained		DOLERITE	PORPHYRITE	PORPHYRY
Medium- or coarse-grained	PERIDOTITE DUNITE	GABBRO	DIORITE (*intermediate*) SYENITE (*alkaline*)	GRANITE

In the description of the common rocks that follows, we group them into two series:

(1) *Granite and its relatives* including syenite, diorite, porphyry and porphyrite, and andesite.

(2) *Basic rocks and their relatives* including the basic and ultrabasic types; trachyte and rhyolite, which are alkaline and acid in composition, are also dealt with in this group, although some rhyolites may be related to granite.

The reason for this grouping is provided by the way in which igneous rocks are distributed. It is found that the rocks of the two groups often (though not always) occur in different geological environments, granite and its relatives in *mobile belts* and basic rocks in *stable continental areas* and *oceanic regions*. This difference, we believe, is connected with the fact that the *primary magma* from which each group is derived originates at a different level in the earth. The primary basic magma comes from deep down, from the sima or mantle (pp. 11–12); the magmas giving the granite family come from the sialic layer of the crust which usually only grows hot enough to liquefy in the mobile belts.

CHEMICAL COMPOSITIONS OF GRANITE AND BASALT

(For mineral composition see Fig. 4.9)

	Granite (acid plutonic) *Average of 546 analyses from Daly*	*Basalt* (basic volcanic) *Average of 198 analyses from Daly*
SiO_2	70·8	49·9
Al_2O_3	14·6	16·0
Fe_2O_3	1·6	5·4
FeO	1·8	6·5
MgO	0·9	6·3
CaO	2·0	9·1
Na_2O	3·5	3·2
K_2O	4·1	1·5
TiO_2	0·4	1·4
P_2O_5	0·2	0·4
MnO	0·1	0·3
	100·0	100·0
Density	2·67	3·0

7. Granite and its Relatives

Granite. One of the two commonest igneous rocks, familiar to everyone as a building or monumental stone, is granite. It is a coarse-grained, light-coloured rock, usually white, pink or grey, consisting essentially of quartz, alkali-feldspars and micas all of which can be recognised in hand specimen (Fig. 4.10). The minerals seldom form well-defined crystals but appear as irregular grains. *Porphyritic granites* contain feldspar crystals very much larger than the groundmass. *Graphic granites* show an intergrowth of quartz and feldspar, attributed to simultaneous crystallisation of the two minerals, which gives interlocking patterns resembling ancient cuneiform writing.

Closely associated with granites in the field are dykes and veins of very coarse-grained pale rocks known as *pegmatites* which represent the last portions of the granitic magma to consolidate. In these portions are concentrated volatiles containing the rarer elements of the magma. Pegmatites therefore often carry, in addition to the common quartz, alkali-feldspars and micas, large crystals of uncommon minerals such as tourmaline, beryl and topaz (Fig. 4.10).

Granite

Pegmatite

Fig. 4.10

Fig. 4.11
Migmatite: the host-rock is darker and finer-grained than the granitic partner

Granite, as its coarse texture indicates, crystallises below the surface in intrusive bodies often 5 miles or more in diameter. Many of these bodies are clearly emplaced by the mechanisms of forceful or permitted intrusion (p. 45). In addition, enormous granitic bodies, known as *batholiths*, may sometimes be formed *in situ* by melting of a considerable volume of crustal material. In the deep parts of mobile belts, and around some granitic intrusions, granitic material becomes intimately mixed with the country-rocks, giving the impression that it has soaked into these rocks. Mixed rocks of this kind (Fig. 4.11) are known as *migmatites* (Greek *migma*, mixture).

Diorite and syenite. The other coarse-grained rocks of this group are much less common than granite. *Diorite* is a darker rock, containing an amphibole or pyroxene together with a plagioclase feldspar. *Syenite* resembles granite but contains little or no quartz; its dark mineral may be biotite or an amphibole or pyroxene. It is said to be an *alkaline rock* because its feldspars are sodic or potassic varieties rather than the calcic plagioclase.

Porphyry and porphyrite. Minor intrusions of acid or intermediate composition are represented by the *porphyries* and *porphyrites*. As their names suggest, these rocks are usually porphyritic and contain large crystals of feldspar or quartz in a fine-grained groundmass. *Quartz-porphyries* show phenocrysts of quartz and orthoclase (often with obvious crystal form) in a fine groundmass of granitic minerals. *Porphyries* have phenocrysts of orthoclase and *porphyrites* of plagioclase, together with some of hornblende, augite or biotite.

Andesite. Andesites are fine-grained volcanic rocks of intermediate composition and are the most common lavas erupted by volcanoes in mobile belts. They may be derived from melted sial or from basic magma contaminated by sial. They are darkish, dull-looking rocks often containing phenocrysts of intermediate plagioclase, augite or hornblende set in a groundmass too fine-grained to reveal its crystalline texture to the naked eye; with a lens or microscope, this groundmass is seen to be made of minute grains of plagioclase together with augite, hornblende or biotite (Fig. 4.12).

8. Basic Rocks and their Relatives

We now describe the second series of igneous rocks which is derived, most probably, from a *primary basic magma* generated by melting of parts of the sima or the mantle (pp. 11–12). The straightforward crystallisation of basic magma under differing crustal conditions gives three very common rocks—*basalts* formed at or near the surface, *dolerites* formed in minor intrusions and

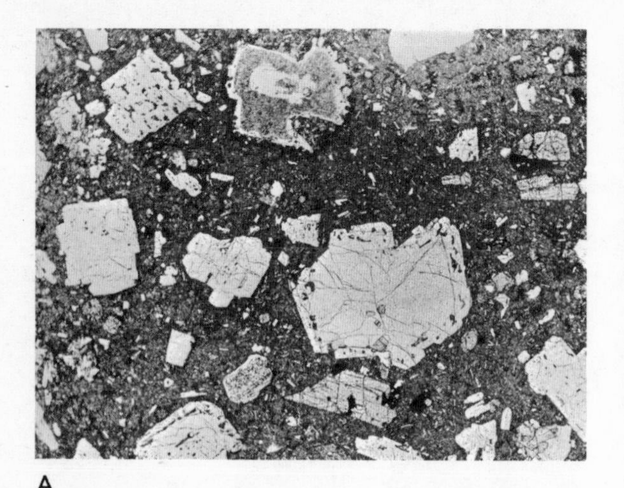

Fig. 4.12 *Lavas*

A. Porphyritic andesite: thin slice showing phenocrysts of plagioclase feldspar
B. Trachyte showing trachytic texture

gabbros produced in large intrusive bodies (Fig. 4.13). With these basic rocks are often associated small amounts of ultrabasic, intermediate or acid igneous rocks which are derived either from modified fractions of the basic magma or from sialic material melted by contact with the very hot basic magma, or from the digestion of sial by basic magma (see andesites above).

Basalt, dolerite and gabbro. *Basalt* (see Table, p. 51), occurring as lavas and small intrusions, is a dark heavy dull-looking rock, often porphyritic and occasionally amygdaloidal (Fig. 4.13A). The essential minerals are anorthite-rich plagioclase feldspar and augite. Olivine-basalts contain olivine in addition. The basalts are fine-grained, occasionally show a little glass and often

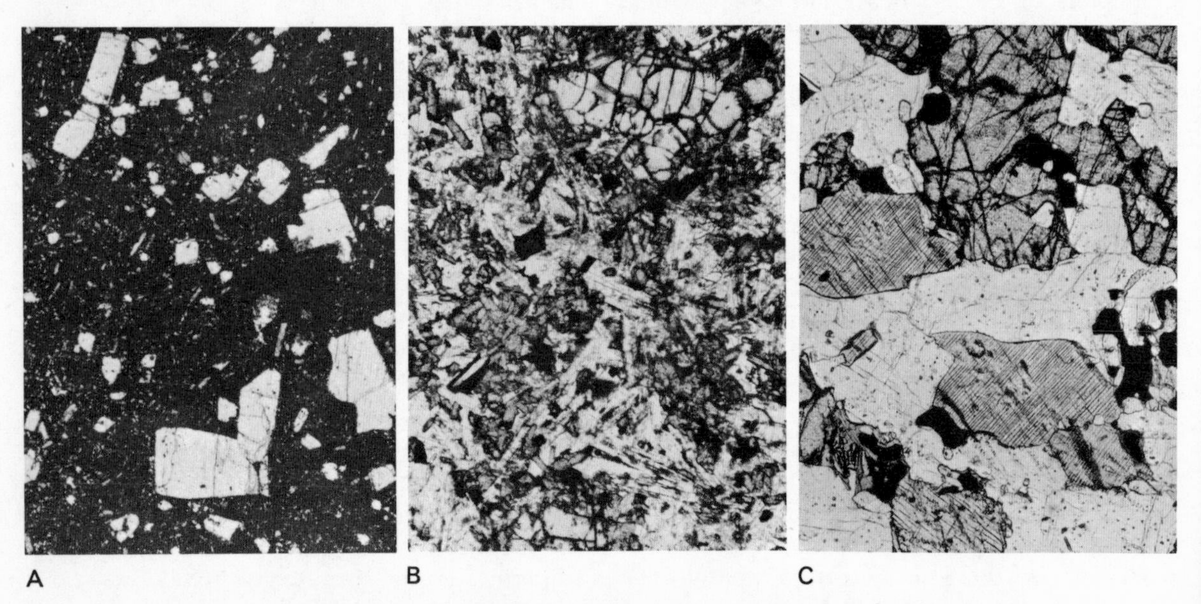

Fig. 4.13 *Basic igneous rocks: three thin slices at the same magnification showing characteristic texture of basalt* (A), *dolerite* (B) *and gabbro* (C)

carry phenocrysts of feldspar, olivine or augite. *Dolerite* (Fig. 4.13B) is closely related to basalt and forms dykes and sills. It is a dark heavy rock in which the essential constituents, plagioclase and augite, sometimes accompanied by olivine, can be distinguished with a hand-lens. Its texture is holocrystalline and sometimes porphyritic; *ophitic texture* is common. *Gabbro* (Fig. 4.13C) is the plutonic equivalent of basalt and dolerite and accordingly is often coarser in grain and never porphyritic. It contains basic plagioclase and augite, with olivine in the olivine-gabbros. A variety called *troctolite* is composed of plagioclase and olivine. Gabbro often occurs in very thick intrusive sheets, such as the colossal *Bushveld complex* of South Africa.

When a basic magma cools, the early-formed silicate minerals, olivine and pyroxene, sink owing to their high specific gravities. In this way accumulations of such minerals may pile up on the floor of the intrusions to give *ultrabasic rocks*. At the base of thick gabbro sheets, layers of *dunite* (olivine rock) or *peridotite* (olivine rock with some pyroxene) are often found. These rocks are dark and heavy, and of rather uniform grain-size: they are often converted into *serpentine*.

Trachyte and rhyolite. The two other extrusive rocks, *trachyte* and *rhyolite*, that we consider with this family are less abundant than basalt and are not themselves basic in composition. They are closely associated with basalts and in some regions have most likely been produced from the primary basaltic magma as a residuum after the removal of the heavy early-formed minerals. Since these early-formed minerals contain much magnesium and calcium, the residual portion of the magma becomes enriched in silica and alkalies, so that the rocks formed from it are *alkaline or acid* in composition. Thus a primary basic magma can be *differentiated* to give ultrabasic, alkaline and acid rocks as well as basic rocks. But some rhyolites may have been formed by the melting of granitic or sialic material.

Trachytes are light-coloured alkaline rocks composed essentially of an alkali feldspar and a dark mineral, augite, hornblende or biotite. They are commonly porphyritic and may have a partly glassy groundmass which often shows a *flow texture* produced by the alignment of tiny feldspar laths—this texture is often called the *trachytic texture* (Fig. 4.12B).

Rhyolites are acid lava-rocks, often glassy since acid magma is viscous and crystallisation is therefore impeded; they are sometimes associated with welded tuffs formed by explosive disruption of the same viscous magma. Quartz and alkali feldspars occur as phenocrysts in a glassy groundmass which shows good *flow-structure* due to colour-banding (Fig. 4.14). Entirely glassy varieties are *obsidian*, a shiny black rock, and *pitchstone*, a resinous-looking variety. The groundmass of non-glassy varieties is formed by minute grains of quartz and feldspar—this is the *felsitic texture* and rocks showing it are called *felsites*.

Fig. 4.14
Flow-banded rhyolite showing contortions produced by flow of viscous magma

5

SURFACE PROCESSES: 1

Introduction, weathering, groundwater

This chapter makes a start on the study of geological processes taking place at the earth's surface, which provides a basis for a uniformitarian approach to the problems of sedimentary rocks. The surface is seen to be shaped by the processes of denudation and deposition through the agencies of water, ice, wind and gravity. The mechanical and chemical processes of weathering are discussed, and the residual deposits produced by them described. The movements of groundwater are seen to lead to the formation of springs and artesian basins. The chemical effects of groundwater above and below the water-table are dealt with and finally its mechanical effects in facilitating soil-creep, landslides and other movements of surface material are mentioned.

1. Introduction

At the surface of the solid earth or *lithosphere* the rocks are exposed to the influences of air and water and, since they are not equally supported on all sides, to the effects of gravity. These three forces, working separately or together, modify and break up rocks already in existence and transport and re-assemble the debris to produce new sediments. These processes going on at the earth's surface are of special importance to the geologist, firstly, because they provide a chance to examine the making of one class of rocks, the *sedimentary rocks*, and, secondly, because they are responsible for moulding the forms of the land-surface and the sea-floor. By applying the doctrine of uniformitarianism, we can therefore use observations of surface-processes to interpret the history of ancient sedimentary rocks and to understand the evolution of *land forms*.

The surface of the lithosphere has acquired its topography as a result of two operations: the first is *denudation* or *erosion* or wearing-away, the second is *deposition* or building-up. The two processes are really complementary, since the debris produced by denudation provides most of the material available for deposition. In broad terms, denudation is dominant on the lands and deposition in the sea.

The *agents of erosion* are rain, rivers, wind, waves and ice which, acting in conjunction with the force of gravity, break down and remove the outer parts of the rocks, thus continually exposing new surfaces to attack. The eroded material is carried away by *transporting agents*—usually water or wind again in some form—and when these agents cease to act it comes to rest. The *agents of deposition* are therefore essentially the same as the agents of transport. The connection between transport and deposition can easily be confirmed by watching the movements of grit and dust on a road surface after heavy rain. The rain, gathering into trickles and rills, washes the loose grit towards the lowest points of the road. Temporary stoppages may occur where the water enters a puddle but, sooner or later, the transported material arrives at the hollows in the road or the gutters alongside and is spread out as small beds of sediment. The process of deposition is known as *sedimentation*.

No surface agent works by itself but, for the purposes of description, it is easier to take them one by one and to consider the effects produced by each agent in turn. We begin by studying how the rocks react to changes of temperature, and to frost and rain—that is, how they *weather*. Next we deal with the action of water which seeps down beneath the surface to provide *groundwater*. Surface waters of one kind or another provide *rivers*, *lakes* and *seas* and, in different conditions, *snowfields*, *glaciers* and *ice-sheets*. These agents are dealt with in turn and finally moving air, or *wind*, is considered. Each surface-agency produces different erosional effects and finally brings about the deposition of sediments of different facies.

2. Weathering

Inspection of ancient buildings or of gravestones in an old churchyard will show that stones decay when exposed to the weather (Fig. 5.1). Natural rocks behave in the same way and the rotted material, if it is not removed by transporting agents, may extend down to depths of tens or even hundreds of feet. A new road-cutting will often reveal an upward transition from fresh solid rock through partly disintegrated material into the surface soil. This kind of reaction which takes place where rocks meet the atmosphere constitutes the process of *weathering*. It represents a *response to a change of environment* according to the principle discussed in Chapter 1: the minerals and textures of many rocks were developed at high temperatures or at considerable depths and, consequently, are not in equilibrium under the conditions prevailing at the surface. The changes brought about by weathering tend to re-establish equilibrium in the physical and chemical environment in which they take place. By this means, new minerals arranged in new ways are produced.

The breakdown of old minerals and development of new ones involve chemical reactions and constitute the process of *chemical weathering*. In addition, the surface-agents may bring about *mechanical weathering* by which the rock is simply disintegrated into smaller and smaller fragments without much chemical change. In most climatic conditions the processes of chemical and mechanical weathering co-operate—disintegration increases the surface area of material exposed to the atmosphere and so speeds up chemical weathering, while chemical reactions frequently produce bulky new minerals whose development helps to disrupt the rock mechanically. But in some climates one or other process is dominant and the characters of the weathered products are therefore dependent on climatic conditions.

Fig. 5.1 *Chemical weathering: a weathered stone figure, no more than a hundred years old, on Bristol Cathedral*

Fig. 5.2 *The screes of Wastwater*

3. Mechanical Weathering

In considering the mechanical breaking-up of the surface rocks we shall leave on one side the part played by the active geological agents such as wind, rivers, the sea or glaciers which are dealt with in later chapters, and concentrate for now on the disintegration resulting from (*a*) change of temperature, (*b*) the ice-wedge, (*c*) the action of plants and animals and (*d*) the crystallisation of salts.

Effects of change of temperature. The coefficients of expansion differ in different minerals and, also, in different directions within any one mineral (except, of course, for those of the Cubic System). When coarsely crystalline rocks such as granites are exposed to rapid changes of temperature, the expansion and contraction of the minerals set up complex strains that are considered to be sufficient to develop intergranular cracks that may cause the rock to crumble. It may be as well to state that such a process has not been experimentally demonstrated.

The ice-wedge. Ice floats in water showing that water expands and becomes less dense on freezing. As a consequence, water freezing in the cracks and pores of a rock will exert great pressures, perhaps as much as a ton to the square inch, on the confining surfaces. Fragments are rifted off and grains forced apart. This frost action, referred to as the *ice-wedge*, is a powerful agent of weathering on high mountains and is responsible for their jagged outlines. The blocks and fragments wedged-off slide down the mountain slopes to accumulate as *scree*, great piles of unstable debris illustrated by the screes of Wastwater in the Lake District (Fig. 5.2). Building stones for use in cold climates have to be tested for their behaviour under repeated freezing and thawing.

Action of plants and animals. Plants cause mechanical weathering by the widening of crevices by their growing roots, as can be observed in hedgerow ditches and old walls. Burrowing animals loosen and mix the soil; earthworms may seem trivial as agents of weathering, but Darwin showed that an enormous amount of work is done by these creatures. The excavations of man—mines, quarries, road and railway cuttings—are important causes of mechanical weathering. On the other hand, a thick plant cover often provides protection against disintegration by binding particles together and keeping off rain and wind.

The crystallisation of salts. This process may take place in fractures and between grains in desert regions where saline waters rise to the surface. The growth of such salts, like the growth of ice crystals already mentioned, exerts pressure on the rock and may assist disintegration.

4. Chemical Weathering

Rain falling through the air takes up oxygen and carbon dioxide, CO_2, as well as other less important atmospheric gases. On reaching the ground it takes up material produced by the decay of plant and animal matter and so forms weak organic acids. As rain-water soaks through the surface layers of disintegrated rock all these substances in solution can react with the rock-minerals to bring about chemical changes. The chief processes involved in this *chemical weathering* are solution, hydration (formation of compounds containing water or OH), oxidation, reduction and the formation of carbonates. We can illustrate some of their effects by considering the weathering, first, of igneous rocks and, secondly, of limestones.

Chemical weathering of igneous rocks. Since the raw materials of the sediments are derived ultimately from igneous rocks it is essential to understand how these rocks react to weathering. For this purpose, we can consider their constituent minerals as belonging to three groups—*quartz*, the *feldspars* and the *ferromagnesian minerals*, the last-named including the FeMg-bearing micas, amphiboles, pyroxenes and olivines. Each group reacts to weathering in a different way. *Quartz*, with a compact structure and stable composition, goes through the weathering process unchanged. It is mechanically broken up and provides the principal constituent of sand. The *feldspars*, silicates of aluminium, the alkalies and calcium, are chemically more complex and break down relatively easily. The strong bases, soda, potash and lime, are converted into carbonates and go into solution, whilst the alumina and silica combine with water to give hydrated aluminium silicates, the *clay-minerals*—these being the new minerals stable under the conditions of weathering (p. 7). The *ferromagnesian* minerals are the least resistant to chemical change. They are converted into clay-minerals, colloidal silica and carbonates of magnesium, calcium or iron. The iron carbonate so formed is oxidised to give the red oxide hematite which itself is hydrated to form the hydroxide limonite.

In summary, then, weathering of igneous rocks results in the formation of resistant quartz, colloidal silica, clay-minerals and iron hydroxides together with carbonates of the alkalies and alkaline earths which are carried off in solution.

Weathering does not take place regularly from above downwards. Many igneous rocks are divided into large rectangular pieces by secondary fractures or *joints* that provide passages along which watery solutions can penetrate far below the surface. Chemical weathering then works inwards through the joint-blocks and, as reaction is especially rapid at the corners, the blocks are rounded-off to produce great spheres. The final product of *spheroidal weathering* of this kind is a deep mass of rotted rock in which are set unmodified spherical cores surrounded by shells of partly altered material (Fig. 5.3).

The more general process of *exfoliation* produces a similar effect on a larger scale. Rocks near the surface develop concentric fractures which separate them into parallel slabs like gigantic onion-skins. The flaking-off of these slabs leaves a curious landscape of *dome-shaped hills* or *inselbergen* (German, 'island hills') such as are seen around Rio de Janeiro and in Bihar,

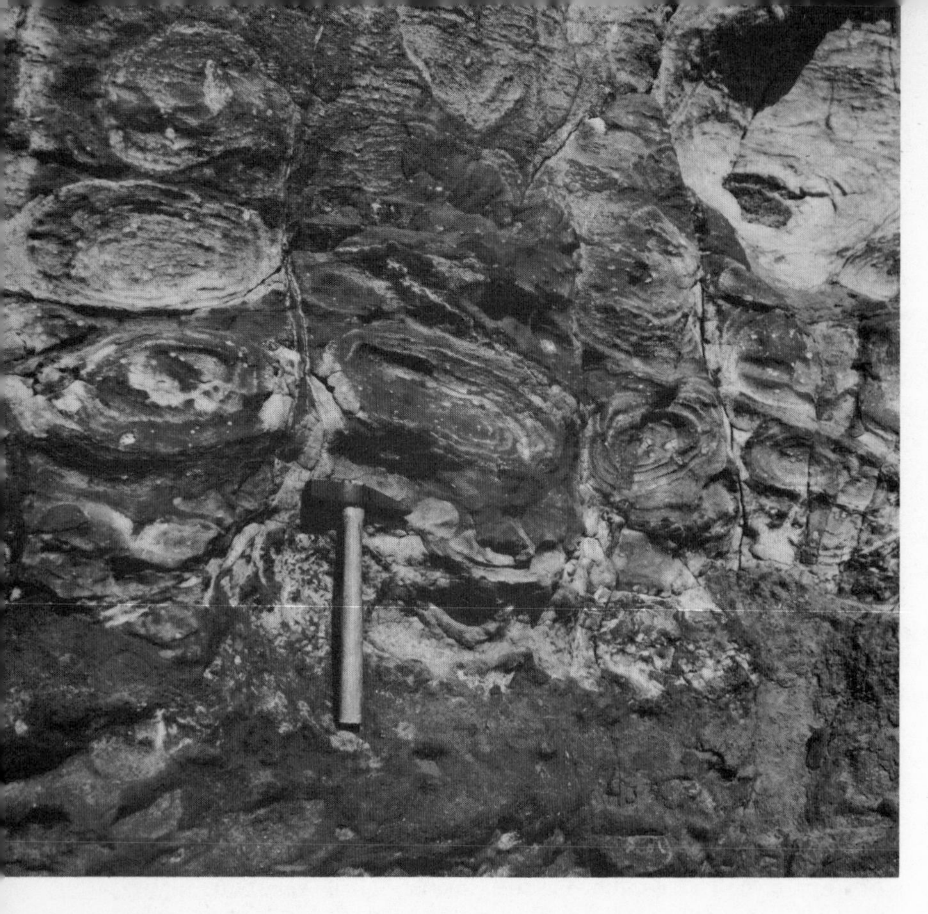

Fig. 5.3 *Spheroidal weathering of a basic rock, Irish Harp, Antrim*

India. The causes of exfoliation include chemical weathering, combined with the effects of temperature changes and the relief of pressure as overlying rocks are removed.

Weathering of limestone. We now examine the weathering of the common sedimentary rock *limestone* which is made dominantly of the mineral calcite, calcium carbonate, $CaCO_3$. Rainwater, charged with carbon dioxide from the air, reacts with limestone to form the soluble bicarbonate, $Ca(HCO_3)_2$. As solution goes on, the rock surfaces become etched and fretted; joint-fractures traversing the rock are widened and deepened (Fig. 5.4) to produce *grikes*, deep gashes arranged roughly at right angles and separating flat paving-stone-like portions of the original surface known as *clints*. Solution is especially active at the intersection of joints and solution-holes, called *swallow-holes*, *pots* or *dolinas* may be formed there. Surface streams may cascade into such pots and by erosion or solution make underground courses for themselves: these courses are enlarged into *caves*. Limestone regions may be honeycombed with such caves, which provide a system of underground drainage: the surface waters are drawn into this system leaving *dry valleys* behind them.

These land-forms developed in limestone regions are so characteristic as to have received a special name, *Karst topography*. They are well developed in the Karst region of Yugoslavia and in the Pennine area of West Yorkshire.

Belts of weathering

Over the world as a whole, *regional variations* in the style of weathering are *controlled by the climate*, especially by the rainfall and the temperature; and these regional variations provide one of the reasons why the scenery of tropical or desert lands looks so unfamiliar to people of temperate countries. *Chemical weathering* takes place most readily in hot, humid climates; it is slowed down by low temperatures and almost stopped by lack of water. *Mechanical weathering* is favoured in regions of rapid temperature-change and of frost action and therefore predominates in sub-polar regions, in deserts and in mountain areas. On this basis, we can recognise the *belts of weathering* of the present day as:

Sub-polar: mechanical weathering, largely frost-shattering.
Temperate: mechanical and chemical weathering acting together.
Desert: mechanical weathering, mainly due to temperature-change.
Tropical: deep chemical weathering, favoured by high rainfall and high temperatures.

5. Residual Deposits

If the weathered covering remains substantially where it was formed it makes a *residual deposit*—the residue of the parent rock. The nature of most residual deposits will be apparent from what has been said about weathering and the student should work out for himself what deposits to expect in different climates. We can consider here two special types, *laterite–bauxite* and *soils.*

Laterite and *bauxite* are formed by chemical decay carried to extremes in the humid tropics. They are clays enriched in iron hydroxide and aluminium hydroxide respectively. They form valuable mineral deposits, bauxite being the only ore of aluminium.

Soils are the most valuable and widespread of the residual deposits. They are formed by the prolonged activity of organisms, organic material, water and air in the upper parts of the weathered zone. Weathering releases from the rock-forming minerals essential elements in forms which can be taken up by plant roots; by degrees, the growth and decay of plants add organic matter, bacteria break this down into new compounds and the general activities of plants, animals and bacteria loosen and lighten the material to produce the crumbly texture of soil. The nature of the soil depends fundamentally on the climate, especially on the rainfall and temperature range—this is shown in the arrangement of the *soil belts* of the world.

Undisturbed soil shows a series of layers constituting the *soil-profile*—take advantage of

Fig. 5.4 *Grikes in Carboniferous Limestone, Yorkshire*

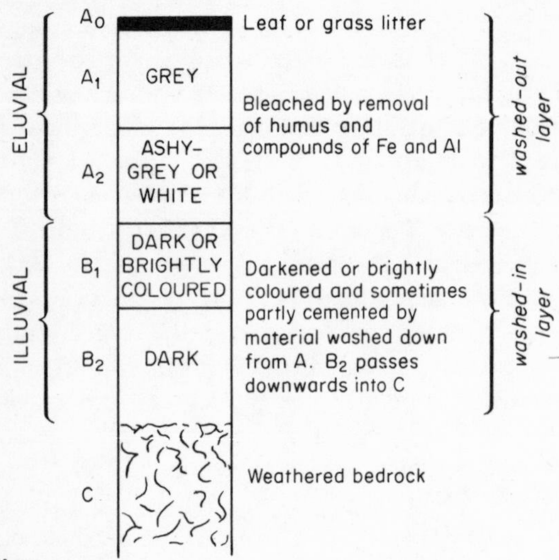

Fig. 5.5
Soil profile of a type common in temperate regions

road-cuttings to examine soil-profiles. In a moist temperate climate like Britain's, three layers are visible in a vertical section (Fig. 5.5). The upper layer, the *A* or *eluvial* (washed out) horizon, is light-coloured, the humus, clay and iron and aluminium compounds having been carried down into the underlying *B* or *illuvial* (washed in) layer. This *B* layer is brightly coloured by iron hydroxides or darkened by humus. The *C* layer below is made simply of the weathered underlying rock. Ancient soils may be represented by the *seat-earths* which underlie coal-seams (p. 208).

6. The Geological Work of Underground Water

In the remainder of this chapter and for the next few chapters we shall be considering the geological work of one particular surface-agent, *water* in all its many forms. This water, whether as rain, streams, ice or the sea, is constantly on the move and its importance is largely due to its mobility. Its migrations in and near the surface are presented schematically in Fig. 5.6.

The migration of groundwater

Rainfall. The rain that falls upon the earth's surface can behave in three ways; some runs off into streams and rivers, some sinks into the ground and some is returned to the atmosphere by evaporation. These three portions of the rain-water are called the *run-off*, the *percolation* and the *evaporation-loss*. The ratio between them depends on a number of obvious factors that the student should assemble for himself. Of the total rainfall, perhaps a half is lost by evaporation, and the remainder is about equally shared between the run-off and percolation. What happens to the run-off is dealt with in the next chapter—our present concern is with the percolation, the portion that enters the ground as *groundwater*.

Porosity and permeability. Two properties of the surface rocks control the movement and storage of groundwater and, incidentally, of other

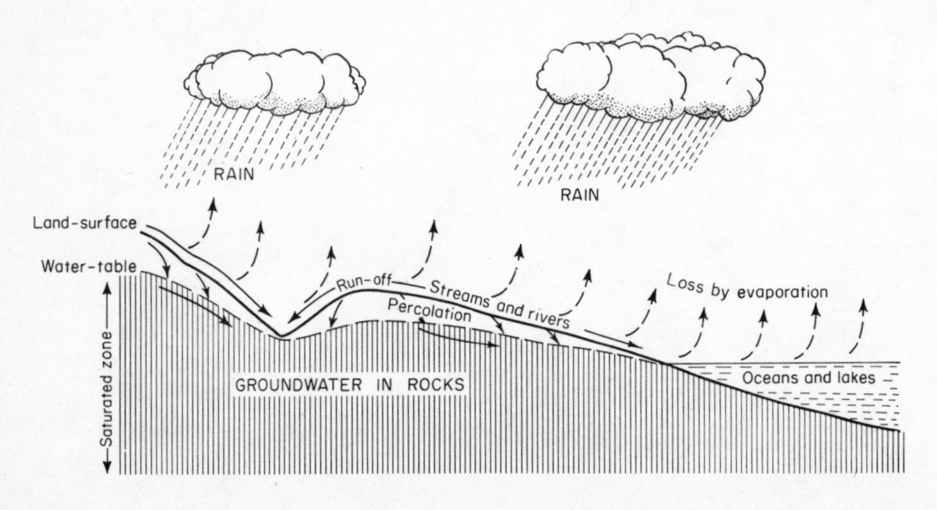

Fig. 5.6
The circulation of water in the outer part of the earth

liquids such as oil. These are *porosity* and *permeability*, by no means the same thing. *Porosity* is the ratio of the spaces (the *voids*), not occupied by solid rock to the bulk volume of the sample; an average porosity is 10%. The voids include minute spaces between grains as well as various kinds of fissures and crevices in the rock. *Permeability* is the capacity of a rock for transmitting fluids; it depends on the size and shape of the pore-voids and the presence or absence of fissures such as joints and other parting planes. For example, chalk has only minute pores, but transmits water quickly along its joints and therefore has a high porosity and a high permeability; clay has a high porosity but, because its pores are minute and unconnected, its permeability is very small. These differences can be appreciated by comparing the look of different parts of the country after prolonged rain—land with a permeable sandy foundation is quickly drained, while clay lands become waterlogged.

So far as the movement of water or oil is concerned, therefore, there are two kinds of rocks—*pervious* and *impervious*. *Pervious rocks* allow water to pass through them and are exemplified by gravels, sands and sandstones and by jointed limestones and other fissured rocks. *Impervious rocks* act as barriers to the movement of groundwater; they include clays, shales, slates and compact unjointed rocks. Any bed carrying and transmitting water is termed an *aquifer*.

The water-table. At some level below the surface the rocks are saturated with water seeping down from above. The upper surface of this saturated zone is called the *water-table*. Its position can be found by observing the level at which water stands in wells or borings (which can be thought of as extra-large voids) and if this level is recorded for many wells it is possible to make a map of the water-table. The topography of the water-table varies with that of the land—in flat country it is flat, in hilly country it undulates but with more gentle curves than the land-surface (Fig. 5.7). The height of the water-table (that is, its nearness to the surface) varies with the rainfall, with a time-lag dependent on the permeability of the surface rocks. In some places and at some times of the year the water-table meets the surface and we then find swamps, springs, rivers or lakes.

Groundwater carries out different activities above and below the water-table. Above the water-table, it is mobile and chemically active: this is the *zone of alteration* or *aeration*. Below it, the water moves only sluggishly and begins to deposit substances held in solution: this is the

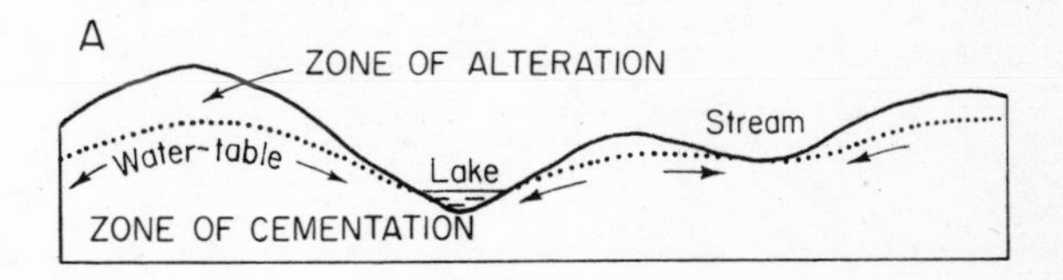

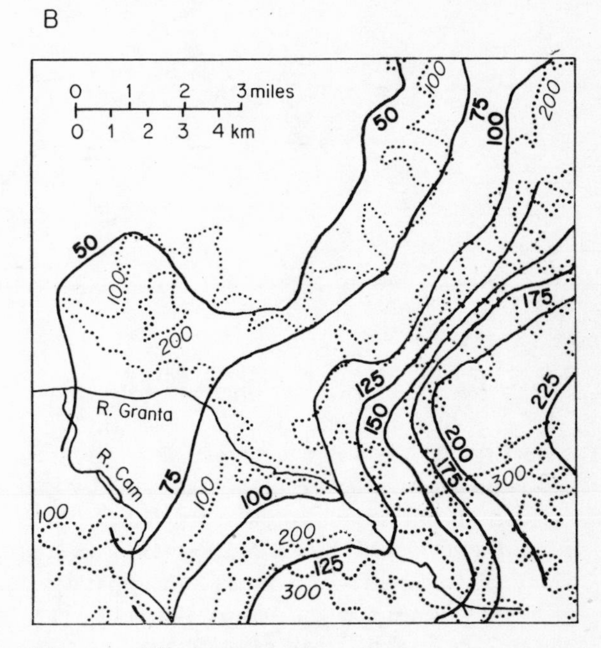

Fig. 5.7 *The water-table*

A. Diagram showing the relationship between the water-table and the topography

B. Contoured map of the water-table (bold lines) in the Chalk south-east of Cambridge showing relationship with topographical contours (dotted lines). Heights in feet (after Bernard Smith)

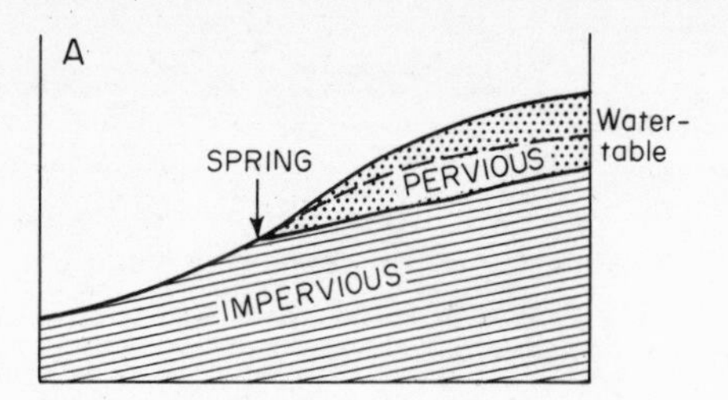

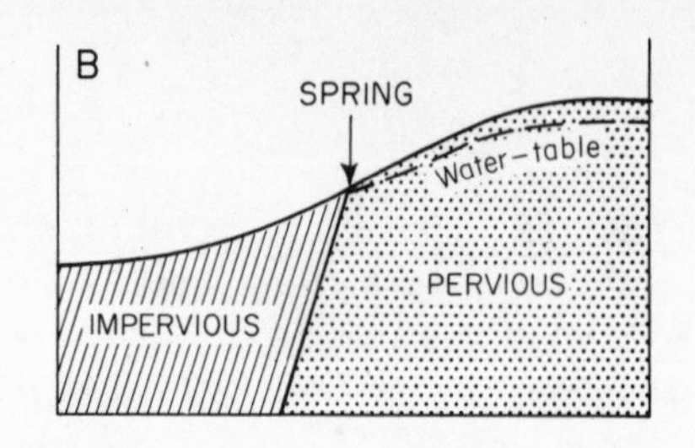

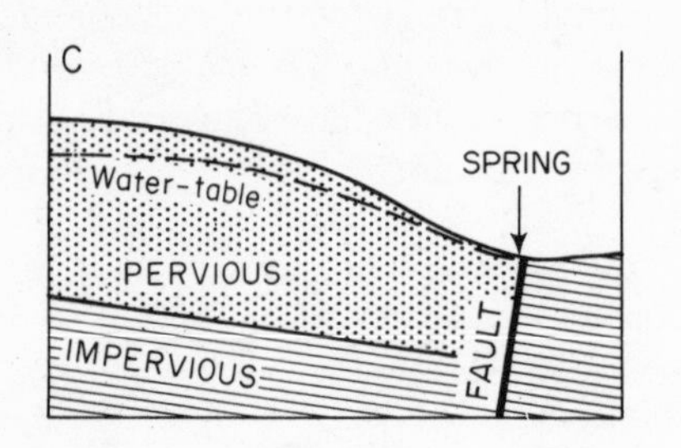

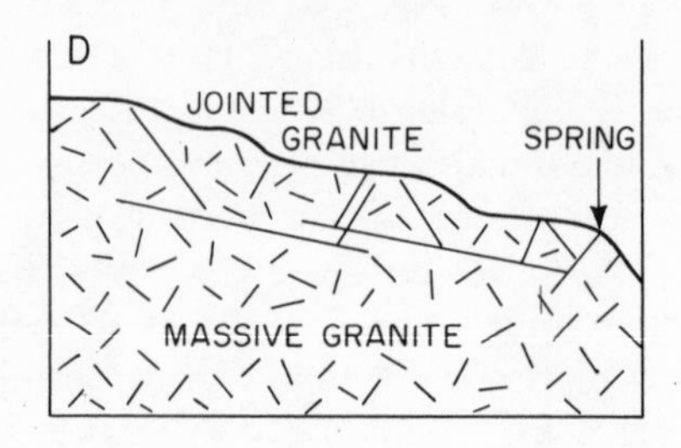

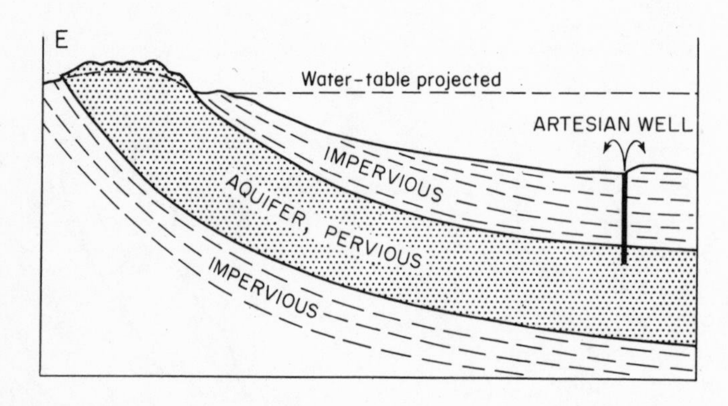

Fig. 5.8
Springs
A. Stratum-spring
B. Overflow-spring
C. Fault-spring
D. Joint-spring
E. Conditions for development of artesian wells

zone of cementation where pores and fissures are filled in with solid material.

Springs. When water contained in a pervious bed is prevented from seeping downward by the presence of impervious rocks in its path it flows out at the surface of the ground to make a *spring*. Many arrangements of pervious and impervious rocks give rise to springs; several types, depending on the interbedding of pervious and impervious strata and on the effects of tilting, folding, faulting and jointing are illustrated in Fig. 5.8 and the reader should be able to think of others.

Artesian conditions. If an aquifer lying between two less permeable beds is tilted, the water in the lower parts of the aquifer is put under a head of hydraulic pressure and an *artesian slope* is established. If a well is put down, this water spouts up to give an *artesian* or *flowing well* (Fig. 5.8). Two opposed artesian slopes unite to make an *artesian basin*. London is built on an artesian basin in which the main aquifers are the jointed and fissured Chalk and the porous Greensand, each sandwiched between impervious layers of clay (Fig. 5.9). In former times, wells put down to the chalk through the clay tapped water under hydraulic pressure; the fountains of Trafalgar Square were supplied by this means. Nowadays, however, London uses so much water that the water-table has been lowered, water has to be pumped up and dirty Thames water is being drawn into the wells: much of the city's supply now comes from rivers and reservoirs.

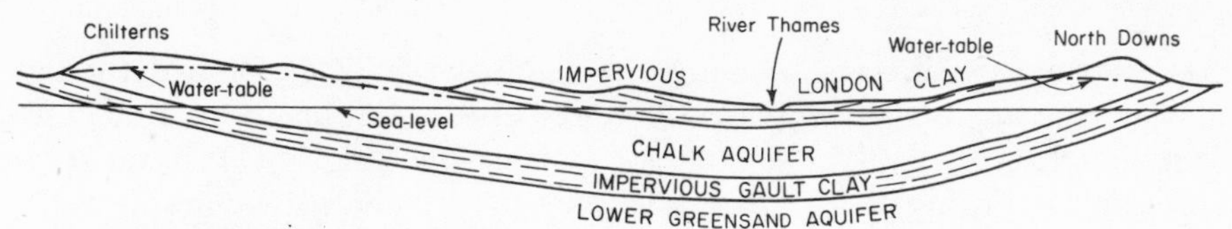

Fig. 5.9 *The artesian basin of London: length of section about forty-five miles, vertical scale exaggerated*

Chemical action of groundwater

The chemical effects of groundwater on the rocks through which it passes can be grouped under three headings—*solution*, *replacement* and *deposition*.

Solution takes place especially above the water-table in the zone of alteration. We have dealt with its effects on limestones and igneous rocks in the section on weathering (pp. 59–60). Widened joints, caves and underground stream-courses are the most spectacular results of solution. The groundwater itself becomes charged with such salts as bicarbonates, sulphates and chlorides of calcium, sodium or potassium: in other words, it becomes *hard* and produces a scummy deposit with soap. *Temporary* hardness is due to the presence of bicarbonate and can be removed by boiling (with the production of a scaly deposit in kettles and pipes); *permanent* hardness is due to sulphates and cannot be got rid of without chemical treatment. Groundwater may be useless for drinking or irrigation purposes if it carries a high content of sulphates or sodium chloride obtained by solution from the rocks of the aquifer. Thus, the water obtained from the great Australian artesian basin is just drinkable but cannot be used for irrigation.

Replacement is achieved by the exchange of certain substances carried by the groundwater for others which it extracts from the rocks. Thus fossil shells made of calcite, $CaCO_3$, may be replaced by silica, SiO_2, or by pyrites, FeS_2, or

Fig. 5.10 *Stalactites and stalagmites in a limestone cave*

siderite, $FeCO_3$. Silicified fossil trees are sometimes found in which the woody tissue is perfectly replaced by opal, $SiO_2.Aq$.

Deposition takes place mainly in the zone of cementation. In the caves of limestone regions, deposition of calcite may be brought about by the breaking-up of the soluble bicarbonate, $Ca(HCO_3)_2$, on loss of carbon dioxide. Successive coatings of calcite left by water dripping from the roof give pendent *stalactites* and surplus solutions falling to the floor cause *stalagmites* to grow upwards, the two often meeting to give pillars of calcite (Fig. 5.10). Calcite is also often deposited as *tufa* or *travertine* from springs in limestone regions and forms encrustations on objects placed in the spring waters—these are the so-called 'petrifying springs'. *Chalybeate* springs deposit hydrated iron oxides or *ochres*. Around the orifices of *geysers* in volcanic regions (p. 49) thick deposits of silica may be built up—this material is known as *siliceous sinter*.

In some conditions, groundwater may take up material evenly distributed through a rock and re-deposit it around some kind of nucleus to form a *concretion* (Fig. 5.11). Examples of concretions are pyrites (FeS_2) nodules in chalk, and impure calcareous *septaria* in clays, as in the London Clay. *Flint* ($SiO_2.Aq$) occurs as nodular concretions in the Chalk of the south of England.

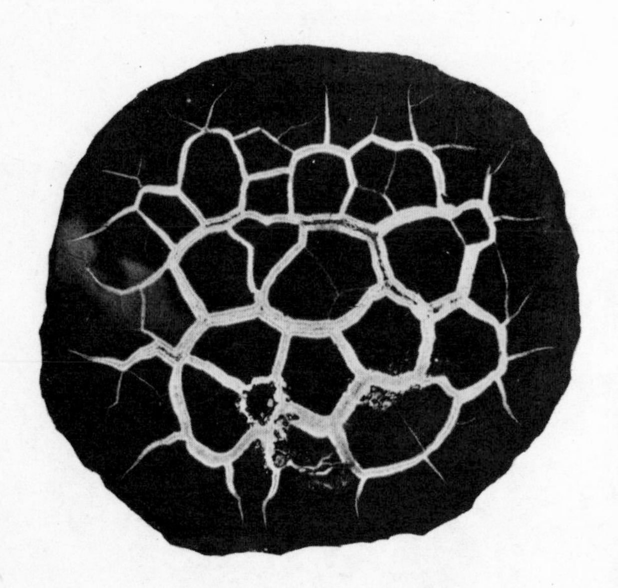

Fig. 5.11 *Concretion: a septarian nodule made of impure calcium carbonate cut to show calcite veins filling shrinkage cracks*

An important operation going on below the water-table is *cementation*, the deposition of mineral matter between the grains of the rocks. The durability of many rocks depends as much on the cement as on the constituent grains. Common cementing materials are *silica*, *calcite* and *iron oxides*. The deposition of a cement decreases the pore-space and so may control the amount, the localisation and the movement of liquids such as water and oil in the rocks.

Mechanical action of groundwater: mass-movements

Rocks saturated by groundwater may be sufficiently lubricated to move or break up under the influence of gravity. Some of these movements are very slow, some extremely rapid. The *slow mass-movements* are exemplified by various forms of *creep*. Loosened weathered material moves slowly downhill by *soil-creep*, indicated by tilted fence-posts and especially by trees which, in an endeavour to keep themselves upright, often show a knee or bend convex down the slope (Fig. 5.12A). By such downhill movements suitably placed beds of partly weathered bedrock may be turned over in the direction of creep; this is called *terminal curvature*. Weathered material obtained, for example, from a mineral vein high up the slope moves down lower, a movement that the prospector has to allow for. *Solifluction* is the slow flowage from higher to lower ground of masses of rock-waste saturated by water; by this process, particularly effective in polar regions of permanently frozen ground, alternating layers, of sand and gravel for example, may become completely disordered and intermixed.

The *rapid mass-movements* of rocks under gravity are of two kinds, *flows*, in which continuous

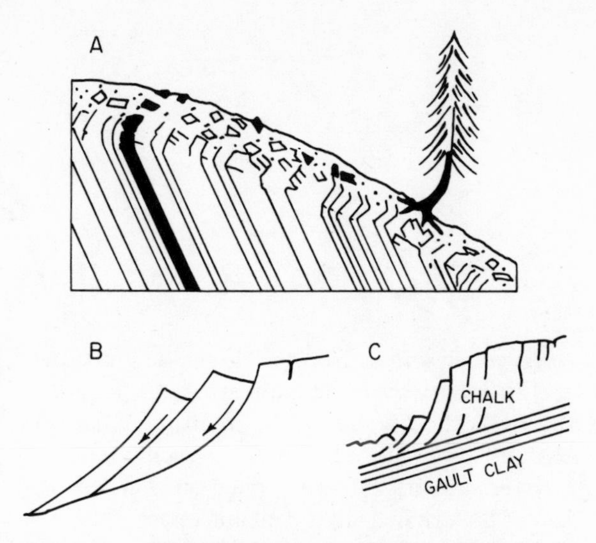

Fig. 5.12 *Mass-movements under gravity*
A. Hill-creep
B. Shearing slide
C. Bedding-plane landslip

deformation takes place throughout the moving mass, and *slides* which move as a whole on a definite surface. *Mud flows* are examples of the first group and need no description. The *slides* take place on curved shear surfaces (*shearing slides*) or on bedding, joint or fault planes in the rock. The conditions are illustrated in Fig. 5.12B. *Landslips* controlled by the structure of the rocks occur when a potential plane of movement is inclined towards a valley or cliff. Bedding-planes, especially when they separate a porous bed from an underlying impervious one, are especially favoured. Rapid mass-movements break up the orderly structure of the rocks and give rise to very distinctive land-forms characterised by chaotic small knolls and hollows. The rock masses affected have of course been moved from their original positions and may give the geologist a wrong idea of the fundamental structure if they are not recognised as landslip material.

6

SURFACE PROCESSES: 2

Rivers and lakes

This chapter deals with the effects of running water as an agent of erosion, transport and deposition. It describes the development of river-valleys, the sediments of rivers, deltas and lakes, and the salt deposits formed by evaporation in regions of interior drainage.

1. Rain, Streams and Rivers

In this chapter we are concerned with the second fraction of the water falling on the earth, the *run-off* which produces streams and rivers (Fig. 5.6); the proportion of the run-off which falls as snow and moves as ice behaves in such a different way from the remainder that we consider it in a separate chapter (Chapter 8).

Rain-action. Rain running down a slope washes the finest weathering products to the foot of the slope where they accumulate as *rain-wash*. During torrential rains such as occur in the tropics, this process may cause *sheet-erosion* and strip great quantities of good soil from the surface. Freak effects of rain-action are seen when the rain washes over a slope of soft material containing sporadic blocks of hard rock. These blocks protect columns of material beneath them and so give rise to *earth-pillars* as in the Findhorn valley of Scotland.

Stream-development. The flowing rain-water is guided into depressions in the slope and its erosive power is concentrated in these to produce *rills*; master rills develop by the flooding into them of water from the smaller rills, so that in course of time a small *drainage-system* is established. This consists of the main stream, flowing directly down the slope, and its *tributaries* which join it from the side. The junction of each tributary with the main stream is *accordant*, that is, without a change of gradient. As stream-erosion proceeds, the bed of each stream is lowered and its valley widened so that less and less of the original slope remains intact.

2. Erosion by Rivers

Stream-flow. The force which makes rivers flow is the pull of gravity. The component of gravity in the direction of flow increases with the slope or *gradient* of the river. Opposed to this force is the internal friction of the water and, especially, the frictional resistance at the floor and walls of the channel. Slowly moving water moves by *laminar flow* in straight paths within laminae parallel to the river-bed. In fast rivers, eddies are set up and the water then moves irregularly by *turbulent flow*. When a river comes to a bend, the velocity of the water increases on the outside of the curve and decreases on the inside. The water is therefore thrown against the outside and its powers of erosion are concentrated there. Deposition may take place in the slack water on the inside of the curve (Fig. 6.1).

Erosion by rivers. Fast-moving river-water may act like water from a hose and sluice away loose weathered material; but for cutting into harder rock it must be armed with particles of sand and gravel. With these it wears away its bed by a vertical filing action known as *corrasion*. The rate of corrasion is controlled by the velocity and mass of the particles used as tools, which in turn depend on the volume of water and the gradient, and also on the hardness or softness of the bed-rock. *Potholes* are produced in the bed of the

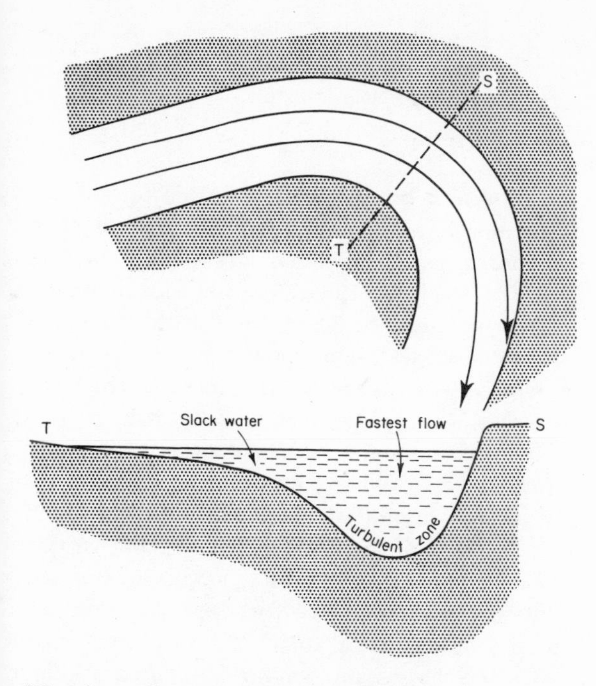

Fig. 6.1 *Stream energy at a bend*

river by the swirling action of pebbles trapped in an initial hollow (Fig. 6.2). They may form deep holes many feet across and adjacent potholes may join up to produce a deep narrow gorge. Since pot-hole formation and corrasion take place only within the river-bed, their effect is to cut down a narrow trench no wider than the river itself.

Base-level. Since water cannot continue to flow where there is no slope, there is a limit below which a river cannot deepen its bed. The ultimate *base-level* is the ocean into which the river must flow. *Temporary base-levels* are provided by lakes, resistant rock-masses and any barriers which hold up the downward passage of the water.

Shapes of river-valleys. The vertical down-cutting due to corrasion is only one of the processes shaping river-valleys. It is accompanied by *broadening* of the valley by weathering, gullying and collapse of the walls. The cross-section of a typical river-valley is therefore V-shaped. The form of the V depends largely on the character of the rocks traversed; open valleys are produced in soft rocks, narrow valleys or gorges in resistant rocks. Still other forms are seen where deposition plays a part in shaping the valley.

When layers of rocks of differing toughness are suitably arranged across a river *rapids* and *waterfalls* result. A strong thick bed may hold up erosion, as exemplified by the Niagara Falls shown in section in Fig. 6.3. Here the hard limestone bed which produces the falls is continually undermined by the erosion of the soft shales below it and a gorge 7 miles long has been formed as the waterfall has retreated upstream.

Spectacular *canyons* are formed under certain conditions—when the valley-walls are stable and

Fig. 6.2 *Potholes in an African river*

Chapter six

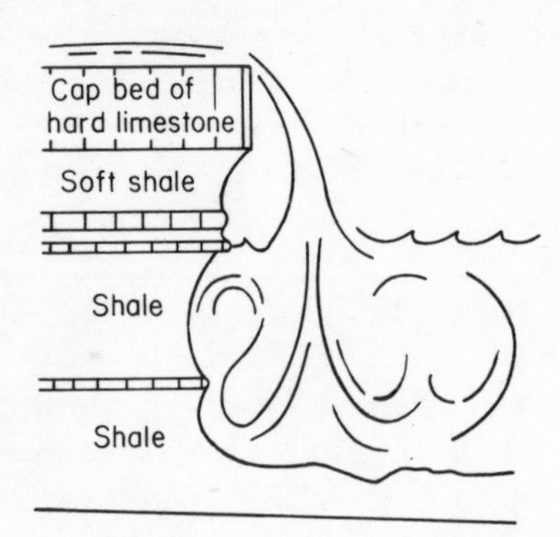

Fig. 6.3 *The Niagara Falls*

a powerful body of water cuts down rapidly after elevation of the land-surface. The ideal example is the Grand Canyon of the Colorado River, 200 miles long, 12 miles across and a mile or more deep.

3. Transport and Deposition

Transport by rivers. The material transported downstream by the river is called its *load*. It consists of two main parts, the first carried in solution, the second as solid particles. The *solution-load* is composed of the soluble products of chemical weathering, such as colloidal silica, hydroxides of aluminium and iron, and carbonates, sulphates and chlorides of calcium, magnesium, sodium and potassium. This solution-load is carried on to the sea or to an inland lake. The *mechanical load* of solid particles is transported in three ways. It may be carried above the floor in *suspension* by turbulent flow, it may proceed by *saltation* in a series of short leaps as the grains get up temporarily into a water layer with higher velocity or, finally, it may be moved by *rolling along* the river-bed. The transported grains undergo abrasion and are therefore reduced in size and partly rounded.

By these operations, the load becomes roughly sorted according to size into river-gravel, river-sand and river-muds which are dropped as separate fractions when the river slows down. The transporting power of a river is considered to vary as the fifth or sixth power of its velocity; by doubling its velocity the transporting power may be increased some thirty- to sixty-fold. Rivers in flood may thus perform immense geological work.

Deposition by rivers. Deposition of portions of the mechanical load of a river takes place when its energy is lowered—by reduction in the gradient (and consequently the speed) or in the volume of water, or by increase in the frictional force opposed to the movement of the water. We can say, in brief, that as the velocity of the river decreases, some of the load comes down.

Deposition due to a sudden decrease in velocity occurs typically where a stream flowing swiftly down a steep mountain slope debouches on to the flatter ground of the main valley or plain. In this *piedmont* area the stream quickly lays down badly sorted material in wedge-shaped layers tapering downstream, to build up fan-like units with steep slopes and well-defined fronts (Fig. 6.5). Such *fans* are well developed in the Po valley of Italy where streams coming from the Alps to the north have built up great deposits and pushed the main river south. Isolated conspicuous fans are called *alluvial cones*; in these, the coarser material lies near the mountain, the finer at the outer edge of the cone.

We can distinguish two classes of river or *fluvial deposits*. The first consists of the *channel-deposits*, formed in the bed of the river itself. These are irregular lenses of gravel, sand, silt and clay, conspicuous among them being the *gravel-bars* formed in the slack water on the inside of bends (Fig. 6.1). The second class of fluvial

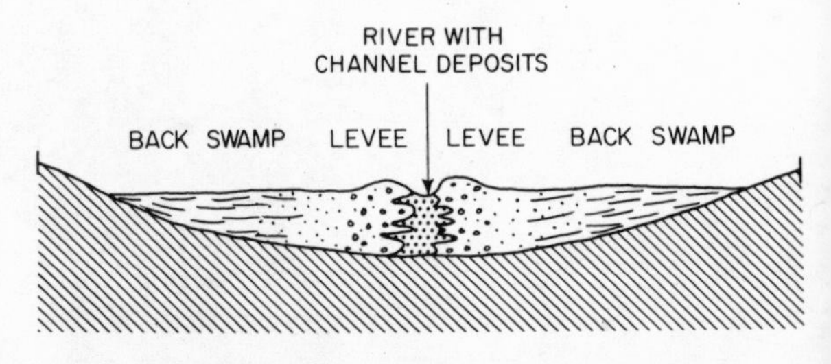

Fig. 6.4 *Deposits in the flood-plain of a river*

Fig. 6.5 *Alluvial fan making a small delta in a Norwegian lake*

deposits includes those formed over the *flood-plain* of the river, that is, the area covered by its floodwater. In this area the velocity of the water is lowered and the load is deposited. The coarsest material, coming to rest nearest the river-bed, may build up natural embankments or *levees* along the sides (Fig. 6.4); behind these levees are the *back swamps* in which the finer fraction of the load is deposited. Over the centuries successive floods build up considerable thicknesses of *alluvium* or *alluvial deposits*.

Alluvial terraces. Along many rivers, the alluvium forms a series of terraces like a succession of great steps rising up the valley sides. The surface of each *alluvial terrace* marks the position of the flood-plain at an earlier stage in the history of the valley and the terraces themselves are relics of earlier flood-plains which have escaped erosion during later down-cutting: the highest is clearly the oldest and was formed when the river flowed at its highest level (Fig. 6.6). The arrangement of terraces is often due to the spasmodic *rejuvenation* of the river by uplift of the land. Each upward movement starts the river on a new phase of active down-cutting and each terrace is the record of an intervening stage of stability and deposition.

Meanders. A bend in the river tends to increase in amplitude by erosion on its outer side and deposition on its inner side (Fig. 6.1). Where the river moves through a flood-plain of its own deposits, this process transforms open bends into tight loops called *meanders* after the River Meander of Troy. Since the main force of the water is concentrated on the downstream side of the loop, each meander tends to migrate slowly downstream. As the loop tightens, it is likely to be *cut-off* by a new channel breaking through across the neck; the abandoned part remains filled with water or marsh as an *ox-bow lake*. Once a river has begun to meander, the momentum of the

Fig. 6.6 *River terraces, Canterbury Plains, New Zealand*

water produces a succession of loops one in front of the other. By this means a *meander-belt* is developed which in course of time swings back and forth across the flood-plain slowly widening the valley (Fig. 6.7).

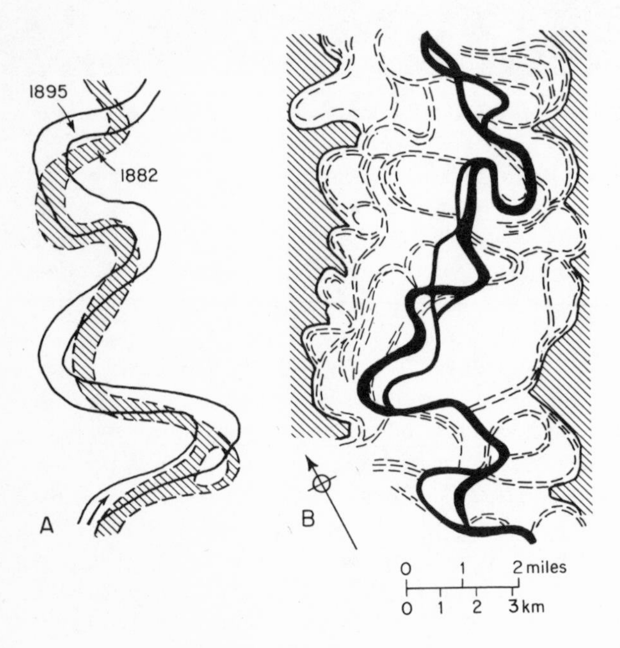

Fig. 6.7 *Meanders*

A. Downstream migration of meanders in the Mississippi

B. The river channel (black) and abandoned meanders of the Rhine

General characters of fluvial deposits. Conditions in a river and over its flood-plain are so inconstant that fluvial deposits are exceedingly variable and include sediments of gravel, sand and mud grades (see p. 107). Elongated strings of coarse material represent the channel and levee deposits and more regular beds of finer muds those of the flood-plain. We may point out that a bed of river gravel has a ribbon-like form instead of the sheet-like form shown by most beds of sediment (p. 108). This is a natural result of its mode of deposition and is a primary character which may help to identify ancient fluvial deposits. Organic remains carried by rivers are usually so battered that they stand little chance of being fossilised.

4. Deltas

Deposition in the flood-plain of a river is often only temporary and, in the end, most of the mechanical load is carried on to the sea or to a lake. Here, the velocity of the water is checked, and much of the load is deposited in a △-shaped area with its apex pointing upstream (Fig. 6.5). This *delta* (the name comes from the Greek letter Δ) is built outward into the sea or lake by the successive tipping of the load in that direction. The part of the delta which stands above water-level is a continuation of the flood-plain and is built up by deposition from bifurcating *distributaries* running across it. The hidden, subaqueous, outer part of the delta forms a great bulge on the sea or lake floor, as revealed by the submarine contours of the classic Nile delta shown in Fig. 6.8.

Deltaic sediments. A whole delta, as distinct from the sub-aerial portion visible on a map, will clearly be a combination of sediments of fluvial and of marine or lake facies. The upper and inner parts, continuous with the flood-plain of the river, are made of coarse irregular beds of sand laid down in the distributaries, and more regular beds of mud and silt laid down between them. The lower and outer parts are formed in shallow water and will contain marine fossils where the delta is built out into the sea. Between the two extremes, sediments of fluvial and marine types may alternate according to local advances and retreats of the sea.

In some deltas, three sets of beds can be distinguished as shown in Fig. 6.9. These are the *foreset*, *bottomset* and *topset* beds. The *foreset beds* are made of the coarser material which is dumped quickly in fairly thick beds on the outer slope of the delta; each bed is inclined at the angle of this slope. The *bottomset beds* are laid down beyond the foreset beds and consist of thin, nearly horizontal laminae of the finer material which settles more slowly; and the *topset beds*, again roughly horizontal, are the fluvial sediments of the delta plain. In course of time, as the delta is built forward into the water, younger topset and foreset beds advance over the older bottomset beds. The foreset beds are inclined in the direction in which the delta is advancing.

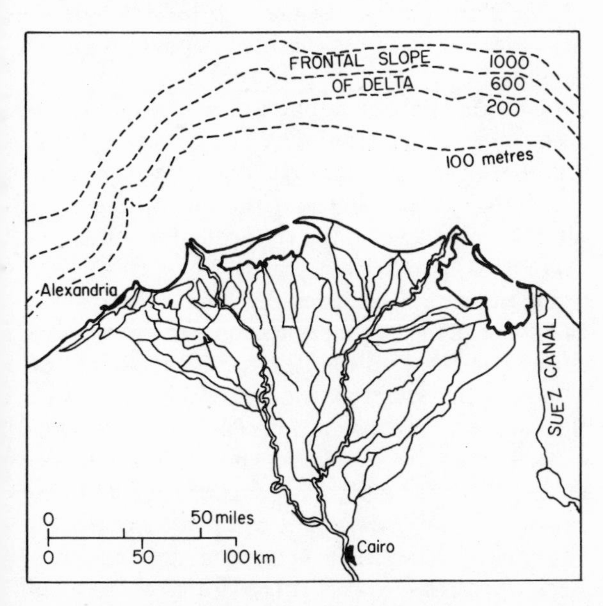

Fig. 6.8 *The sub-aerial and submarine parts of the Nile delta*

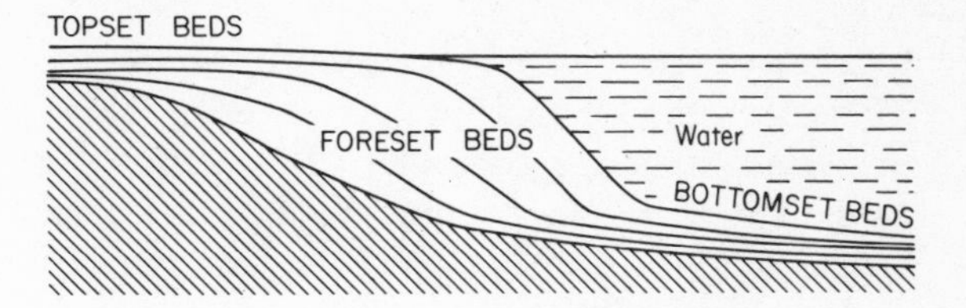

Fig. 6.9 *Foreset, topset and bottomset beds*

The deposits of a large delta have in cross-section the shape of a lens tapering towards both the land and the water. Borings in modern deltas reveal that this lens may be extraordinarily thick, often many thousands of feet, and that the facies of the deposits is fluvial and shallow-water throughout. It is clear that the floor of the delta region must have sagged to make room for all this material and that subsidence must have gone on at about the same rate as deposition, so that the surface of the deposits always remained near water-level. The enormous bulk of the deltas of big rivers shows how effective these rivers must be as agents of erosion and transport.

5. The Fluvial Cycle

The *fluvial cycle* is the name given to the series of progressive changes in land-forms which results from the development of a river system. All the processes of erosion and deposition connected with rivers combine to bring about these changes, and in areas of moderate rainfall they are among the dominant factors shaping the landscape.

The river-profile. If the longitudinal profile of a normal river from source to mouth is constructed by plotting to scale its height above sea-level at a large number of points, it will be found to make a more or less smooth curve, concave upwards, steeper towards the source and gentler towards the base-level at the mouth (Fig. 6.10). Tributaries show similar profiles, meeting that of the main stream accordantly. *Irregularities in profile* are related to the presence of lakes or of beds of hard rock which provide temporary base-levels—the profile flattens off on approaching these features and then steepens again below them. All irregularities, even those so great as the Niagara Falls, will ultimately be smoothed out if the river has time enough; lakes will be filled in with sediments, rapids and waterfalls eroded away, local steep stretches will be preferentially eroded and gentle stretches will receive deposits to build them up. When all irregularities are removed and the profile is smooth, the river is said to be *in equilibrium* with the factors controlling its development and its profile is a *profile of equilibrium.* Given time a river would finally wear down its valley to the level of its mouth, a condition known as *base-levelled.* This condition is never quite achieved, but we see an approach to it in the *peneplain* ('almost a plain') where slow rivers with low gradients meander in wide alluvial plains through a broad rolling country of gentle slopes and low relief.

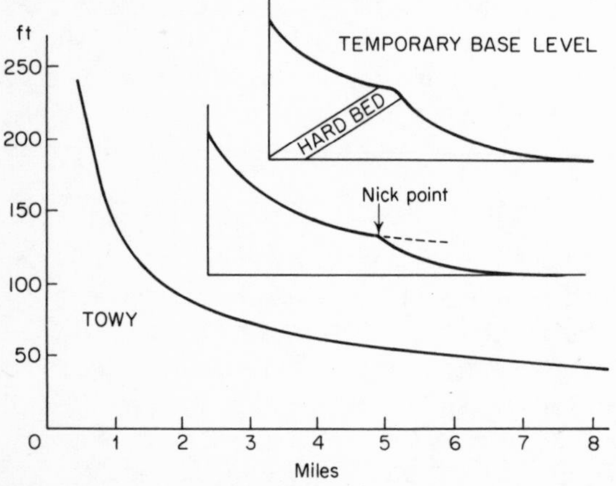

Fig. 6.10

River profiles: equilibrium profile of the River Towy (after O. T. Jones) with (above) *a temporary base-level provided by a hard bed in the river-valley and* (centre) *a nick-point at the junction of profiles related to different base-levels*

Stages of the fluvial cycle. If we consider the history of a drainage system from the time when it is first established on a newly uplifted land-surface, we can distinguish the three stages of youth, maturity and old age in its development. The stage of *youth* is characterised by steep gradients, with consequent vigorous erosion giving rise to gorges or to narrow, steep-sided valleys. Waterfalls and rapids are common and the river-profiles are not in equilibrium. The fast-flowing streams have little need to deposit their load and there is therefore no flood-plain. In the stage of *maturity*, the gradient is moderate and the valley both deeper and wider. The stream is in equilibrium, a flood-plain is beginning to appear and meandering takes place within it. In *old age*, the gradient is still lower, the valley is wide and open and lateral cutting by the migrating meander-belt is the main erosional process. A broad flood-plain is formed and the well-developed meander-belt wanders slowly across it. Irregularities in the profile are largely smoothed out, the profile is one of equilibrium and the condition of base-levelling or peneplanation is almost achieved.

Accidents to the fluvial cycle. The simple sequence of changes outlined above may be modified or interrupted by various causes. One type of modification is due to the *structure of the rocks* into which the river-system is cutting. Another results from the *effect of earth-movements* which by raising or lowering the land-surface may alter the base-level of the rivers and force them to establish a new equilibrium.

Influence of bedrock. A *consequent river* is the consequence of the formation of an inclined surface, such as an uptilted coastal plain or the slope of a new volcano, and it flows directly down this surface. As erosion proceeds, new drainage-channels may be formed depending on the weaknesses of the lower layers of rock that begin to be exposed—such streams are called *subsequent.* Erosion by a subsequent stream flowing on a weak layer between harder layers may dig a valley running almost perpendicular to the consequent streams.

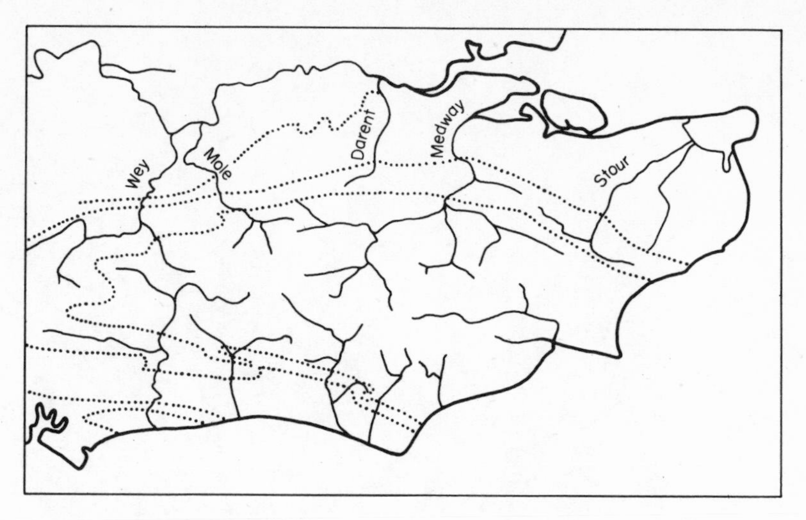

Fig. 6.11 *River-system of the Weald*

The main rivers were developed as consequent streams on the dome-shaped Wealden anticline (see pp. 164–5) and many of their tributaries are subsequent streams parallel to the outcrops of soft strata (geological boundaries shown by dotted lines). A strong tributary of the Medway has captured the headwaters of the Darent

All the consequent streams initiated on a slope do not grow equally along parallel straight lines. Minor inequalities of surface, of bedrock or of slope, lead to the increase in size of certain streams which then enlarge their basins by *capturing* other streams (Fig. 6.11). Tributaries of active rivers cut into the basins of their neighbours and *behead* them. Inequalities in the rocks of the valleys especially help in this *piracy*—the stream flowing in soft, easily eroded rocks captures those that are compelled to erode hard rocks, as is illustrated by the rivers of Yorkshire and of the Weald (Fig. 6.11). A series of consequent rivers initiated on a certain layer of rock may in course of time remove this layer to reveal another set of rocks arranged in a quite different pattern. Such a *superimposed drainage* system is illustrated by the rivers of the Lake District (Fig. 6.12).

Fig. 6.12

Superimposed drainage in the Lake District

A system of radial rivers was developed on a dome of Carboniferous limestone (stippled) and when erosion removed the limestone from the centre of the dome, the radial drainage was superimposed on underlying rocks whose main structures run WSW–ENE (dotted lines).

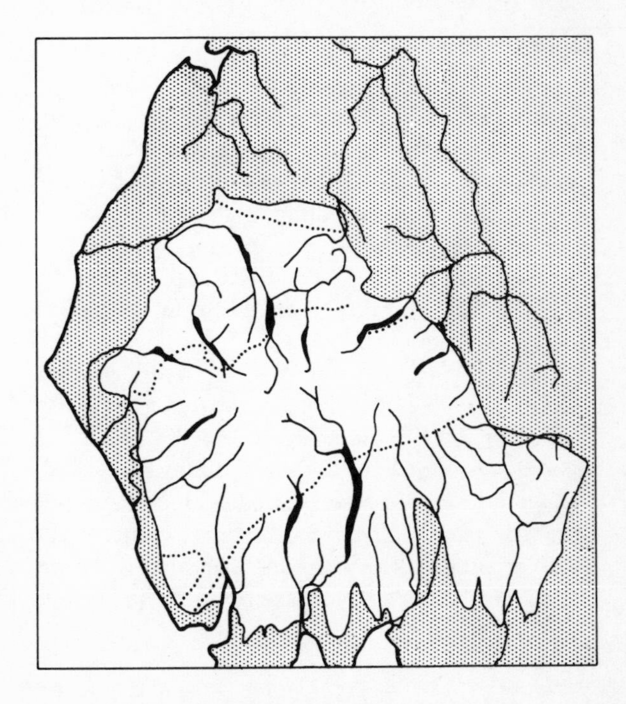

Influence of earth-movements. Drastic interference with the equilibrium of a river-system results from regional elevation or depression of the land. Elevation results in *rejuvenation* of the river which then begins active erosion again and digs down in its flood-plain to produce *entrenched meanders* on the sites of the existing

Fig. 6.13
The Parallel Roads of Glen Roy

meanders. The Grand Canyon of the Colorado already mentioned is a result of rejuvenation of a river-system. Depression of the land results in the *drowning* of a river-system and the dismemberment of its components. The pattern of river-erosion is found to continue on the sea-floor and deep inlets of the sea fill the old river-valleys. A drowned topography of this kind is seen in south-west Ireland.

6. Lake-processes

Lakes are a sign of youth in a river-system and are doomed to be filled up by *deposition*; they act in fact as gigantic settling-tanks. Most lakes are parts of river-systems draining ultimately to the sea and therefore have outlets, but some, the *salt lakes*, are situated in areas of *interior drainage* and have no outlet. These salt-lakes accumulate not only the mechanical but also the solution-load of the rivers feeding them.

Lakes occupy basins of many different origins. Firstly they may result from *crustal movements*. The Rift-lakes of East Africa and the Dead Sea in the Jordan valley (Fig. 14.7) lie in rift valleys (p. 173), produced by depression of a narrow zone between parallel fractures. Broad warping of the surface produces depressions occupied by wide, relatively shallow lakes exemplified by Lough Neagh in Ireland and Lake Victoria in East Africa. Small lakes occupy craters of volcanoes (*crater lakes*) or cut-off meanders (*oxbow lakes*), or *hollows in glacial deposits* or *rock-basins* excavated by *glacial erosion*. Many lakes are held up by a *barrier* of some kind formed across a river-valley—a lava stream, a landslip, a ridge of glacial deposits, a bar formed by wave-action or a man-made dam. One of the most interesting of these barriers is a glacier. The Marjelen See in Switzerland is held up in a tributary valley by the Aletsch Glacier that occupies the main valley. During the Ice Age (p. 222) exactly similar temporary lakes were formed in the glaciated regions of that time. One such lake filled Glen Roy in western Scotland and the various levels of the water are still recorded by old shore-lines seen one above the other on the valley-side as narrow terraces—the famous *Parallel Roads of Glen Roy* (Fig. 6.13). Surplus water from these lakes escaped by *glacial overflows* or spillways which now appear as trenches across ridges.

Sedimentation in lakes: lacustrine deposits

Much of the mechanical load of the rivers is deposited in lakes. *Deltas* of gravel, sand and clay are built out by the incoming rivers and the flood-plain deposits are extended by this means until, given time, the site of the lake becomes a level alluvial plain. Much *organic material* is incorporated in lacustrine deposits: drifted vegetable matter, siliceous tests of microscopic diatoms (p. 145), calcareous shells and remains of algae accumulate to give *lake-marl*. *Bog iron-ore*, hydrated iron oxide, is formed in some shallow lakes.

Many lake-processes vary with the seasons because they depend on the volume of water in the rivers and the activity of organisms. The most celebrated type of *annual lamination* is that seen in the banded clays which were deposited in temporary lakes during the retreat of the ice-sheets which covered north-west Europe and northern America in the geologically recent Ice Age (p. 222). In these banded clays, or *varved clays*, each year's deposit consists of a summer layer of pale sandy material passing gradually up into a thinner winter layer of fine dark clay (Fig. 6.14). Over 20,000 of these varves have been counted and they record a rare example of geological processes which can be measured in terms of single years.

Deposition in salt lakes

Under conditions of interior drainage where rivers have no outlet to the sea, not only the mechanical load but also the chemical load is trapped. As the rivers continue to bring in their solution-load and evaporation proceeds from the lake surface, the waters of the lakes are turned into

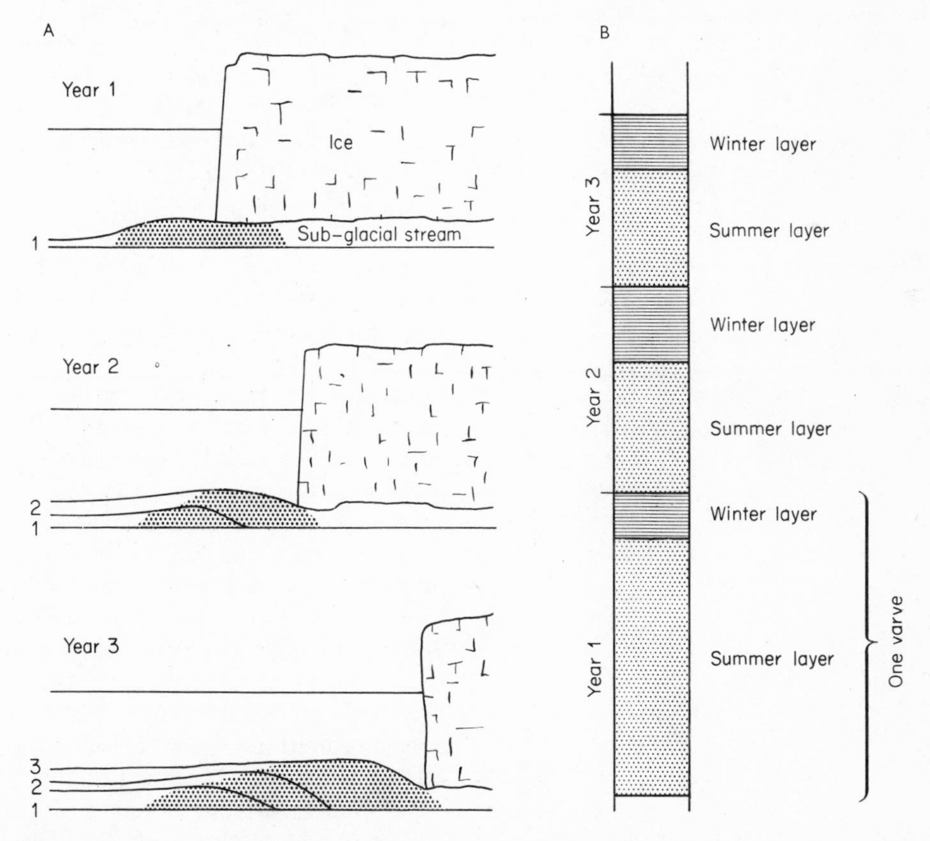

Fig. 6.14 *Varves*
A. Delta and varved sediments built up year by year by a sub-glacial stream entering a lake
B. Summer and winter layers in three successive varves

concentrated salt-solutions and eventually precipitate various salts to form the *saline residues* or *evaporates*. Salt lakes may be rich in sodium chloride, sodium sulphate, alkali carbonates and other soluble salts, depending on the rocks of the drainage-basin.

These *terrestrial salt lakes* are numerous in the warm-arid climatic belts of today. They can be exemplified by the salt lakes of the western United States, such as the Great Salt Lake of Utah; in shallow parts of this lake, rock-salt, NaCl, is being deposited, whilst sodium sulphate is thrown down in winter. Other saline lakes, as in Nevada, are precipitating sodium carbonates. Analyses of the waters of salt lakes reveal, in addition to the substances so far mentioned, salts of potassium, magnesium and bromine. If such lakes were to dry up completely these *bittern salts* would be deposited—such deposits, however, are rare.

Lakes formed by the shutting off of a portion of sea-water as a consequence of earth-movements begin as sea-water containing about 3·5% of dissolved matter, the greater part of which is common salt, NaCl. When sea-water is evaporated hydrous calcium sulphate (the mineral gypsum) is deposited first, then sodium chloride (rock-salt): the remaining mother liquor is enriched in the bittern salts such as magnesium sulphate and chloride, potassium chloride and sodium bromide. In terms of minerals, this order of deposition is that of increasing solubility, but it is modified in nature by many factors. To illustrate the succession of deposition we may use the famous Stassfurt salt deposits of Germany—from the top down these are:

Shales, sandstones and clays
    Rock-salt, NaCl, variable thickness
    Anhydrite, $CaSO_4$, 30–80 metres
Salt-Clay, 5–10 metres
    Bittern minerals
    Older rock-salt and anhydrite, interlayered
    Anhydrite and gypsum

The very soluble bittern minerals have possibly been preserved in these deposits by the blowing over them of the Salt-Clay to provide a seal against leaching.

Ancient salt-deposits are often thousands of feet thick, a fact that poses a considerable problem when it is remembered that sea-water contains only 3·5% of salts. In some instances, it may be possible that a mechanism similar to that shown in the Gulf of Karabugas in the Caspian may have operated (Fig. 6.15). The Gulf has a mean depth of only 15 metres and is connected with the Caspian Sea by a narrow strait no more than a metre or two in depth. Each day water carrying some 350,000 tons of salt enters the Gulf whence no return is possible, and accordingly the saline content of the Gulf is steadily increasing as water evaporates from its surface. If present conditions continue, rock-salt will begin to be deposited in a couple of centuries.

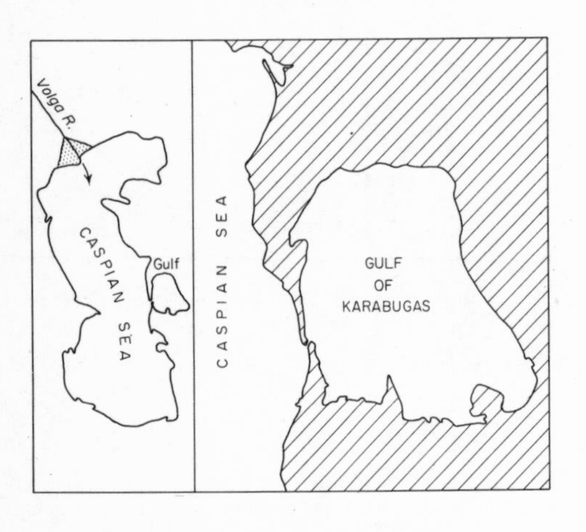

Fig. 6.15 *The Gulf of Karabugas: maps* (above) *and section*

7

SURFACE PROCESSES: 3

The sea

This chapter deals with the sea as an agent of erosion, transport and deposition. Special attention is paid to the processes of marine deposition and to the sediments, mechanical, organic or chemical in origin, which are laid down in the epicontinental seas; we emphasise this topic because the commonest types of ancient sedimentary rocks are of marine epicontinental facies and an understanding of marine environments is necessary for their interpretation. In addition, the chapter deals with the composition of sea-water and with the shaping of coastlines by marine erosion and deposition.

1. Importance of Marine Processes

What goes on in the seas and oceans is of immense importance to the geologist because the commonest types of sedimentary rocks with which he has to deal in his task of interpreting earth-history are those deposited in the seas of past ages. As we have seen already, such rocks are now widely exposed in the interiors of continents and even in the highest parts of mountain ranges where they have been uplifted by earth-movements.

Ocean waters may act as agents of erosion or transport or deposition. Some of their activities can be observed directly along the shore-lines where land and sea meet, but the more important processes taking place on the sea-floor itself are hidden from direct observation. The techniques of *marine geology* by which these processes are studied tend to be difficult and expensive, but nevertheless in recent years great advances have been made by means of the collection of samples and cores from the sea-bed, by underwater photography, and by echo-sounding and other geophysical methods.

The main features of the sea-floor have already been mentioned (p. 13). It will be remembered that the *continental shelf* fringing the continents is limited outward by the *continental slope* beyond which are the deep *ocean basins*. We shall contrast the *shallow* or *epicontinental seas* covering the continental shelves with the *deep seas* over the continental slopes and the *abyssal seas* of the true oceanic basins. Almost all the ancient marine sediments now exposed in continental areas were deposited in epicontinental seas or on the continental slopes.

2. Composition of Sea-water

Each river delivers its solution-load of dissolved salts to the sea and from these contributions, continued over hundreds of millions of years and concentrated by the evaporation of water from the surface, comes the familiar saltiness of the sea. Not all the ingredients of the solution-load, however, remain in the water. Some, for example calcium carbonate, are *extracted* by marine organisms to make shells and skeletons. Other elements are *adsorbed* by particles of clay on the sea-floor and become locked up in the sediments forming there. These differences are reflected in the chemical composition of sea-water. The igneous rocks which provide the ultimate source of the solution-load in rivers contain roughly equal amounts of sodium and potassium; but because the potassium is largely adsorbed by clays, sodium becomes the dominant metal element of sea-water. All elements are probably present in minute proportion in the sea and their total amounts must, in view of the enormous volumes of the oceans, be large. It has been estimated, for example, that 9,000,000 tons of gold are present in the sea which may yet prove an important source of ores.

The composition of the dissolved matter in sea-water is given in the following table:

COMPOSITION OF MATERIAL IN SOLUTION IN SEA-WATER (DITTMAR)

	In 1,000 gms. sea-water	*% of all salts*		
$NaCl$	27·213	77·758	Na	30·593
$MgCl_2$	3·807	10·878	Mg	3·725
$MgSO_4$	1·658	4·737	Ca	1·197
$CaSO_4$	1·260	3·600	K	1·106
K_2SO_4	0·863	2·465	Cl	55·292
$CaCO_3$, etc.	0·123	0·345	Br	0·188
$MgBr_2$	0·076	0·217	SO_3	7·692
			CO_3	0·207
	35·000	100·000		100·000

This composition of the salts is very constant, but their *concentration*, which determines the *salinity* of the sea, is more variable, depending on the size of the ocean basin, the contribution by rivers (small seas receiving much fresh water have low salinity) and the climate (small seas concentrated by rapid evaporation have high salinity). The range of salinity, expressed as grams per 1,000 grams of water, is shown by the following figures: Red Sea, 38·8, Persian Gulf, 36·7, Mediterranean, 37–39, Bering Sea, 30·8, Black Sea, 18·0, Baltic Sea, 7·8.

3. Movements of Sea-waters

Sea-waters are moved in four ways, by currents, tides and waves affecting water alone and by turbidity currents in which the water is mixed with sediment. *Currents* of convection type are produced as a result of differences in density of the water due to variations in temperature and salinity. They produce relatively slow but very extensive movements. Other currents are produced by frictional drag of wind across the surface of the sea. *Tides* result from the oscillation of masses of water under the attraction of the moon. In the open sea the rise of the tides is small; but strong currents may be generated near coast-lines and, in funnel-shaped estuaries such as that of the Mersey, tidal waters may be piled up daily to heights of 40 feet or more.

Waves, the most spectacular form of moving sea-water, result from the pressure and drag of wind on the sea surface. In the open sea, each particle of water in a symmetrical wave moves in a circular orbit whose diameter equals the height of the wave. When a wave reaches shallow water it becomes taller and when its height is roughly equal to the depth of the water it topples over or *breaks* and moves forward as a *breaker*. This produces a second, smaller wave which in turn advances and breaks, the process being repeated until the shore is reached. The force of breaking waves can be enormous, over 6,000 lbs. per square foot being recorded during a gale, and their capacity for erosion and transport of material is correspondingly great.

Turbidity currents. Very fine particles of mud, when stirred up with water, can stay in suspension for a considerable time and if the water carrying them is moving by turbulent flow they may drift almost indefinitely without settling out. Water filled with such particles is cloudy or turbid. Because of their higher density and viscosity, bodies of turbid water will flow at quite high velocities down a slope beneath the clear water of a sea or lake and are capable of transporting large particles. Such *turbidity currents* may act as agents of erosion, transport and deposition in deep waters not reached by waves or normal currents.

Turbidity currents may be generated in nature by a number of different mechanisms. A fast-flowing river carrying a large load in suspension may deliver turbid water to a lake or sea and this water then flows on along the floor as a turbidity current—such conditions are found where summer melt-water comes down into a lake from a glacier. Secondly, violent storms may stir up mud from the sea- or lake-floor. Thirdly, earthquake shocks may similarly stir up sediment. Lastly, and perhaps most important, loose sediments which have been piled up rapidly near the shore may become unstable and begin to *slump* or slide towards lower levels on the sea-floor. In doing so, the material becomes churned up with water again and generates turbidity currents.

4. Marine Erosion

It is possible to find evidence of the wearing-away of the land by marine agents on almost any coast and this *coastal erosion* naturally forms our first topic. But erosion also goes on on the sea-floor even beneath the waters of deep oceans—this *bottom-erosion* has particularly spectacular results in the production of *submarine canyons* which form our second topic.

Coastal erosion. Most coastal erosion is of a mechanical kind, brought about by the force of the moving water and the battering action of rock-fragments carried by it (Fig. 7.1). Mechanical erosion is greatest where and when the movements of water are most powerful, that is, along the parts of the coast most exposed to currents and waves, and in periods of storm. The force of breaking waves may be sufficient in itself to wrench off rock fragments. The sudden compression and expansion of air in crevices when a wave breaks against them may loosen and suck out such fragments; crevices are widened and ultimately passages or caves are opened up which may connect with the land-surface by *blow-holes*. The fragments torn from the rocks provide tools by which further erosion is carried out. The battering of these fragments cuts into the coastal rocks with a kind of horizontal sawing action—a form of *corrasion* (p. 68)—whose effect is to produce a wall-like *cliff* facing the water.

Continual marine erosion near the base of the cliff leads to *undercutting* and is followed by collapse of the upper parts under the influence of

Fig. 7.1
Marine erosion in Aberdeenshire, showing the influence of lines of weakness in the rocks

Chapter seven

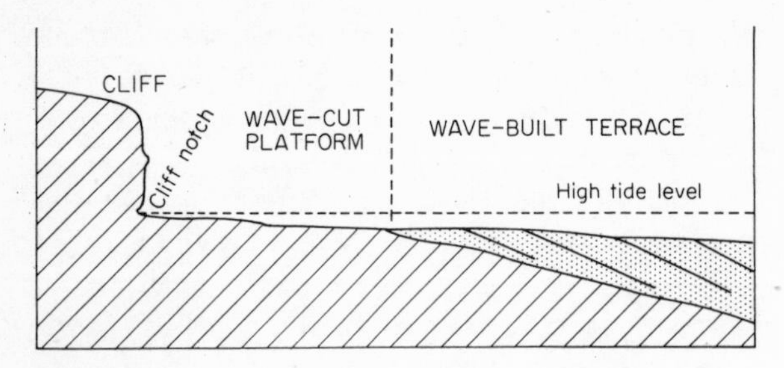

Fig. 7.2
Wave-cut platform and wave-built terrace

gravity and the sub-aerial agents of erosion. In this way, the whole cliff-face recedes, leaving behind it a *wave-cut platform* near water-level. The battered and rounded material used as tools —boulders, pebbles and sand—is spread out seaward and extends the wave-cut platform in this direction as a *wave-built terrace* (Fig. 7.2).

Details of the shaping of the coast by marine erosion depend largely on the properties and structure of the coastal rocks. Bands of hard resistant rocks project as *headlands*, while softer rocks are more rapidly worn back to give *bays*. Intense localised erosion may take place along planes or zones of weakness such as bedding-planes, joints or faults. A variety of coastal forms (Fig. 7.1) is produced by penetration of the sea along such planes—they include *sea-caves*, *blow-holes* (see above), and *geos* (a Norse word for narrow steep-sided inlets, used in Scotland and Shetland). Where erosion follows planes roughly parallel to the shore-line, a column of rock or *sea-stack* may be left standing free of the cliff or connected to it by a natural *arch*. Where planes of weakness slope down towards the sea, *cliff-landslides* may result.

Submarine canyons. The continental margin (p. 13) is often scored by great chasms known as *submarine canyons* which begin near the outer edge of the continental shelf and continue down the continental slope till they reach depths of up to 2,000 metres. These canyons are steepest near their heads, where they cut into the continental slope; here, their walls may be almost vertical and their floors may be incised to depths of 1,000 metres. They may be joined, without change of gradient, by tributary canyons resembling the tributaries of sub-aerial rivers. An example of a submarine canyon is provided by the Hudson Canyon off the Atlantic coast of the northern United States (Fig. 7.3).

There has been much discussion about the origin of submarine canyons. They are undoubtedly erosional features and the fact that many are cut through fairly well-consolidated sediments (a canyon off California is even cut in shattered granite) shows that a powerful agent of erosion must have been at work. Yet the majority of canyons lie at depths below sea-level at which waves and currents of ordinary types would not be expected to have much power. The explanation now generally accepted is that the canyons were eroded by turbidity currents.

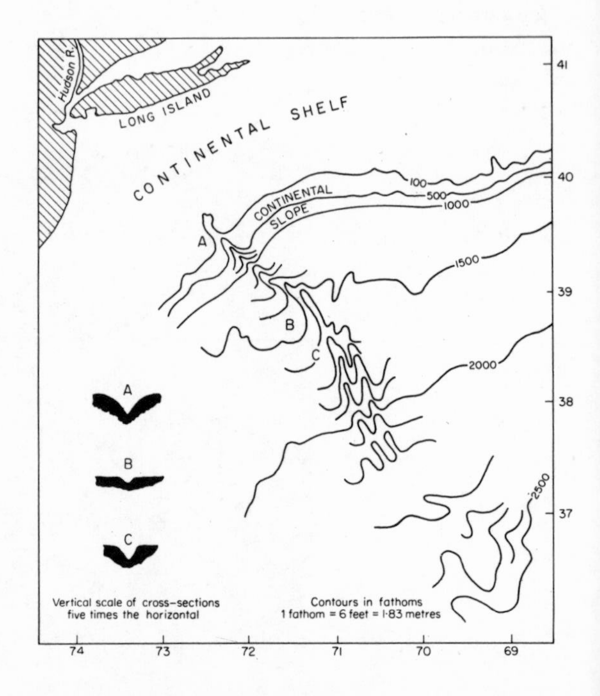

Fig. 7.3 *The Hudson submarine canyon off the Atlantic coast of North America: contour map and profiles (based on Heezen and others)*

Submarine canyons and turbidity currents. There is evidence to suggest that many canyons were formed at about the time of the geologically recent Ice Age (p. 222); their walls, in some instances, are cut through sedimentary rocks known from their characteristic fossils to have been deposited immediately before this Ice Age and the floors may show a thin cover of sediments laid down after it. During the Ice Age, a large volume of water was locked up in solid form in ice-sheets and as a result, the general sea-level was lowered, possibly by about 300 feet. Such a lowering of sea-level would not be great enough to expose the sea-floor in which the canyons are cut to sub-aerial erosion, but it may have been sufficient to bring great areas of recently deposited marine sediments within reach of wave-action. These sediments could therefore be repeatedly stirred up during storms with the production of numerous turbidity currents. These currents were guided into local depressions of the sea-bottom and their erosive action was thus concentrated along narrow paths. The deep canyons were dug as the turbidity currents gathered speed on reaching the steeper gradients of the continental slope.

5. Transport in the Sea

Moving sea-water acts as an agent of transport not only for the products of marine erosion but also for the very much more bulky products of weathering and erosion which are brought to the sea from the lands. All this material, ranging from large boulders to minute clay-particles, is moved about by waves and currents, dumped many times as the velocity of the currents decreases and, at last, deposited beyond the reach of normal wave or current action. During these complex movements the large particles are battered and rounded by constant collisions with each other. This rounding process is often well illustrated by the fragments of broken glass one finds on the sea shore. In addition to rounding, the particles are *sorted* according to size and density. Incoming *breakers* near the shore-line are powerful agents which can transport both large and small fragments towards the shore. The less powerful *backwash* of each wave can only carry out again the finer particles, so that the coarse material remains stranded near the shore-line. Sand grains may be dropped in shallow water while the finest particles of all may be passed out into deep water. One common effect of *wave-action* is thus to form a zone of coarse deposits high up the beach passing outward into finer deposits.

Waves which meet the shore-line obliquely transport material parallel to the coast. This *longshore drift* is often demonstrated by the occurrence of pebbles of a distinctive rock spread out along the shore-line on one side of the outcrop of this rock in the cliffs. Attempts are sometimes made to check longshore drift by building widely spaced low walls or *groynes* perpendicular to the shore-line; the moving material then piles up against the side of the groynes facing the direction from which it is being transported.

The effects of transport and sorting by *currents* depend largely on their speed. Where tidal currents are concerned, a to-and-fro movement may be produced by the *flood* and *ebb* (that is, the incoming and outgoing tide) and the net movement of material depends on which current is the stronger. Currents flowing over mixed coarse and fine material may have a *winnowing* action, removing the finer particles and leaving behind a residual or *lag* deposit of coarse material. Particles of clay size may remain in suspension for so long that they are carried far afield. Final deposition is often brought about by *flocculation*, the clotting together of clay particles to form larger granules.

The role of *turbidity currents* as agents of transport has already been mentioned. The sorting action of these currents is of a special kind. As the turbidity flow finally slows down its whole load, including both coarse and fine material, begins to settle out. The larger and denser fragments fall more rapidly than the fine particles and so come to rest at the bottom of the deposit with finer material on top of them. A bed laid down by a turbidity current is therefore *graded* from coarse at the base to fine at the top (see Fig. 10.2).

Chapter seven

6. Marine Deposition

The material deposited in the sea originated in three ways. Most of it consists of particles derived from the weathering and erosion of older rocks. This *detrital* quota is supplied from the land by rivers, glaciers, ice-sheets and winds and from the coast by marine erosion. Secondly, *material of organic origin* makes some very important types of marine sediment. It is derived mainly from the protective shells and skeletons of marine animals and plants and the bulk of its substance has been extracted from solution in sea-water by these organisms. Finally, *chemical deposits* are formed directly by precipitation of substances from solution in sea-water; we have already met examples of these deposits in the evaporates of marine salt lakes (p. 78). The components of marine sediments which are derived directly from the land are called *terrigenous*, those derived from the sea itself are called *pelagic*. Most marine deposits contain both kinds of components, terrigenous material predominating near the lands and pelagic material in abyssal deposits.

Environments of marine deposition. The seas are so vast and the range of conditions within them so great that there are many different environments of marine deposition and, consequently, marine sediments of many facies. The variable factors include the depth of the water, its distance from land, its temperature and salinity, its fauna and flora and the extent to which it is disturbed by waves and currents. We may make a first classification of marine environments according to depth of water which, in a general way, increases with distance from land (Fig. 7.4). From the coast outwards, four environments can be distinguished; they are those of the *littoral*, *neritic*, *bathyal* and *abyssal* zones and are dealt with in this order below. The first two are, broadly, the environments of the *epicontinental seas*.

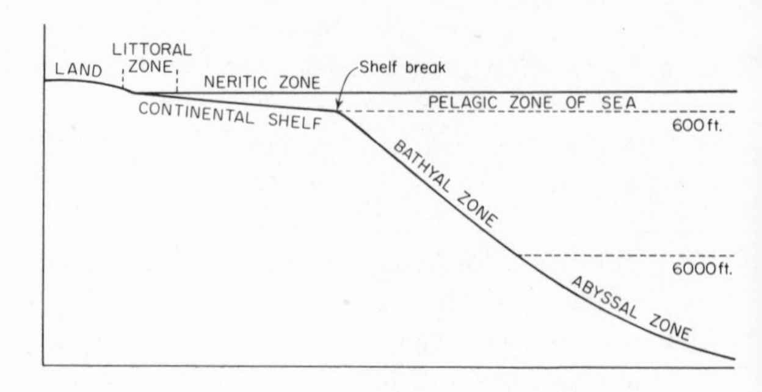

Fig. 7.4 *Environments of marine deposition*

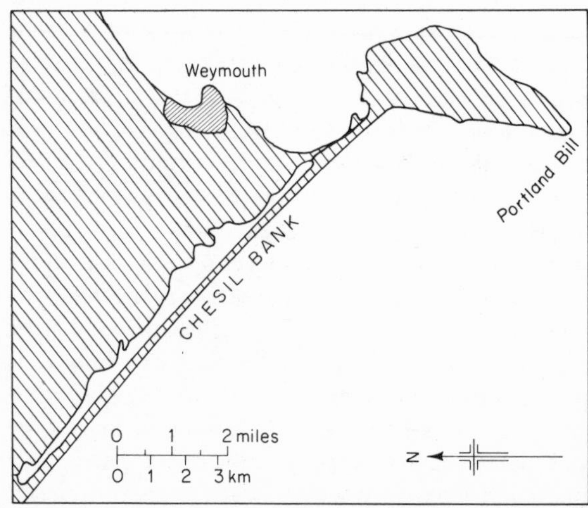

Fig. 7.5 *Chesil Bank, southern England: a shingle bar cutting off a lagoon*

7. Littoral Deposits

The *littoral zone* lies along the shore-line itself between high tide mark and low tide mark. Littoral deposits can be examined at low tide on beaches, mud-flats, estuaries, and so forth. As everyone knows, they are very variable, ranging from bouldery or pebbly deposits to sands and muds. Coarse material predominates, but muddy sediments may accumulate in estuaries and in places sheltered from waves and currents. The general effect of littoral deposition is to simplify the coastline by filling in or cutting-off indentations (Fig. 7.5).

Beach deposits usually consist of pebbly and sandy material sorted by wave action. They contain few organic remains other than broken and worn shell fragments. The pebbles are made largely of resistant rock-types and are usually partly rounded by abrasion. They are generally concentrated high up on the beach and may make a *storm beach* only reached by the waves in times of rough weather. The pebble or *shingle* deposits grade seawards into sands and even muds sloping towards the sea at a low angle. The surface is often moulded into *ripple-marks*, small, closely spaced ridges produced by current and wave action. Behind the beach itself there may be *dunes* of sand blown back by wind from the drying beach sand.

Banks and bars of sand and shingle may be built up offshore, where the gradient of the sea-floor is low, by the rotating action of breakers. These structures may finally be built up above normal high tide level, forming barriers to the further action of the waves. The development of a bar may thus shut off a sheltered *lagoon* which begins to fill up with fine-grained muddy sediments (Fig. 7.5). A bar developed between two headlands in such a way as to cut off the bay between them is a *baymouth bar* or *ayre* (Shetland). A bar connecting an island with the mainland is a *tombolo*. Longshore drift may build out a *spit* where tidal currents flow past a headland, and the washing of waves round the end of the spit may produce a *hook*.

The last types of littoral deposits which must be mentioned are *estuarine deposits* formed where a river mouth widens out to open into the sea. Here the conditions are unusual. The estuary is protected from vigorous wave action, though it may be swept by strong tidal currents. The mixing of fresh and salt water tends to cause flocculation of the clay particles carried in suspension. These two factors together lead to the deposition of fine-grained material and typical estuarine deposits are *clays or silts* often showing a close lamination, and containing a good deal of decaying organic matter.

8. Neritic Deposits

Detrital sediments. The neritic deposits are those laid down below low tide mark on the continental shelf. They are variable both in character and distribution because there are many local differences in the strength of waves and currents, the supply of detrital material and the activities of organisms. Maps of the sea-floor

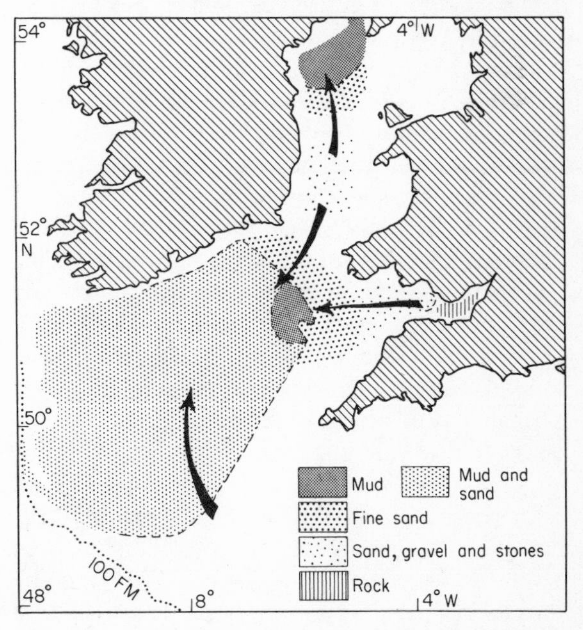

Fig. 7.6 *The distribution of neritic sediments on the floor of the Irish Sea and English Channel (after Stride). The arrows show directions of decreasing grain-size*

around southern Britain, for example, show that mud, sand and gravel are accumulating in different areas, the coarseness of the sediment being controlled largely by the strength of the tidal currents; areas swept by unusually fast currents are floored by bare rock and are not accumulating any sediments (Fig. 7.6).

Perhaps the most widespread neritic sediments forming today are fairly coarse *sands*, well sorted by current action. The surfaces of such sands may be ripple-marked and may also be moulded into wave-like *megaripples* several feet in height by the action of currents flowing over them. They often show the kind of bedding known as *current-bedding* which is described on page 108. A special type of marine sand is *greensand* coloured by the green mineral *glauconite* which is a hydrated silicate of iron and potassium.

Neritic *muds* are found opposite the mouths of large rivers such as the Amazon and in places sheltered from strong currents. Flocculation of iron compounds brought in by rivers leads to the deposition of *siderite*, iron carbonate, $FeCO_3$, and *chamosite*, hydrated iron aluminium silicate. Valuable iron-ores, such as those of the English Midlands, are produced by these minerals in ancient neritic sediments.

Limestones. Sediments which are made largely of carbonates—the many varieties of *limestones* and *dolomites*—are derived from material held in solution in sea-water. This material is extracted from solution by marine organisms or, more rarely, precipitated directly. Because the shallow seas are well-lit and well-aerated, they support a large population of animals and plants, many of which protect themselves from waves and from each other by building shells or skeletons of calcium carbonate. These calcareous skeletons provide the material for limestones of organic origin.

These *limestones of organic origin* can be formed in several ways. Empty shells and skeletal fragments are treated by waves and currents like detrital particles, that is, they are rounded and broken up, sorted according to size and deposited when current velocities decrease. The resulting sediments are *shell-banks* made of large fragments, *shell-sands* of sand-sized particles, and *lime-muds*. In addition, organic limestones may be built without the action of waves and currents by the growth of generations of fixed or sedentary organisms with calcareous skeletons. These are *reef-limestones* and the most important organisms which build them are *corals* and *algae*, members of the group of primitive plants to which the seaweeds also belong. The important features of reef-limestones are, firstly, their primary shapes—they form short, rather thick beds or lenses or bun-shaped masses corresponding in extent with the living reef—and, secondly, the preservation of at least some of the organic remains *in the position of growth* on the site occupied in life by the organism. *Coral reefs* have some special features which we discuss in a moment.

In some warm and shallow seas and lagoons where terrigenous material is scarce, calcium carbonate is precipitated directly from sea-water as a *chemical sediment*. These chemical limestones naturally do not show the organic structures seen in shelly or reef-limestones. They sometimes take the form of *oolites*, so called because they are made up of innumerable little spherical particles resembling fish-roe (Greek, *oion*, egg). Oolitic limestones, which can be seen forming today in the Bahamas, arise from the deposition of successive coatings of calcium carbonate around tiny nuclei as these are trundled to and fro between tide-marks.

Coral reefs

Coral reefs are interesting to the geologist because they are formed in a restricted and specialised environment and ancient coral reefs may therefore provide records of the past distribution of this type of environment. The *reef-building corals* (p. 126) and the algae and other animals associated with them will, at the present day, grow only in clean well-lit water only a few metres in depth and at temperatures of 25–30°C. Consequently, modern coral reefs are restricted to a belt extending some 30° north and south of the equator and are most often developed in the

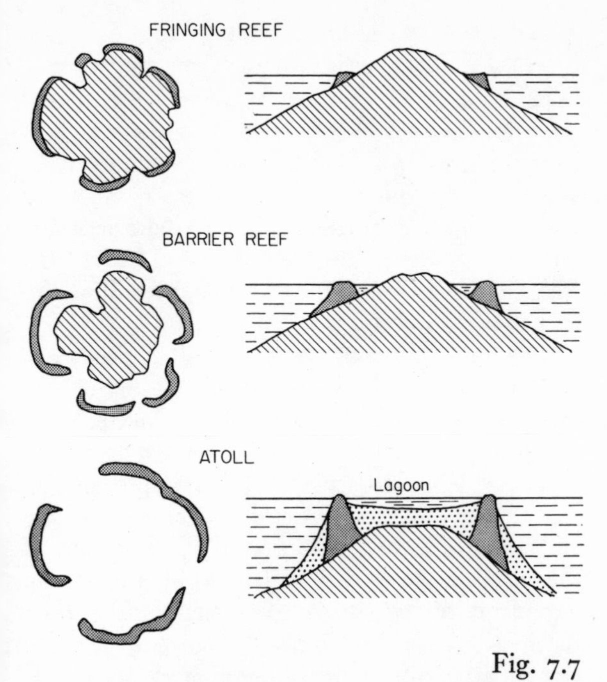

Fig. 7·7
Coral reefs: diagrammatic maps and sections

paths of warm westerly equatorial currents. Furthermore, since the organisms are particular about the depth of the water, reef formation can only start from a pre-existing foundation and it will be stopped by rapid sinking or elevation of this foundation.

Coral reefs of the present day occur in three forms—as *fringing reefs*, *barrier reefs* and *atolls* (Fig. 7·7). *Fringing reefs* are attached directly to the rocky shore. *Barrier reefs* stand on the sea-floor and are separated from the shore by a belt of calm shallow water. The *Great Barrier Reef* of Australia, chief example of this type, extends for a thousand miles at distances of 20 to 100 miles off the Queensland coast. *Atolls* are small, irregular ring-shaped reefs which are not attached to any visible land and occur especially in the Pacific Ocean. The ring of reef-material encloses a sheltered *lagoon* averaging perhaps 45 metres in depth. The outer wall of the reef is made by steep slopes of reef-debris falling rapidly to the ocean depths.

The formation of fringing and barrier reefs is easy to understand, since the shallow sea-floor off the coast provides a suitable basis from which coral growth can start. The origin of atolls is a little more complicated since they often occur within deep oceans. Charles Darwin thought that fringing reefs could evolve into barrier reefs and finally into atolls as a result of slow subsidence of the sea-floor. Continued upward growth of the corals kept pace with subsidence—thus the corals could keep their heads just below the sea-level, but the land around which they grew was swamped and finally drowned. A rival to this *subsidence theory* of Darwin's was proposed by John Murray who thought that atolls might be built up without changes of relative sea-level if the coral ocean already contained a large number of submarine hills with tops not far below sea-level from which coral growth could start. It is now known that submarine hills or *seamounts*, mostly representing old volcanoes, do exist in many parts of the Pacific and, moreover, borings and geophysical surveys have shown that in some atolls coral rock is underlain by basic volcanic rocks. As we have already seen, there was probably a general lowering of sea-level during the recent Ice Age and during this time many volcanic islands may have had their summits planed off by marine erosion. When the sea-level gradually rose again with the melting of the ice-sheets, reef-building corals which had colonised the stumps of these islands were able to grow upwards, keeping pace with the rising water, to build barrier-reefs at the margins of the platforms; further submergence converted the reefs into atolls. The formation of atolls can thus be best explained by this *glacial control theory*.

9. Bathyal or Slope Deposits

Muddy deposits. The continental slope and adjacent regions receive two main types of deposit. The finest terrigenous material drifted out in suspension by surface currents may settle to give *marine muds*. Such muds may be blue, green, red or yellow, and are coloured largely by glauconite (p. 86) or by other iron compounds. They may, in some regions, be rich in volcanic ash or in fine coral debris. The organic remains entombed in deep-water marine muds include skeletons of free-swimming creatures and, more important, those of minute organisms which in life float passively in the surface layers of the sea as the *plankton*; *foraminifera* which make their skeletons of calcium carbonate and *radiolaria* with skeletons of silica are the two most common groups whose remains rain down into the mud.

Turbidites. The second widespread type of bathyal sediments are *turbidites*, the deposits of turbidity currents flowing down the continental slope. The characteristic turbidites make *graded beds* in which coarse sediment passes gradually up into finer material (p. 107; see also Fig. 10.1). The composition of a turbidite depends a good deal on that of the source area. For example, if the sediment stirred up to make a turbidity current is rich in volcanic material, the graded bed which settles out from the current will also be rich in this material. Shallow-water fossils contained in the original sediment may be incorporated in the turbidite derived from it. Turbidites are sometimes laid down in delta-like masses around the mouths of submarine canyons and they may also be spread out more widely over the deep sea floor.

Deposits of restricted deep-sea basins. Small deep-water basins may be developed by various mechanisms quite close to land and these basins are often cut off from the seas of the open ocean by a submarine ridge at which the sea-floor rises to shallow depths. In such small deep basins, exemplified by the Black Sea and by some Norwegian fjords, the bottom waters may become stagnant and devoid of oxygen. As a result, the organic matter which falls to the sea-floor does not decay completely. Organic material contributes as much as 30% to the total bulk of sediment (mostly mud) and in the reducing conditions produced by lack of oxygen *hydrocarbons* are formed from it. Sulphides of iron and other metals are also produced and colour the mud black. This black mud, rich in organic matter, is called *sapropel* and is believed to be the parent material of *petroleum*.

10. Abyssal or Pelagic Deposits

The characters of the typical abyssal deposits depend on their remoteness from land. These deposits receive so little terrigenous material that contributions from other, somewhat surprising sources form most of the total bulk. As these contributions are small, abyssal deposits accumulate very slowly; it is estimated that the Red Clay (see below) increases by only about a millimetre in a thousand years.

The chief components of abyssal deposits are organic remains derived from planktonic organisms, volcanic dust and the finest particles of terrigenous material. The composition of the deposits depends largely on the depth of the water. With increasing depth calcium carbonate and finally silica become soluble in sea-water; in the deepest seas the skeletons of planktonic organisms are therefore dissolved before they reach the floor. Two classes of abyssal deposits can thus be distinguished—the *oozes*, largely of organic origin, and the *Red Clay*.

The *oozes* are of two main types, calcareous and siliceous. *Calcareous ooze*, which accumulates down to depths of some 3,600 metres, contains some 30% of skeletal material from foraminifera and pteropods (pteropods are gastropods which are adapted for pelagic life) enclosed in a slimy mass of colloidal clay particles. *Globigerina ooze*, in which the skeletons of foraminifera called *Globigerina* predominate, covers 50 million square miles of the ocean floor at an average depth of 3,600 metres. *Siliceous ooze* derives its material from the silica-skeletons of radiolaria and microscopic pelagic plants known as diatoms. As silica is less soluble in sea-water than calcium carbonate it may accumulate down to depths of over 5,000 metres. *Radiolarian ooze* covers some 5 million square miles in the tropical Pacific; *diatom ooze* is most widespread in polar regions, especially around Antarctica.

The *Red Clay* accumulates, with extreme slowness (see above), at depths greater than about 4,000 metres, covering half the floor of the Pacific Ocean. It consists of very fine particles of clay-minerals mixed with the finest volcanic ash and pieces of waterlogged pumice. It is coloured chocolate or red by manganese or iron compounds and nodules of manganese oxide are sometimes so abundant as to constitute a possible source of the metal. Because of the slow rate of deposition, some very strange components, which in ordinary sediments would be so much diluted as to be unnoticed, appear quite conspicuous. These include particles from meteorites and the earbones of whales and teeth of sharks (both made of very insoluble bony material). Finally, Red Clay is many times more radioactive than the igneous rocks of the continents, possibly because radioactive elements in sea-water are adsorbed on the clay particles. Altogether, it is a very queer deposit indeed.

8

SURFACE PROCESSES: 4

Ice-masses

This chapter deals with the geological work of glaciers and ice-sheets. It considers the effects of erosion, transport and deposition by ice and includes a short account of the ice-sheets of Greenland and Antarctica. The effects of the recent Ice Age and of ice-ages in earlier periods of geological history are briefly mentioned.

1. Glaciation

At the present day, as much as 10% of the land surface is covered by ice in the form of glaciers and ice-sheets. The main ice-sheets are of course situated in the polar regions of the Arctic and Antarctic, but smaller glaciers occur in the high mountain ranges of temperate and even tropical latitudes. The effects of *glaciation*, the processes associated with ice-action, are of special interest to geologists because the earth has recently passed through an *ice-age* during which glaciation was abnormally widespread. The ice-sheets of this *Great Ice Age* covered much of northern Europe and America and, as the effects they produced are still very conspicuous, it is important to know how they worked.

On uniformitarian principles, it would be excellent to interpret the effects of the Great Ice Age and of earlier ice-ages by reference to the work of present-day ice-masses. But glaciers and ice-sheets are awkward subjects to investigate and the rate at which they work is, by human standards, very slow. In practice, therefore, most of our knowledge of ice-action has come from observations of regions which were glaciated during the Ice Age and from which the ice has long since melted away. We shall refer to these regions a good deal in the next few pages and shall deal more systematically with the Ice Age in Chapter 17.

2. The Formation of Ice-masses

In high latitudes and at high altitudes snow collects in the *snowfields* above the *snow-line.* Where the snowfall is adequate and the summer temperatures low, the thickness of snow gradually increases and its condition changes. Snow falls as flakes made of skeletal ice-crystals belonging to the hexagonal system. Under pressure in the snowfield it is gradually converted into a more compact aggregate of ice pellets looking opaque and white as a result of the presence of air-bubbles. This material is called *névé* or *firn.* When the snow becomes more than about 100 feet thick, the névé at its base is transformed by further compaction and recrystallisation into *glacier-ice* formed by solid interlocking grains, free from air and therefore translucent and blue-grey in colour; the same transformation takes place when a snowball is made unfairly hard.

As snow piles up, its weight begins to force the excess ice outward and it begins to flow very slowly under the influence of gravity. In mountain regions, ice moves down the valleys as tongue-like *glaciers* which continue to advance until the supply of ice is equalled by the wastage due to melting. In high latitudes, the ice may cover all the land to make an *ice-sheet* (Fig. 8.1). On reaching the sea, *icebergs* may break off and float away, a process known as *calving*.

We can think of any ice-mass as being made up of two parts, one of *accumulation*, where the supply of new snow exceeds wastage due to melting and evaporation, and one of *ablation* where wastage exceeds supply. The amount of ablation depends on climate, altitude and exposure. At the present day, most glaciers and ice-sheets are retreating because ablation is greater than accumulation; in these conditions the layer of ice becomes thinner, the snout retreats and finally the ice-body deteriorates into separate, stagnant masses no longer connected with the source of supply.

Melting in the region of ablation produces abundant *melt-waters* which flow out from the ice-bodies to make streams, rivers and lakes. *Superglacial streams* flow over the ice itself when the surface melts in summer. These streams may plunge down cracks in the ice to form well-like hollows or *moulins* and may then continue beneath the ice as *subglacial streams* to join the other melt-waters near the snout of the glacier.

Fig. 8.1 *The margin of the Greenland ice-sheet*

3. Movement of Ice

It is difficult to find out how a glacier moves, though it is fairly easy to demonstrate—for example by showing that lines of stakes driven into the ice are gradually displaced—that it does move. Moreover, although glacier-ice is brittle and has considerable rigidity, a glacier can wind its way round bends in its valley as if it were flowing like a viscous liquid. Observation shows that we can distinguish two zones within glaciers (Fig. 8.2):

(1) an upper *zone of fracture*, up to 200 feet thick, in which the ice is brittle and opens under tension to give deep cracks called *crevasses*.

(2) a lower *zone of flow* in which the ice behaves as a viscous material.

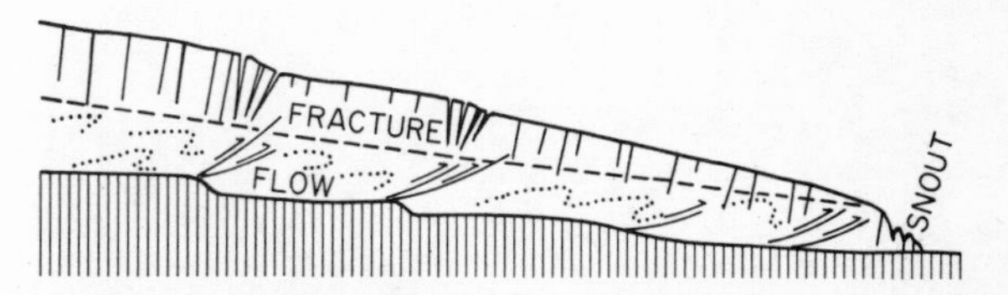

Fig. 8.2 *The zones of fracture and flow in a valley-glacier. Crevasses and other fractures develop above steps in the valley-floor*

Movement seems to take place in two ways. In the first, rigid blocks of ice are pushed forward passively along surfaces of movement within the ice—this kind of motion is demonstrated by the displacement of layers of dust or *dirt-bands* incorporated in the ice. In the second way, movement of the glacier is achieved by movements of

Chapter eight

the individual ice-grains within it. These grains are thought to rotate until certain planes of the crystal structure come into a particular orientation; after this stage has been reached, further movement takes place by *gliding* within the atomic lattice of the ice-crystals themselves. The sum of all the rotations and gliding of individual grains produces an effect of flow in the ice-mass as a whole. As would be expected, this *quasi-viscous creep* takes place mainly in the zone of flow. Passive displacement on fracture-planes in the ice occurs in the zone of fracture and near the rock-floor, especially where its gradient is steep. Steps in the valley-floor produce *ice-falls* marked by close-set crevasses (Fig. 8.2).

The motive force of glacier flow is *gravity*—the ice moves, as a whole, from high levels to low levels. But because the solid ice-mass can transmit the motive impulse from the source area, the base of a glacier can ride up and down over irregular surfaces and its work is not limited by a base-level as river-erosion is; consequently, ice-masses can scoop out hollows which are below sea-level and can cause *overdeepening* of a valley or lake-floor.

The velocity of ice-movement depends on the gradient, the thickness and the temperature of the ice, increasing as these increase. Alpine glaciers move 80–150 metres a year, some Greenland glaciers as much as 33 metres in a day—these can be seen to move—and the very cold Antarctic ice only a few millimetres a day. Movement is not uniform. In valley-glaciers, friction against the walls and floor results in *differential flow*, by which the central and upper parts move faster than the marginal and basal parts.

4. Classification of Ice-masses

We can recognise two types of ice-masses, formed in different circumstances and behaving in different ways. These are (1) the *mountain glaciers* which fill the valleys of a mountain area and are dominated by rocky mountain-slopes, and (2) the *ice-sheets* which virtually cover up the land and form enormous low domes of ice, as in Greenland.

5. Mountain Glaciers

Corrie-glaciers. In mountainous regions such as Snowdonia, the Scottish Highlands or the Alps which were glaciated during the Great Ice Age, one often sees, scooped out of the sides of the mountains, great armchair-shaped hollows with steep back walls often thousands of feet high (Fig. 8.3). These hollows are called *corries* (Scotland), *cwms* (Wales) or *cirques* (France). They often contain a small *corrie lake* held up by a rock-barrier across the front or by a pile of glacial debris, and they often make the head of a present-day river-valley. It is believed that when the climate was colder each corrie was filled with ice in the form of a *corrie-glacier*.

Corries appear to begin their existence as accidental hollows in the mountain-side where snow lies throughout the summer. These hollows are deepened and extended by frost action until they become occupied by small glaciers capable of erosive work. Where a glacier leaves the snow-

Fig. 8.3 *Corrie: Glaslyn, a cwm carved in the side of Snowdon, North Wales*

Fig. 8.4
Roches moutonnées, Iceland

field there is often developed a great crevasse called the *bergschrund* which seems to play a part in developing the characteristic steep back wall of the corrie—it has been discovered that the bergschrund reaches down between the rockwall of the mountain and the glacier-ice. The rock-wall is frost-rifted and blocks broken from it are tilted towards the glacier. In winter the berg-schrund fills up with snow which freezes on to the loose blocks and in the following summer, as the bergschrund opens, these blocks are pulled away from the mountain by the moving ice. Thus, the *plucking-out* of blocks causes the corrie to eat its way back into the mountain.

Valley-glaciers. Where snow is plentiful, ice moves out far beyond the corrie, following the course of pre-existing river-valleys to produce *valley-glaciers*. These glaciers act as agents of erosion, transport and deposition and their operations greatly modify the valleys taken over from the rivers.

Erosion by valley-glaciers. The sole of the glacier is shod with frozen-in blocks of all sizes which act as abrasives to wear away the valley-floor. Small crags over which ice has passed are smoothed, rounded and streamlined by friction in this way. They are called *roches moutonnées* (French *mouton*, sheep) from their resemblance to the backs of a flock of sheep (Fig. 8.4). It is often possible to tell from their shape which way the ice which moulded them was travelling—the upstream sides are rounded off, the downstream sides steepened by the *plucking* of blocks which got loosened and frozen into the ice. Large crags standing in the path of the ice provide protection for softer rocks on their downstream side, so that a tail is preserved there. This *crag-and-tail* structure is well seen in Edinburgh—the Castle stands upon the crag and the tail on its lee side runs down to Holyrood.

Abrasion by glaciers produces deep scratches known as *glacial striae* on the rocks of the floor which run in the general line of ice movement and provide a record of this movement. *Glacial polish* is produced on some soft rocks, especially lime-stones. Weak places are gouged out and a *fluting* or *grooving* of the surface may result. The stones carried in the glacier are themselves worn, scratched and polished. They often acquire a characteristic sub-angular shape—that of a 'faceted flat-iron' from rubbing against the valley-floor while firmly embedded in the ice. Constant abrasion produces a quantity of finely powdered *rock-flour*.

Fig. 8.5 *A valley-glacier and its deposits, Iceland: hummocky glacial moraine is seen in the foreground and fluvioglacial gravels in the valley-bottom*

Transport by valley-glaciers. A glacier makes a very different agent of transport from a river. It acts as a kind of conveyor-belt, carrying along fragments of all sizes with little sorting, and when it melts the material is dumped wherever it happens to be (Fig. 8.5). Melt-waters derived from the ice wash away the finest material, leaving behind a *glacial moraine* of mixed sandy, clayey and bouldery material. The total load carried by a glacier may be enormous and the largest individual blocks may weigh many tons.

Valley-glaciers receive scree and other debris from the rocky mountain-slopes above them. This material, lying on the surface of the ice near the valley sides, makes a moving scree called a *lateral moraine*. When two valley-glaciers join one another, the lateral moraines on their inner sides unite and move on as a single *median moraine*. Many rock fragments become enclosed in the ice by falling down crevasses, and these are carried along as *englacial* material. Some reach the floor and join debris picked up from below to make the *subglacial* material. When the ice finally melts, the morainic material is let down on to the ground; the lateral moraines build up ridges parallel to the length of the valley, while the englacial material collapses on top of the subglacial material to give an irregular cover of *ground-moraine*. *Terminal moraines* are built up around the snout of the glacier as crescentic ridges roughly perpendicular to the length of the valley.

The effect of a glacier on its valley. The broad ice-stream, winding down a valley taken over from a pre-existing river, modifies the valley to suit its own means of progression and so leaves evidence of glaciation which can be recognised long after the ice has melted away. We have seen that a normal river-valley has a V-shaped cross-section. The glacier widens the bottom of the valley by lateral plucking and by direct abrasion until the *cross-section becomes U-shaped* (Fig. 8.6). At the same time, kinks in the valley are straightened out by the wearing-away of spurs projecting from the valley sides. The *truncated* or *bevelled* remains of these spurs can often still be recognised. Since glacial erosion depends largely on the volume of the ice, the main valleys down which glaciers flow are *over-deepened* and over-widened in comparison with tributary valleys. The accordant junctions of the tributaries are therefore destroyed and on melting of the ice their valleys are left as *hanging valleys* whose streams must join the main river by cascading down the steep walls cut by the glacier.

Finally, the effects of glaciation destroy the profile of equilibrium established by the river and produce irregularities of profile which lead after melting to the development of many *glacial lakes*. Some of these lakes occupy *rock-basins* gouged out by the ice. The floors of such lakes may descend far below sea-level and could not possibly have been eroded by rivers, whose action is related to base-level. Loch Morar in western Scotland is over 1,000 feet deep, although its surface is less than 50 feet above sea-level and many of the *fjords* of Norway are similarly over-deepened. Over-deepening is favoured by the occurrence of belts of soft or shattered rock, and a good example of this control is provided by the Great Glen of Scotland which contains Loch Ness and Loch Lochy and is eroded along the line of a big earth-fracture. Modifications of the smooth river-profile are also produced by the dumping of moraine, and many glacial lakes are held up behind morainic barriers.

Piedmont glaciers. In mountain regions of high snowfall, glaciers may protrude beyond the narrow valleys on to low, flatter ground—the *piedmont*—at the foot of the mountains. Here, several glaciers may unite to make a great 'lake' of ice, the *piedmont glacier*. The standard example is the Malaspina Glacier of Alaska on the coastal plain below the Mount St. Elias ranges. This glacier is almost stationary and the morainic material it carries supports extensive forests; occasional movements of the ice beneath lead to the engulfing of trees and soil by the glacier.

Fig. 8.6

A glaciated valley: Glen Rosa, Arran, showing U-shaped profile, truncated spurs and, in the distance, the hanging valley of a tributary

Chapter eight

6. Ice-caps and Ice-sheets

These ice-bodies, in contrast to the corrie, valley and piedmont glaciers, cover up most of the land-surface and are not dominated by rocky mountain-slopes. *Ice-caps* are relatively small. They cover high plateaux in Iceland, Norway and Spitzbergen, their formation depending as much on altitude as latitude. During the Great Ice Age, ice-caps covered the mountain-ranges of the Alps and Caucasus.

Ice-sheets are very much larger, covering large parts of continents and having areas of millions of square miles. The upper surface of the ice makes a great dome from whose high parts ice moves outward under gravity. Isolated mountain peaks projecting through the ice are known as *nunataks*. The two ice-sheets of the present day are those of Greenland and Antarctica.

The *Greenland Ice-sheet* (Fig. 8.1) makes a flat dome fringed by a narrow marginal ribbon of high mountains through which the ice moves to the sea in great glaciers. Measurements by geophysical methods, including observations on artificially produced shock-waves, have shown that in the central parts of Greenland the base of the ice goes down to well below sea-level (Fig. 8.7). It seems as if the enormous weight of the ice-sheet has depressed the floor beneath it, so that the ice now fills a great basin. This depression of land beneath the ice is an example of a local isostatic adjustment (p. 13).

The *Antarctic Ice-sheet* is many times larger even than that of Greenland and has an area of 5 million square miles, one and a half times the area of the United States. Measurements of ice-thickness by the Commonwealth Transantarctic Expedition of 1958 have shown that a rugged mountainous land lies buried beneath the ice (Fig. 8.7). A remarkable feature is the development of enormous expanses of *shelf-ice* extending out to sea from the continental ice-sheet. The

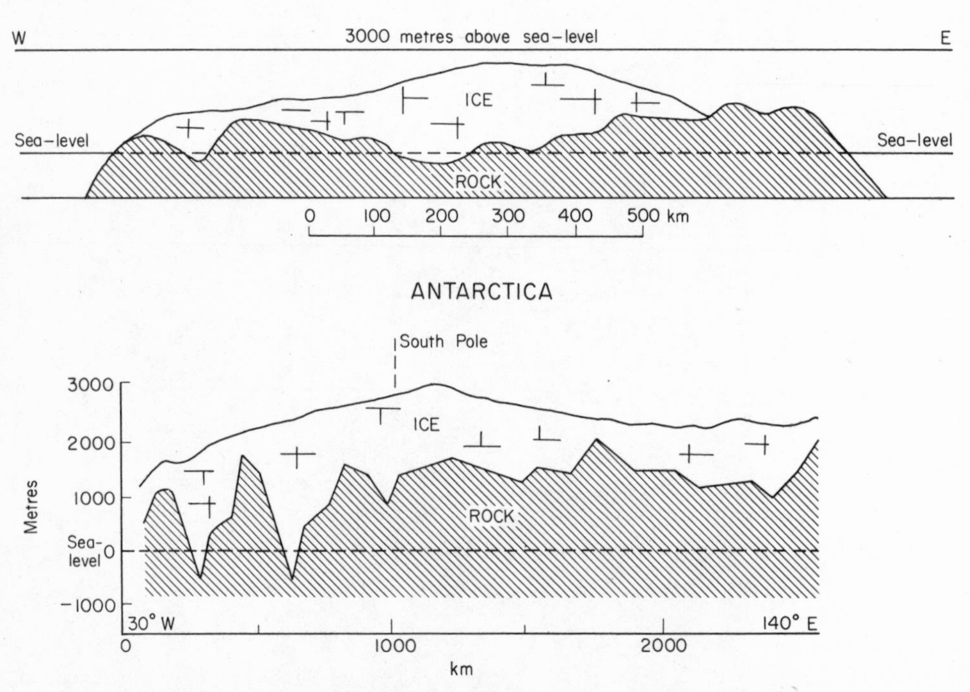

Fig. 8.7 *The Greenland and Antarctic ice-sheets: profiles showing the thicknesses of ice determined by geophysical measurements*

Ross Shelf-Ice is some 500 miles long and up to 300 miles wide. It may be as much as 1,000 feet thick and the ice-cliffs at its seaward margin are up to 280 feet in height. This shelf-ice is afloat, and has been formed by the pushing out over deep water of the inland ice with its cover of névé.

7. Deposition by Ice-masses

Glacial drift. The general term *glacial drift* is used for the material deposited by ice-sheets and glaciers, the word drift having been first applied to the superficial debris left after the Great Ice Age in the mistaken belief that this debris was a product of the Flood.

The *lateral moraines* of valley-glaciers sometimes survive as ridges or shelves on the valley-sides, but these and the *median moraines* tend to be largely destroyed by later stream-action. *Terminal moraines* are often preserved as crescentic barriers across valleys, breached by outflowing streams; sometimes several moraines one behind the other mark positions where the glacier snout remained stationary for some time in its retreat. *Ground-moraine* formed by slow collapse of the ice during melting is the typical deposit of an ice-sheet. It is composed of two rather different parts, a lower layer of compact rock-flour carrying boulders of all sizes, derived from the sub-glacial material; and an upper part of looser, more porous drift representing englacial and superficial debris from which the finer particles have been washed away by melt-water.

Boulder-clay. The sediment deposited by ice is called *boulder-clay* or *till* (Fig. 8.8). Because the material is not sorted during transport, till is a mixture of rock-flour, sand and boulders, the latter often sub-angular and sometimes scratched. Its composition depends on that of the rocks over or past which the ice has flowed: sandstone bed-rock gives a sandy drift, chalk a chalky drift, and so on. The boulders in till are called *erratics* (Latin *erro*, I wander) and when they are derived from distinctive rocks it is often possible to identify the source from which they came and so to establish the direction of travel of the ice-mass. For example, the boulder-clay on the eastern coasts of Britain contains erratics which can only be matched in the Oslo region of Norway and it is concluded that, during the Great Ice Age, Scandinavian ice spread across the North Sea to reach this country. Conspicuous erratics which are perched in spectacular situations are known as *erratic blocks* (Fig. 8.9)—in Wales and Scotland, erratic blocks are often

Fig. 8.8 *Boulder-clay or till, Co. Down, Northern Ireland*

Fig. 8.9 *An erratic block of Silurian sandstone resting on Carboniferous Limestone: the Norber, Yorkshire*

Fig. 8.10 *Drumlins in Strangford Lough, Co. Down, Northern Ireland*

scattered high on the flanks of the mountains, showing that the greater part of the land was covered by ice during the Ice Age.

The ground-moraine laid down by ice-sheets forms a very irregular blanket, varying from a few feet to several hundred feet in thickness. It is often piled haphazardly in little hummocks but may instead be moulded into many small streamlined hills, all much the same in height and shape. These hills, known as *drumlins*, appear to have been shaped by the action of clean, moving ice as it rode over pockets of stagnant ice laden with debris. They are elongated in the direction of ice-flow and their sides and crests are smooth and rounded. In mass, they give to the landscape a very recognisable 'basket-of-eggs' appearance (Fig. 8.10).

The blanket of boulder-clay left by ice-sheets of the Great Ice Age has smothered many irregularities of the pre-glacial land-surface. Deep river channels were sometimes filled in, and the displaced rivers were forced to make new post-glacial channels. The *buried channels* of old rivers sometimes prove a considerable nuisance during mining or engineering works when they are encountered unexpectedly. They can be detected in advance by the sinking of many boreholes or by geophysical techniques which depend on differences in physical properties between the boulder-clay and the substratum.

Fluvioglacial deposits. The melting of glaciers and ice-sheets at their margins and in the region of ablation provides melt-waters which carry away some of the load of the ice. Fine rock-flour held in suspension often gives the water a milky blue appearance such as is seen in many Alpine streams. These waters formed in connection with ice-masses lay down *fluvioglacial* *deposits* with the general characters of river and lake sediments. The primary shapes of the deposits are often rather unusual and are decided by the relationships of the streams and the ice (Fig. 8.11). Deposits of superglacial streams appear as lenticular or ribbon-like bodies in the boulder-clay. Streams running between the side of a glacier and its valley-wall build up shelf-like *kame-terraces* on the side of the valley. Streams issuing from the ice-front may make small deltaic cones which build low knolls or ridges; similar forms are produced from the collapse of debris filling crevasses in stagnant ice, and both are known as *kames*. Beyond the terminal moraines of the ice-mass, the escaping melt-waters lay down an *outwash fan* of sands and gravels, the deposits of many wandering, heavily laden streams. It is often pitted by rounded depressions without outlets known as *kettle-holes* which mark the places where the last blocks of ice remained unmelted.

Eskers are winding ridges of sand, gravel and boulders, up to 100 feet high and often tens of miles in length; they have steep sides, narrow flattish tops and look remarkably like railway-embankments. Their origin is reasonably connected with deposition by subglacial streams during the retreat of an ice-sheet—as the ice-margin retreated, the point at which the stream issued from it moved back and so, year by year, the length of the esker was increased. *Beaded eskers* may be formed when the ice-margin is

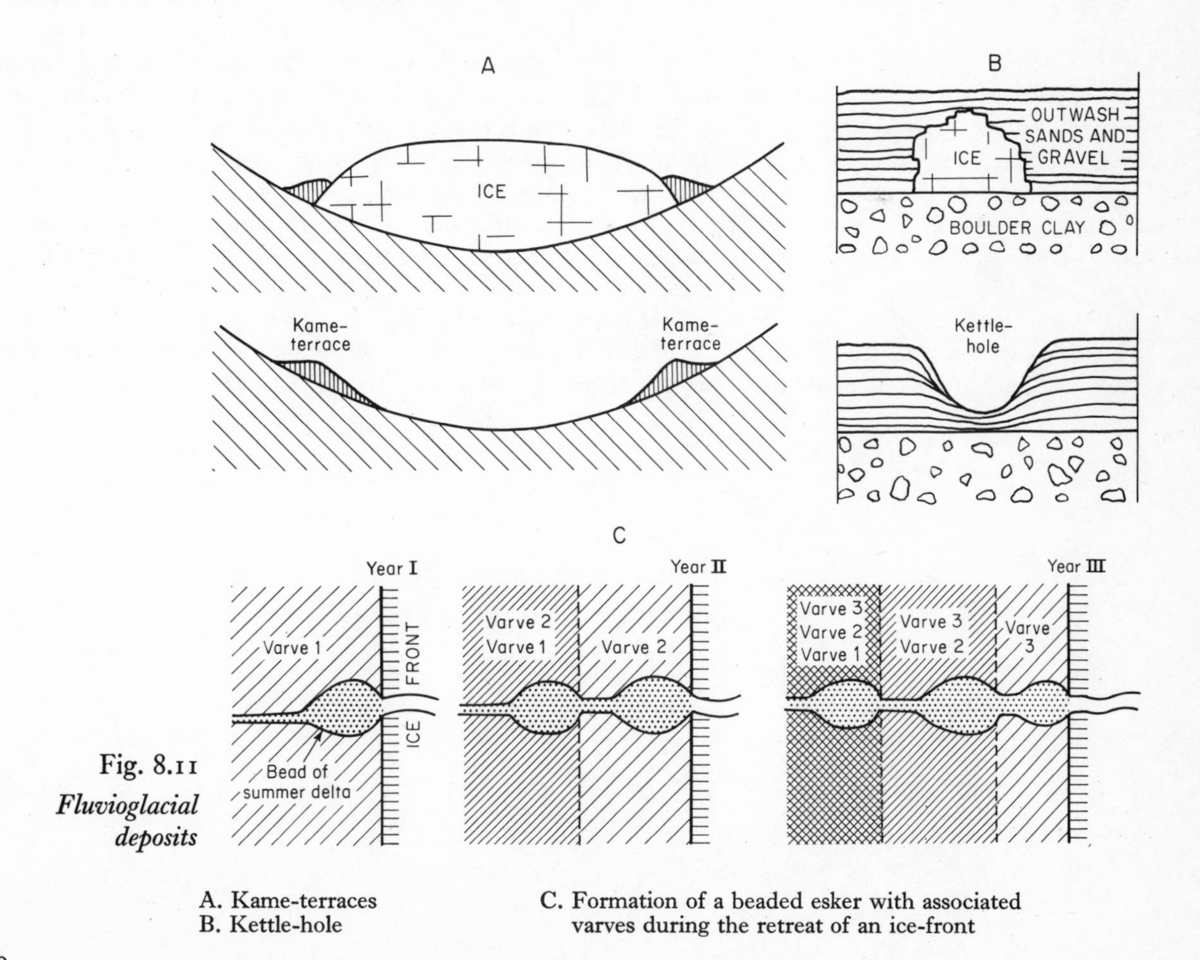

Fig. 8.11
Fluvioglacial deposits

A. Kame-terraces
B. Kettle-hole

C. Formation of a beaded esker with associated varves during the retreat of an ice-front

fringed by standing water; the high, wide beads represent summer deltas laid down when the stream was most active and the lower parts which link them are the winter deposits.

Lakes held up against a retreating ice-front receive other deposits besides those of beaded eskers. The finer material carried into the lake by the glacial streams spreads out beyond the deltas to form thin beds of fine sand and mud. These deposits are the *varved sediments* described on page 77. As we have already seen, each complete bed or *varve* shows a gradation from a coarser summer layer laid down when abundant melt-water entered the lake to a finer winter layer representing the material which settled out when melting stopped during the cold weather (Fig. 6.14).

8. Ice-ages

Ice-ages are not very common geological events and if we did not happen to be living in the aftermath of one, the effects of glaciation might not seem very important. Geologists working in regions covered by ice-sheets of the Great Ice Age, however, find themselves surrounded by evidence of former glaciation in the presence of extensive deposits of boulder-clay and in the glacial sculpturing of the landscape. Moreover, this glaciation had other effects of world-wide importance, such as the general lowering of sea-level caused by the locking-up of so much water in ice-sheets (consequences of this are discussed on pages 83, 87) and the isostatic adjustments that resulted from imposition and removal of the load of ice. We have to take account of all these effects when we try to understand the earth as it is today.

The Great Ice Age is a recent affair, geologically speaking. It took place during the last million years of geological time in the period known as the Pleistocene (p. 119) and the ice vanished from Britain only about 10,000 years ago. Evidence of earlier ice-ages is also abundant. The most famous of these is the *Permo-Carboniferous Glaciation* occurring 300 million years ago. Rocks which represent boulder-clays laid down during this glaciation are found in India, South Africa, South America and Australia, in localities many of which now lie in tropical latitudes. We have mentioned the problem of these deposits in Chapter 2 and have seen that they may have reached their present situations as a result of *continental drift*; at the time of the glaciation the regions affected were grouped in one great continental mass near the south pole and the subsequent break-up and drifting apart of this mass carried the ancient glacial deposits into their present surprising positions (Fig. 2.9).

9

SURFACE PROCESSES: 5

Wind

The wind is an active geological agent in regions where the climate is arid and in this chapter we deal with the distinctive erosional land-forms and sedimentary deposits which are produced by it.

1. Erosion by the Wind

The cover of loose material produced by weathering is liable to be blown away when it is not held together by moisture or bound down by the roots of plants. Wind is therefore an important agent of erosion and transport in regions where the climate is arid or semi-arid. At the present day its activities are most spectacular in the *deserts* of the arid zones lying north and south of the tropics, for example, the Sahara, the Arabian and Gobi deserts in the north, the Kalahari and Australian deserts in the south. In more humid climates, wind erosion acts on a small scale on the drying sand of beaches, rivers or old lake-floors and, most important to man, on soil laid bare by ploughing (p. 229). Through much of geological time there were no land plants to stabilise the weathered material and wind-erosion may then have been of more widespread importance.

Where wind-action is concerned, we must make a distinction between *dust* and *sand*, because the wind treats these two fractions in different ways. Dust particles, less than 0·06 millimetre in diameter, can be carried in suspension high in the air even by winds of average velocities. As a result, such particles may be blown away and deposited far from their source; their mobility is illustrated by the fact that dust from the Sahara often falls in southern Europe and on ships far out in the Atlantic. Sand particles, with average diameter about 0·20 millimetre, on the other hand, are too heavy to remain long in suspension—when moved by the wind they keep low and frequently fall back to earth. The wind is therefore an efficient sorting agent and tends to lay down *dust deposits* and *sand deposits* in separate bodies. The two types together constitute the *windborne* or *eolian deposits*. Pebbles and boulders, of course, are hardly moved at all and in regions of wind-erosion they may be left behind as tight-packed stony layers known as *lag-gravels*. *Deflation* is the term given to the process of blowing-away of loose material. It is not limited by a base-level and may lower the desert-surface below sea-level.

Solid rocks, as distinct from loose particles, can only be eroded by wind as a result of bombardment by the sand-grains which it carries. This natural *sand-blast* is concentrated near ground-level since, as we noted above, sand-grains driven by the wind do not rise high. It acts in a way similar to that of the artificial sand-blast used to etch glass, eroding and polishing the surfaces on which it impinges. Corners are rounded off and soft layers worn away in a manner well illustrated by the pedestal of the Sphinx, which is made of layers of varying hardness. Cleopatra's Needle, now erected on the Embankment in London, shows an etched face which was subjected to centuries of sand-blast in the Egyptian desert; the opposite face on which the column lay is relatively unaffected. Cliffs and rocks standing up above the level of the desert floor become undercut by the sand-blast. Isolated rocks are sometimes left as *rock-mushrooms*

Chapter nine

Fig. 9.1 *Dreikanters on a desert surface, South-West Africa*

perched on a thin, eroded stalk. Loose stones lying on the desert floor are worn away on the windward side until they overbalance and so present a different surface to the sand-blast. In the end, they often develop a characteristic three-sided form rather like a brazil nut; such wind-faceted stones are called *dreikanter* (German, three edges), or *ventifacts* (Fig. 9.1).

The sand-grains which are the agents of all this erosion are themselves affected by it. They become so thoroughly rounded that they are referred to as *millet-seed sand* and they are often polished or frosted by the multitude of impacts.

2. Dust Deposits: Loess

Dust deposits build up outside the desert areas where the particles can be anchored by moisture and vegetation as they settle. An example of such a deposit is provided by the *loess* of China which covers an area the size of France to a thickness of some hundreds of feet; the dust of the loess is derived from the Gobi Desert. It accumulates as a brownish or yellowish deposit made of minute angular particles of quartz, feldspar, calcite and other minerals, and containing occasional snail-shells and bones of land animals. The loess is very uniform and does not show well-defined bedding-planes. Although unconsolidated, it will stand in vertical cliffs tens of feet high because it tends to part along vertical surfaces marked out by extraordinarily long grass-roots.

Dust deposits similar to the loess include the *adobe* of North America and the *brick-earths* of southern England and central Europe. The material of these deposits was derived from the rock-flour produced by ice-abrasion during the Great Ice Age: the deposits are *periglacial*, that is, situated in areas around the margins of the Pleistocene ice-sheets.

3. Sand Deposits

Eolian sand deposits are laid down in or near deserts and in more restricted areas such as the dunes behind beaches in moister regions. Their principal constituent is quartz, the most obdurate of the common minerals; but occasionally they may be made of broken-up shells or even of grains of rock-salt, gypsum or (in polar regions) of snow. Since the environment of deposition is arid and inhospitable, animals and plants are scarce and the deposits consequently contain very few fossils: such fossils as are present are the remains of terrestrial organisms.

Sand-grains driven by the wind move by *saltation* in a succession of short leaps. The impact of descending grains on the loose surface knocks other grains up into the air or drives them forward by a kind of *surface-creep*, so that the superficial layers of sand move gradually down-wind. Where there are small irregularities different parts of the surface receive unequal

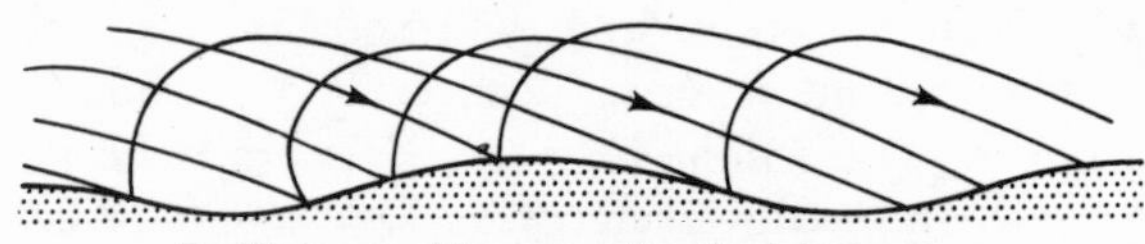

Equilibrium condition: wave-length of ripples = horizontal length of path of sand grains

Fig. 9.2 *The transport of sand-grains by wind (after Bagnold)*

The grains moving forward in short leaps strike the originally irregular surface at varying angles, and a state of equilibrium is not reached until a system of evenly spaced ripples is built up; these ripples then themselves migrate down-wind

numbers of impacts and, by degrees, the irregularities are therefore developed into a rhythmic series of *sand-ripples* which provide a surface in equilibrium with the system of winds blowing across it (Fig. 9.2).

The moving sand eventually accumulates as *sand-drifts* and *sand-dunes*. Conspicuous drifts may be built up in the lee of obstacles acting as wind-breaks or on patches where the surface is damp. Their form depends on the local sand-supply and on the velocity and constancy of the winds. Sometimes the sand is heaped up irregularly to produce a *sand-sea*, but very often it is arranged in armies of dunes, all of much the same shape and orientation. When the wind blows continually from the same direction, it tends to produce crescentic dunes known as *barchans* (Fig. 9.3) arranged with the horns of the crescents pointing away from the wind. Elongated ridges or *seifs* result from the combined action of persistent gentle winds which supply the sand and stronger intermittent winds blowing from another quarter which trim the dunes into ridges.

Sand-dunes have a complicated internal structure which can be seen on cutting into present-day dunes by special methods and also by examining consolidated sandstones representing ancient dunes. Their structure is due to the way in which the dunes accumulate and migrate under the influence of the wind. Wind carrying sand sweeps rapidly over the windward slope but is checked beyond the summit and drops some of its load near the top of the lee slope. The upper part of this slope therefore builds up rapidly and the

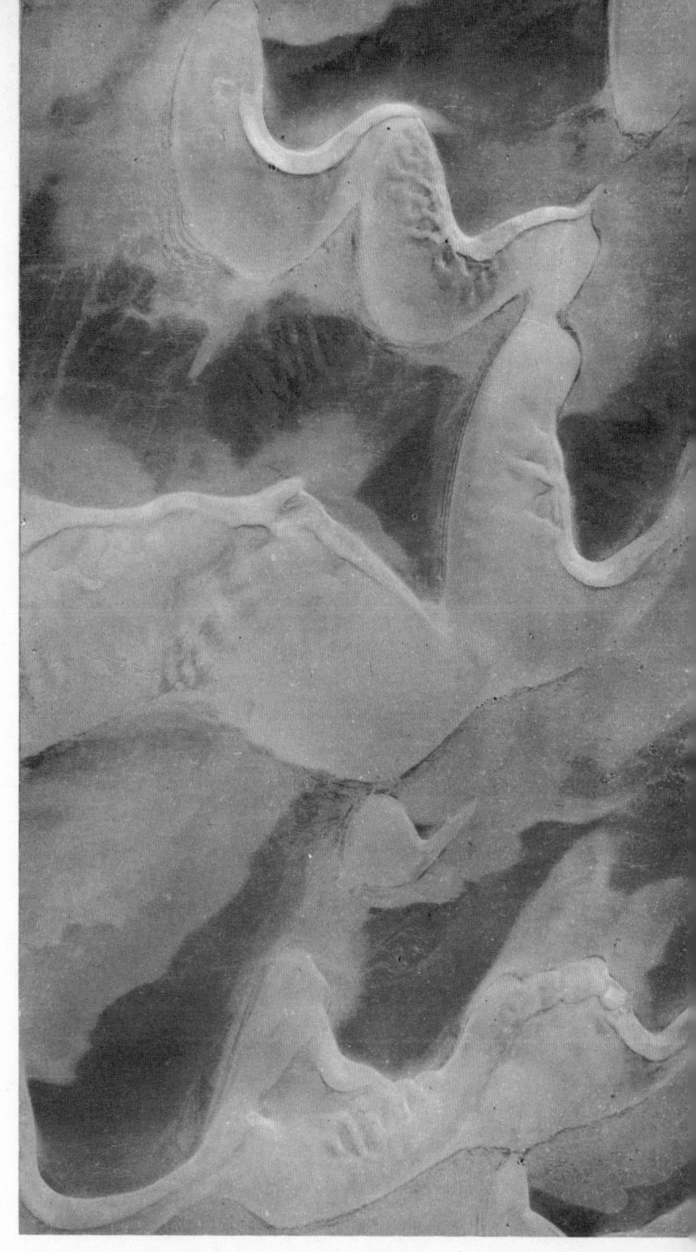

Fig. 9.3 *Barchans in Arabia viewed from the air*

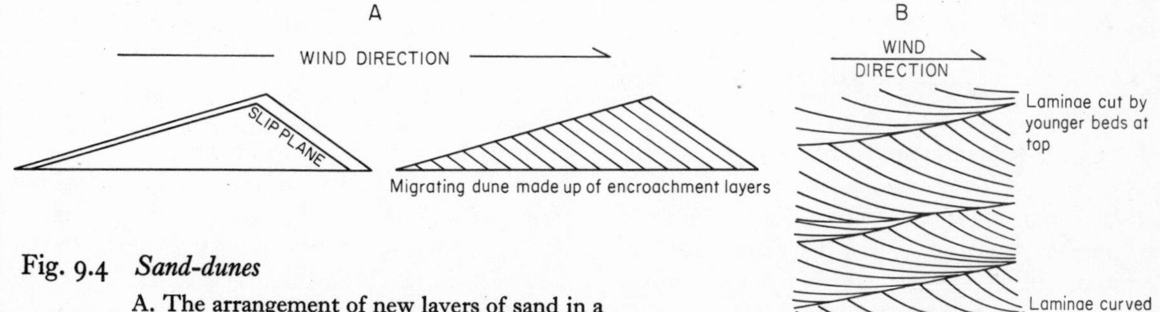

Fig. 9.4 *Sand-dunes*

A. The arrangement of new layers of sand in a stationary dune (*left*) and the structure of a migrating dune (*right*)

B. Dune-bedding, a type of current-bedding

Fig. 9.5 *Dune-bedding in eolian sandstone of Triassic age, Mauchline, Scotland*

slope becomes steeper. When its inclination approaches 34°, the angle of rest of dry sand, the sand on this slope avalanches or shears away on a slip-plane inclined at an angle a degree or two short of 34°. This process is repeated, so that the whole dune moves down-wind as sand is piled up above successive slip-planes (Fig. 9.4).

It will be noticed that the laminae of sand enclosed between slip-planes are inclined to the horizontal at about 32°; the direction in which they slope depends on the direction of the wind. It often happens that in the course of deposition a change of wind either blows away part of the dune or causes it to migrate in a new direction. New sand lamellae will then be laid down with a different orientation and their lowest members will cut at an angle across the set previously deposited. This arrangement, illustrated in Figs. 9.4, 9.5, is an example of *current-bedding* or *false bedding* (p. 108).

4. Land-forms and Sediments of Eolian Origin

The kind of landscapes to be found in regions where wind is a dominant geological agent can be deduced from the descriptions already given and are familiar to most people as the setting of 'Western' films. Where erosion is the main process at work, fine loose material is swept away and a desert-floor of bare polished rock or a surface of lag-gravel is left. Upstanding rocky masses are etched and undercut. Where deflation acts on unconsolidated sediment or weathered material, the wind scoops out wide shallow basins. These may be deepened so far that they intersect the water-table to produce fertile *oases*.

Where sand is accumulating, the landscape is dominated by the pattern of dunes or the confused structure of a sand-sea. The details of the topography are always changing as the sand migrates, but the general impression is one of extreme monotony. These eolian land-forms of arid climates are associated with the forms produced by short-lived torrents which cut deep channels or *wadis* after occasional desert storms.

Eolian deposits are almost as distinctive as eolian land-forms. They are mainly sandstones made up of well-sorted, rounded grains and exhibiting large-scale false bedding produced by the migration of dunes. Pebbles with the characteristic dreikanter form may be present but fossils are scarce. Other types of sediment formed in an arid environment may be interbedded with the eolian sandstones. These include torrential stream deposits and evaporates formed by the drying out of salt lakes; together with the eolian deposits, they define a *desert facies* of sediments.

5. The Triassic Desert: applying the Uniformitarian Method

By applying our knowledge of the desert facies, we can conclude that during the Permian and Triassic geological periods, between 270 and 200 million years ago, the British area formed part of a great desert. Permian and Triassic deposits can be seen along the south Devonshire coast, in the west Midlands and in many other scattered localities (Fig. 9.6). They are red in colour, due to the presence of oxidised iron compounds, and consist largely of sandstones. Many of these are made of millet-seed sand-grains, contain occasional dreikanter, and show very conspicuous dune-bedding (Fig. 9.5). They are interpreted as fossil dunes. The older rock-surface on which they rest is very irregular and is clearly a land-surface showing evidence of wind-erosion. Interbedded with the sandstones are layers of rock-salt (e.g. the salt-deposits of Cheshire) and thin-bedded fine sandstones or mudstones which appear to have been laid down in temporary lakes and occasionally retain impressions of the footprints of reptiles which walked across the dried-out lake. The bones of land-reptiles are almost the only other fossils. No marine fossils occur.

The orientation of the inclined sand-lamellae formed on the lee-slopes of sand-dunes is, as we have seen, decided by the direction in which the wind blows—the lamellae slope, in general, downwards in a down-wind direction. By applying this observation to the Permian and Triassic sandstones and measuring the orientation of the lamellae in dune-bedded rocks, it is possible to work out the directions of the dominant winds during the time, more than 200 million years ago, when they were being deposited. Measurements made in various parts of Britain indicate that there was a fairly constant wind-system blowing from easterly and north-easterly directions (Fig. 9.6). It has been suggested, from comparisons of this wind-system with the wind-patterns of modern deserts, that in Triassic times the British area lay further south and in an orientation which was askew with respect to its present position. According to this suggestion, Britain (and the rest of Europe) must have moved northward and undergone a clockwise rotation of some 36° as a result of continental drift since Triassic times. The proposition fits in fairly well with the conclusions drawn from palaeomagnetic observations which suggest that the continents have moved relative to each other and to the poles (p. 17).

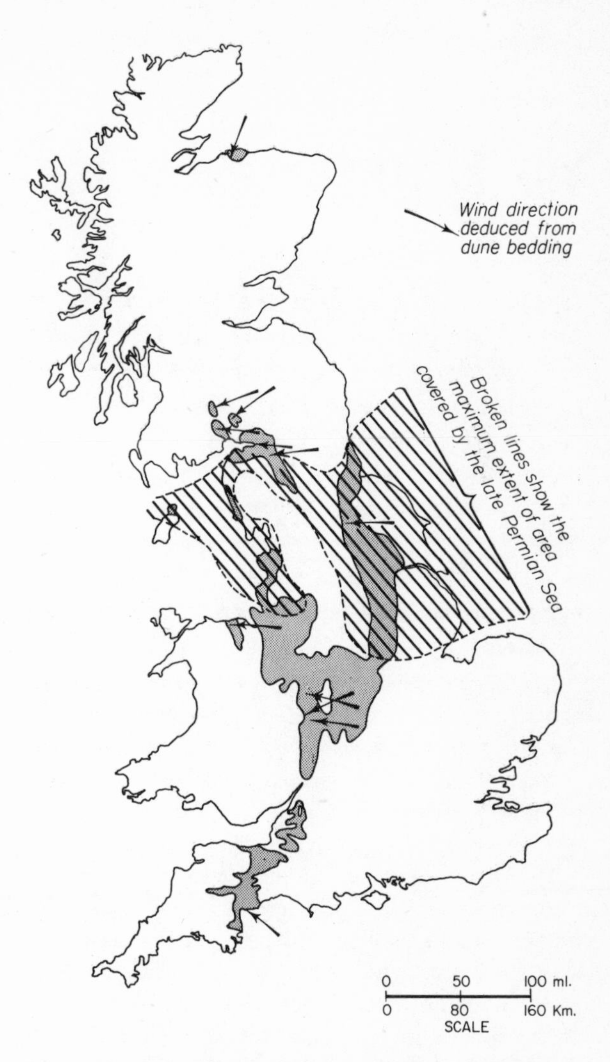

Fig. 9.6 *The Permo-Triassic desert in Britain*

The map shows the principal outcrops of New Red Sandstone (Permo-Triassic), the wind-directions recorded by the arrangement of dune-bedding in these rocks and the maximum extent of the late Permian Sea (see Chapter 17).

10

THE SEDIMENTARY ROCKS

Having now seen how the sediments were formed, we go on in this chapter to consider the sedimentary rocks for which they are the raw materials. We deal with the lithology and geology of the chief sedimentary rocks, that is, their physical and chemical properties and their primary arrangements in space and time.

1. Sediments and Sedimentary Differentiation

We have defined the *sedimentary rocks* as those formed by the accumulation of material at the earth's surface as a result of geological processes going on there. The surface processes which are in action today are held, on our doctrine of uniformitarianism, to have produced the same kinds of sediments in the past: and these ancient sediments were the raw materials of the sedimentary rocks. We should perhaps emphasise at this stage that the acceptance of uniformitarianism does not imply that geological conditions at the earth's surface have always been exactly as they are now—we know, for example, that at certain periods the general temperatures were higher or lower than they are at present and these differences naturally had an effect on the kind of sediments formed in those periods. More important, we know that animals and plants have played a much greater part in the making of sediments during the last fifth of geological time than they did at any earlier time, and that the nature of their activities has changed more than once. On the uniformitarian principle, the operation of different processes would naturally give rise to different results.

We can most easily classify the sediments by recalling what happens to the igneous rocks on weathering (details are given in Chapter 5, section 4). Certain resistant minerals such as quartz survive unchanged and are ultimately incorporated in the new sediments. They provide the *inherited minerals* of these sediments and tend to be concentrated in certain types of sediment (Fig. 10.1). Other igneous minerals, such as the feldspars and ferromagnesian minerals, break down during weathering to give rise to new minerals and to colloidal and dissolved substances. The *new minerals*, chiefly clay-minerals, are concentrated in a second group of sediments and the *colloidal matter*, largely iron hydroxides, in a third. The *substances taken into solution* include calcium and magnesium salts which are precipitated by chemical and organic processes as carbonate rocks, and sodium and potassium salts which may in special circumstances crystallise out to give *evaporates*. Finally, a group of sediments including coal and peat is produced by the piling-up of decaying plant matter.

We see therefore that the products of weathering are sorted out, as is shown diagrammatically in Fig. 10.1, into fairly distinct chemical and geological groups. This natural sorting-out or *sedimentary differentiation* provides us with a classification of sediments into two broad groups:

1. *Detrital sediments* made by the accumulation of *particles* of minerals or rocks, represented by (*a*) the *pebbly rocks* and (*b*) the *sands*, made chiefly of inherited minerals or rocks; and (*c*) the *clays* made chiefly of new minerals.
2. *Chemical–organic sediments* formed by the precipitation of material from solution or by organic processes, represented mainly by the *limestones*, the *evaporates* and the *coals*.

The sediments produced in the ways outlined above go on changing after deposition—they may, for example, be worked over by groundwater carrying salts in solution, and squeezed by the weight of new sediments laid down on top of them. Changes produced by such means as these, called *diagenetic* changes, convert the sediments into consolidated *sedimentary rocks*.

2. Textures and Structures

The textures of *detrital rocks* are decided mainly by the size and shape of the component grains and by the way in which they are arranged.

The *size* of the components may vary from that of boulders down to the finest rock-flour. The accepted divisions of particles of different sizes are as follows:

Gravel, diameter >2 millimetres (usually much greater than 2 millimetres).
Sand, diameter $2-\frac{1}{16}$ millimetres.
Clay, diameter $<\frac{1}{16}$ millimetre.

The Greek words for particles of these three types provide general names for the rocks made from such particles: thus we can distinguish the *psephites* or pebbly rocks, the *psammites* or sandy rocks and the *pelites* or clayey rocks. The degree of *sorting* depends on the way in which the sediment was transported and deposited. *Well-sorted* sediments, in which the particles are all of roughly the same size, are formed when wind, waves or currents have time to work over the material and to separate finer and coarser fractions. *Poorly sorted* sediments, in which fine and coarse material is jumbled together, result from rapid deposition, from deposition by turbidity currents and from glacial action. The *shapes* of grains also depend largely on the origin of the sediment. They are expressed by saying that the grains are *angular*, *sub-angular* or *rounded*. Angular grains cannot have moved far. Well-rounded grains associated with dreikanter pebbles may have been transported by wind, rounded pebbles may have been rolled about on a beach.

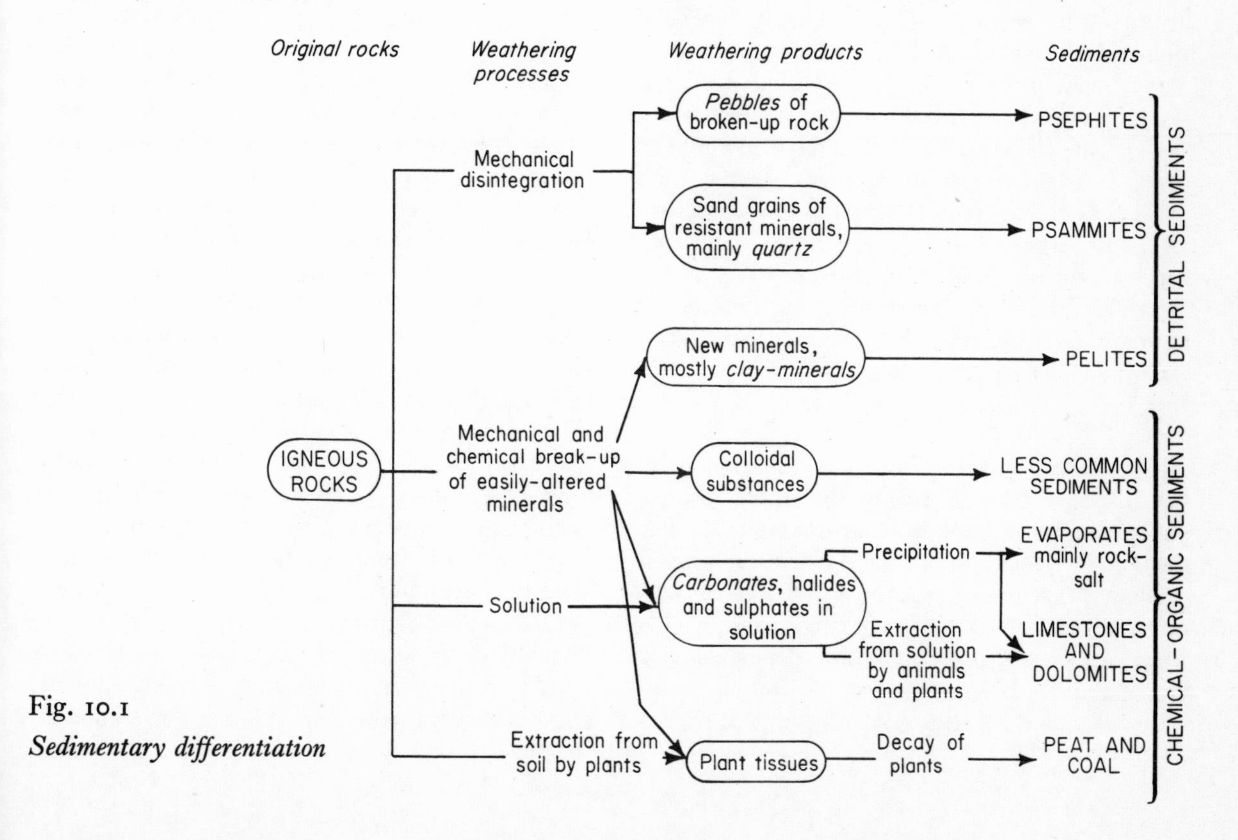

Fig. 10.1
Sedimentary differentiation

Chapter ten

The way in which the constituent grains fit together determines the *fabric* of the sediment. In some rocks, the grains seem to be arranged quite haphazardly. In others, we can see a definite pattern about the arrangement of their longer and shorter dimensions; this type of pattern produces a *preferred orientation* of grains. Flakes of detrital mica or of clay-minerals usually settle flat on the surface, as a sheet of paper will settle when allowed to fall; they therefore have a preferred orientation, with their longer axes parallel to the bedding-planes. Pebbles transported by strong currents in a stream may come to rest with their largest surfaces tilted to slope upstream, each pebble overlapping its downstream neighbour like tiles on a roof. This structure can be used to decide the direction of flow of the current which produced it. In glacial deposits, the *till-fabric* is characterised by the arrangement of a significant number of pebbles with their longest axes lying parallel to the direction of flow of the ice which carried them.

The textures of the *chemical–organic* group of sediments are so varied that it is best to deal with them when we come to describe individual rocks. We can say in the meantime that the textures of organic deposits depend on the characters of the organism concerned in making them—for example coal, derived from plant debris, may preserve some traces of the microscopic structure of plant tissues or spores. Chemical sediments formed by crystallisation of minerals from solution reveal under the microscope a mosaic of grains welded together by growth as ice-crystals are welded in a frozen pond.

3. Bedding

The *structure* of a sediment is determined by the arrangement of the components on a larger scale. The most important structural feature is the *bedding* whose significance has already been discussed (Chapter 1, section 3). A *bed* of sediment (see Fig. 1.2) represents the product of a single act of sedimentation. It is deposited as a flattish, nearly horizontal layer of limited thickness but wide lateral extent and is bounded top and bottom by bedding-planes. Each *bedding-plane* marks a pause in the process of deposition. Before deposition begins again, the surface of the sediment may be hardened, dried out or marked in some way, so that it makes a sharp plane of separation when the next layer is laid down on it. No two beds are ever identical, so that, as sedimentation continues, there is built up a *succession* of individualised beds or strata with a distinct *stratification*. A succession of beds which are all of the same general composition is called a *formation*: examples in Britain are the Chalk, the London Clay and the Millstone Grit.

Bedding (Fig. 10.2) is of four main types, some of which we have met before: because it is so important, we deal with all types here. *Regular bedding*, as its name implies, is shown by a sequence of parallel-sided beds separated by plane bedding-surfaces (see Fig. 1.2). *Current-bedding* or *false bedding* is a more complicated structure produced when layers of sediment are deposited on an inclined surface such as the frontal slope of a small delta (p. 72, Fig. 6.9), the lee slope of a sand-dune (p. 103, Figs. 9.4, 9.5) or the slope of any small shelf or channel in the sea- or river-bed. The depositional laminae laid down on the sloping surface are usually lenticular and gently curved. Slight erosion by the current flowing over them carries away their thin upper parts leaving a surface of erosion truncating the inclined laminae. New laminae are then laid down above this surface in the same pattern as those below, so that a contrast is produced between the asymptotic curvature at the base of each group of laminae and the sharp erosional plane of truncation at their top. We shall see later (p. 179) that this contrast can be used to decide whether disturbed beds are right-way-up or upside-down.

Graded bedding is characterised by a regular change in grain-size within a single bed from coarse at the bottom to fine at the top. We have already encountered two methods by which graded beds are produced: firstly, by the settling through deep water of the mixed collection of detritus carried by a turbidity current (p. 80): and, secondly, by the continuous change in grain-

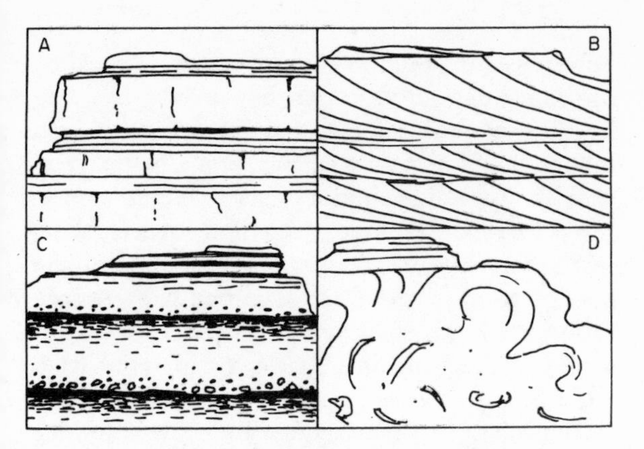

Fig. 10.2 *Four common types of bedding*
A. Regular bedding B. Current-bedding
C. Graded bedding D. Slump structure

size of the detritus available in the area of deposition, as exemplified by the varves laid down in lakes fed by streams which freeze up in winter (p. 77, Fig. 6.14).

The last bedding structure we shall mention is *slump-bedding* or *slump-structure* which is produced soon after deposition by the sliding or *slumping* of layers of wet, incoherent sediment on the sea-floor. During this process, the original bedding is crumpled, and layers of different kinds of sediment become broken up and mixed together. Very frequently, the top of the slumped mass is washed away by currents and the next bed is therefore laid down on a smooth erosion-surface truncating the crumpled laminae of the slumped bed.

4. Sediments into Sedimentary Rocks

As soon as a layer of sediment is laid down, diagenetic processes (p. 107) get to work and convert it into a harder and more coherent sedimentary rock. Some water is squeezed out by the weight of successive new layers (*dewatering*) and its removal leads to *compaction* or closing together of the components. The remaining water, circulating through the rocks, deposits material carried in solution and in exchange takes away other soluble substances. Much of the *salt* trapped in marine sediments is removed by leaching. Other changes have already been considered in connection with the activity of groundwater (p. 66). They include *cementation* by deposition between the grains of a cement of calcite, silica or iron hydrates: the formation of *concretions* by concentration at a few points of material originally spread through the rock and the *replacement* of original components by silica or iron compounds.

In limestones, *recrystallisation* frequently takes place during diagenesis—the calcite grains of shell debris, for example, may go into solution and be replaced by a new mosaic of calcite which does not preserve the original organic structures. Also in limestones, calcite may be replaced by dolomite, $CaMg(CO_3)_2$, by a process of *dolomitisation*. All these changes take place near the earth's surface at ordinary temperatures and pressures.

5. The Common Sedimentary Rocks

We have divided the sedimentary rocks into two classes, the *detrital* and the *chemical–organic*. Subdivision in these two classes can be made by *grain-size* in the detrital class (p. 107) and by *chemical composition* in the chemical–organic class. Proceeding on these lines, we arrive at the classification given in the Table below. The commoner rocks listed in this are selected for further description.

CLASSIFICATION OF SEDIMENTARY ROCKS

A. DETRITAL ROCKS

1. *Pebbly or psephitic:* gravel, conglomerate, breccia, tillite 2. *Sandy or psammitic:* sand, sandstone, arkose, greywacke	*particles consist mainly of inherited minerals*
3. *Clayey or pelitic:* mud, clay, mudstone, shale	*particles mainly new minerals produced during weathering*

B. CHEMICAL–ORGANIC ROCKS

1. *Carbonate:* Lime-mud, shell-sand, limestone, dolomite
2. *Siliceous:* flint, chert
3. *Ferruginous:* iron compounds
4. *Aluminous:* laterite, bauxite
5. *Saline:* evaporates, e.g. rock-salt, gypsum, anhydrite
6. *Carbonaceous:* peat, coal, bitumen

Chapter ten

6. Detrital Sedimentary Rocks

Pebbly or psephitic rocks. The fragments in the coarse detrital rocks are by definition over 2 millimetres in diameter, and in practice they are usually pebbles or boulders very much larger than this: the shingle on a sea-beach is a typical modern rock of this group. The composition, size and shape of the pebbles, and the manner of occurrence of the rocks which they make vary according to the geological circumstances in which they accumulated.

(i) **Conglomerates** are rocks in which the pebbles are *rounded*; the pebbles are close-packed and are bonded together by finer sediment or by a cement of silica, calcium carbonate or iron compounds (Fig. 10.3). In one type of conglomerate, the pebbles are well-sorted, well-rounded and fairly small; they are made predominantly of one or two resistant rocks or minerals such as quartz, quartzite or flint. Such conglomerates are typical of *shore-lines* where the sea advances slowly on a low-lying land and where there is consequently plenty of time for weathering and sorting of the components. At the other extreme are conglomerates made of poorly rounded pebbles of many rock-types and of all sizes. In contrast to the well-sorted shore-line type of conglomerates which usually make only thin layers, these badly sorted conglomerates often make formations hundreds of feet in thickness. They are formed by the rapid deposition, often by torrents building up alluvial fans, of material eroded from high mountains. The Old Red Sandstone of Scotland provides typical examples of such conglomerates.

Fig. 10.3 *Conglomerate*

(ii) *Breccias* differ from conglomerates in being made of *angular* fragments. The material in them may be of many kinds and sizes and has not been transported far enough to become rounded or sorted; indeed, many breccias are derived from *screes* (p. 58). *Slump-breccias* are the result of the mixing-up of sedimentary layers during slumping and consist of fragments of the more coherent beds enclosed in a matrix derived from the remaining sediment.

(iii) *Boulder-clay* or *till*, the bouldery deposit laid down by ice (p. 97, Fig. 8.8), contains relatively few pebbles and boulders, which are often faceted by abrasion and are scattered through an unbedded matrix of badly sorted sand and rock-flour. The hardened rock derived from boulder-clay is known as *tillite*.

Sandy or psammitic rocks. Sand-grains are, by definition, between 2 millimetres and $\frac{1}{16}$ millimetre in diameter. When we speak of *sand* we usually mean quartz-sand, since the resistant mineral quartz is generally by far the commonest component of sediments of this grain-size; however, feldspars, micas, small shell-fragments and many other components may be mixed with quartz or, occasionally, may be concentrated to form sands of unusual composition. Coherent rocks formed by the compaction or cementation of sands are known as *sandstones* (Fig. 10.4). As with the pebbly rocks we can distinguish different types formed in different environments.

Pure *quartz-sands* made of well-sorted and well-rounded grains result from prolonged chemical and mechanical working-over along the margins of shallow seas. These sands are converted into *quartzitic sandstones* or *orthoquartzites* by cementation with silica; more rarely, they may be cemented by calcium carbonate or dolomite. Other types of sandstone, not necessarily formed under the same conditions, are *feldspathic sand-*

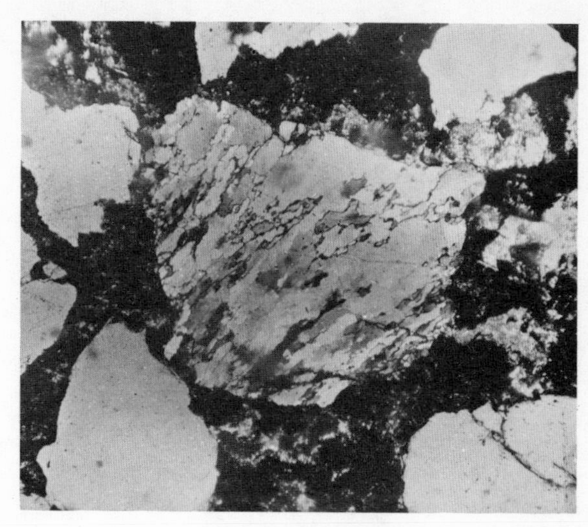

Fig. 10.4 *Psammitic rocks*

Left Quartzitic sandstone made of well-rounded and well-sorted quartz grains with silica cement (some grains appear dark because the thin slice is viewed by polarised light)

Right Arkosic sandstone made of angular, badly sorted particles of quartz, feldspar and metamorphic rocks set in a matrix of silt

stone containing a little feldspar in addition to quartz; *micaceous sandstone* in which mica-flakes are plastered on closely spaced bedding-planes—the rock readily splits along these planes and is accordingly called a *flagstone*; *greensand* coloured by the iron-bearing mineral *glauconite*; and *ferruginous sandstones*, red, yellow or green in colour, which are cemented by iron hydrate.

Two other rather different types of sediment fall within the psammitic group, *arkoses* and *greywackes*. *Arkoses* are made of quartz and feldspar grains, the feldspar often amounting to one-third or more of the whole. The grains are moderately well-sorted, usually rather coarse and are partially rounded and cemented by an iron compound or, more rarely, by calcite. Most arkoses are red, pink or grey in colour. *Current-bedding*, often of a type associated with torrential stream-action, is a characteristic structure. Arkoses are products of rapid deposition which left little time for chemical decay of the feldspars, and they are frequently laid down in alluvial fans and on flood-plains by streams eroding mountainous lands where much granite is exposed. They make thick sequences and are often interbedded with mixed conglomerates.

Greywackes are very different in their origin. They are dark-coloured sandstones made of badly sorted, partially angular grains of many minerals and rocks jumbled together, the finer material packed in between the coarser particles. Coarse greywackes in which the angular grains are easily visible can be called by the old rock-name of *grit*. *Graded bedding* is usually a characteristic structure of greywackes and slump-bedding is also common. The low degree of sorting and rounding and the occurrence of minerals which weather easily suggest that deposition was rapid. The graded bedding, often repeated again and again through thousands of feet of sediment, indicates that the transporting agents were turbidity currents.

Clayey or pelitic rocks. The particles of the clayey rocks, less than $\frac{1}{16}$ millimetre in diameter, are of three origins: (1) clay-minerals formed by chemical weathering and transported to the region of deposition, (2) the finest particles of inherited minerals such as quartz, and (3) minerals such as quartz, white mica, chlorite and iron hydroxide produced on the spot by crystallisation in the new sediment of colloidal material.

Muds when deposited contain a great deal of water which is gradually squeezed out by the weight of material accumulating above them.

Chapter ten

Clay, formed by compaction and partial reconstitution of mud, is still moist and readily absorbs water to become plastic. It has a characteristic smell and forms a structureless substance, white when pure but usually coloured brown, green or red by iron compounds. Further compaction and dewatering gives rise to structureless *mudstone* and finally to *shale* which is hard, brittle and splits readily along closely spaced parting-planes parallel to the bedding. *Black shales* rich in organic matter and in pyrites (FeS_2) are developed from black muds deposited in enclosed deep-water basins as described on page 88. *Fire-clays* make layers immediately beneath coal-seams and are regarded as the old soils ('seat-earths') in which the coal-forests grew. Growth of the abundant plants depleted the soils of alkalies, lime and magnesium, leaving behind a grey, unbedded fire-clay which makes a valuable refractory material. Finally, *marl* is a calcareous clay or shale.

7. Chemical–organic Sedimentary Rocks

Only two of the groups of chemical–organic deposits are described here because the other main types have been dealt with in previous chapters, flint and chert on p. 31, bauxite and laterite on p. 61, and evaporates on pp. 77–8.

Limestone and dolomite. Limestone and dolomite are carbonate rocks, limestone being made of calcite, $CaCO_3$, and dolomite of the mineral of the same name, $CaMg(CO_3)_2$. *Magnesian limestone* contains up to 10% of the dolomite molecule held in solid solution in the calcite. All carbonate rocks may contain some sandy or clayey detrital material. The majority of limestones are organic in origin, types formed by direct precipitation of calcium carbonate being rather rare. In all limestones, a good deal of solution and redeposition goes on during diagenesis (p. 109). *Dolomitisation*, involving the replacement of calcite by dolomite, takes place as a result of the exchange of calcium for magnesium present in sea-water and most dolomite-rock is produced in this way during diagenesis from original limestones.

Organic limestones (Fig. 10.5) vary in character according to the nature of the organisms contributing to them. Very often, debris from one particular type of organism predominates and one can therefore distinguish shelly limestone, coral limestone, crinoidal limestone and foraminiferal limestone—the last type is that which was used in building the Pyramids. *Reef-limestone* is derived from the fixed organisms which make coral reefs (pp. 86–7). As we have seen already, it is built up on the site of the reef as a lumpy mass of rather limited lateral extent: such masses occurring in ancient sediments are called *reef-knolls*. *Algal limestone*, made largely by the growth of algae (p. 46), is prominent in many limestone-formations such as the Carboniferous Limestone of Britain. *Chalk* is a soft white limestone, very fine-grained and composed largely of the minute skeletons of microscopic organisms including foraminifera (especially the species *Globigerina*), radiolaria, sponge spicules and microscopic plants, all set in a hardened calcite-mud.

The most distinctive types of *chemical limestone* are the *oolites* (Fig. 10.5B) already described (p. 86); *pisolite* is a similar type in which the spherical particles are larger, about the size of peas. Oolitic limestones make important building-stones—examples much used in Britain are the *Bath Stone* and *Portland Stone* (used in St. Paul's Cathedral), both even-grained white or creamy rocks. Other chemical limestones include the deposits from calcareous springs known as *tufa* and *travertine*—the latter is used for interior decoration and is a cream-coloured rock, porous and cellular and occasionally beautifully banded.

Dolomite rock is, as has been said, usually of secondary origin and, as a result of the solution and redeposition that goes on during its formation, the primary textures and shapes of fossils are often difficult to recognise. Dolomitisation of calcite involves a reduction of volume and many dolomites become cavernous or collapse into angular fragments welded together again by further crystallisation. The development of large pore-spaces as a result of dolomitisation may make a dolomitised reef-limestone, for example, a potential *reservoir-rock* for oil. In appearance, dolomite rock is usually massive, granular and sugary-looking, white when pure but often coloured yellow or brown by iron carbonate.

The carbonaceous rocks: coals and bitumens. When plants and animals die, their substance is usually acted upon by bacteria and other organisms which break down the organic compounds and convert them into water, carbon dioxide and simple inorganic salts. Under some circumstances, however, lack of oxygen during rapid accumulation may make complete breakdown impossible. In these circumstances, the partially decayed debris is buried and by further diagenetic changes is converted into *carbonaceous rocks* whose main component is organic *carbon* often combined with hydrogen as various *hydrocarbons*. When carbonaceous rocks are brought into a plentiful supply of oxygen and reaction is encouraged by heating them, the processes of breakdown of organic compounds which were suspended long ago during deposition can be brought to completion—in other words, the rocks will *burn* with the production of water and carbon dioxide and with the release of energy in the form of heat. Thus, the carbonaceous rocks are natural *fuels* of great value to man.

The coal series. The coals are bedded carbonaceous rocks formed from accumulations of plant debris. The effects of chemical and bacterial action at the surface, followed by compression due to burial under younger material, convert a layer of loose vegetable matter into a compact *coal seam.* Each stage in the process of change is represented by a different-looking rock, the products of the entire sequence of changes being known as the *coal series* shown in Fig. 10.6. The position of a given rock in the coal series determines its *rank.* The improvement of rank from peat through to anthracite is marked by increase in density, hardness and carbon-content and by decrease in content of moisture and volatile substances; in general, the higher the rank the more valuable the fuel.

Peat is a little-modified accumulation of bog plants which can be seen forming at the present

A

Fig. 10.5 *Limestones*

A. Shelly limestone, the Aymestry Limestone, Shropshire

B. Oolitic limestone, thin slice (× 25)

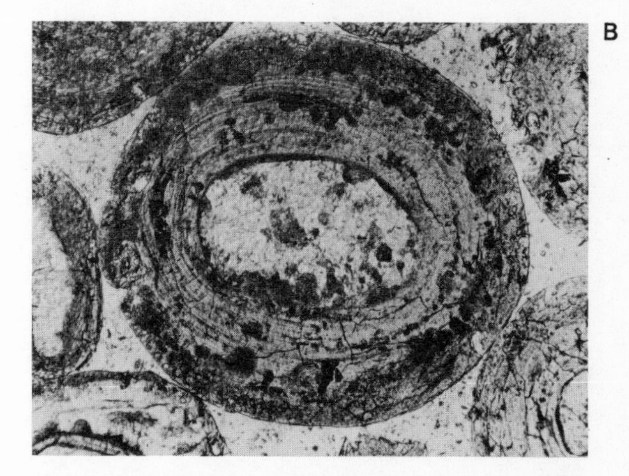

B

day on badly drained moorlands; it is used as a fuel especially in Scotland and Ireland. The vegetable character is very clear, the peat is light and porous and the colour brown. Towards the base of a layer some tens of feet thick, peat becomes somewhat darker, more compact and richer in carbon. *Lignite* or *brown coal* represents a further stage in the transformation. It is more

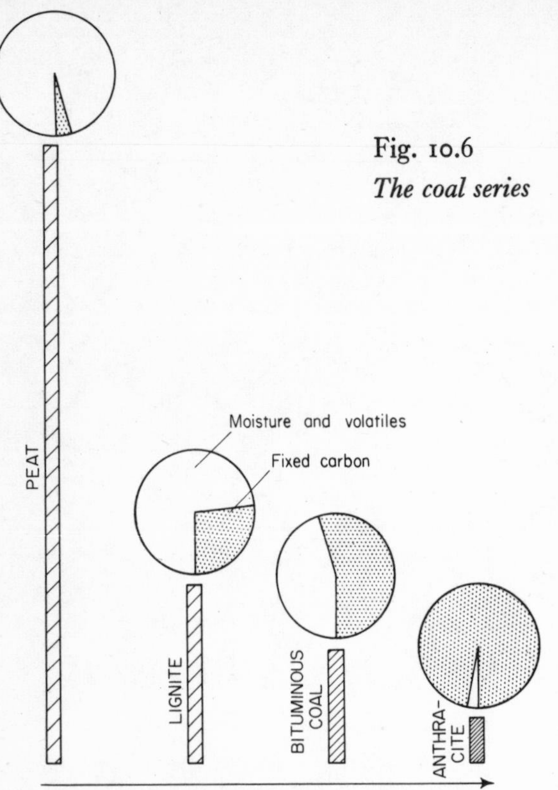

Fig. 10.6
The coal series

compact than peat and brown or black in colour but still clearly reveals its vegetable origin.

Bituminous coal, exemplified by most ordinary house-coals, is black, hard and brittle, and shows only occasional traces of plant matter. Its structure can be seen by examining a large lump of coal—such a lump is usually a rough cube, and is bounded by bedding-planes and by two sets of joints perpendicular to each other and to the bedding. The bedding-planes along which the coal parts are covered by soft, powdery material in which chips of what looks like charcoal can sometimes be seen: this material is called *fusain*. The main part of the lump is made up of other substances, of varying lustre, occurring in layers parallel to the bedding. *Durain* is hard and dull and is made of the resistant parts of plants such as spore-cases. *Clarain* is brighter and is made of a kind of consolidated jelly derived from decomposed vegetable matter, containing fragments of resistant plant-tissue. Finally, *vitrain* makes streaks of bright glassy-looking material resembling the jellified matrix of the clarain layers. A lump of house-coal should be examined and these layers identified.

Anthracite, the highest-ranking coal, is black, unbanded and structureless; it has a brilliant lustre, breaks with a conchoidal fracture and shows little or no trace of original organic structures.

Anthracite and bituminous coal occur in Britain and western Europe as seams in the Carboniferous System, deposited some 350 million years ago. They are derived from the organic debris which accumulated on the floors of vast swampy forests; individual seams, seldom more than 6 feet in thickness, may extend over thousands of square miles. The origin of anthracite is not very clear. Certain seams of bituminous coal pass laterally into anthracite in South Wales and it therefore does not seem that the anthracite has been more deeply buried than the ordinary coal. It is possible that pressures due to earth-movements or even the heat of igneous intrusions may play a part in turning bituminous coal into anthracite.

Most coals are formed from vegetation that accumulated more or less where it grew. There are, however, some coals that are more reasonably considered to be the result of the accumulation of *drifted* vegetable debris. Examples of such coals are the *cannels*. Cannel is a dense, lustrous, greyish-black coal with a conchoidal fracture. It ignites in a candle flame and burns with a smoky candle-like flame—hence its name. It contains a considerable amount of ash and occasionally shows fish scales and other fossil fragments. It occurs in lenticular beds in ordinary bituminous coal seams and represents the fillings of *wash-outs* or contemporary stream-channels in the coal-swamp.

Bitumens. The bitumens are composed of a number of hydrocarbons of which the paraffins, the naphthenes and the benzenes are the chief. They vary from very liquid oils to solid substances such as asphalts and are the parent materials from which petroleum is manufactured.

It is generally agreed that the bitumens are produced from original organic material entrapped in sediments. The *source-rocks* are the *sapropelites*, muds containing much organic matter largely of algal origin, deposited in a reducing environment like that of the present Black Sea (p. 88). This organic material, transformed into the hydrocarbons, migrates under suitable conditions into *reservoir-rocks*, such as porous, jointed or fractured sandstones and limestones, to form *oil-pools* located on favourable geological structures as described later (p. 231).

8. Environments of Deposition

In our examination of the making of modern sediments, we were able to relate the characters of the sediments directly to the environment of deposition, because the agents of deposition could still be seen in action. In studying older sedimentary rocks, we have to reconstruct the environment of deposition from the characters of the rocks themselves because, of course, the land or sea in which they were deposited has long since disappeared. The clues for this reconstruction are provided by the composition and texture of the rock, by the kind of bedding, by the fossil content and by the relationship between any one bed and its neighbours. The sum of all these features decides the *sedimentary facies* from which we may attempt to deduce the conditions under which each rock was formed (p. 3). It may be useful at this point to summarise the information scattered through the last chapters concerning the possible environments of deposition and the kinds of sediments laid down in them. Such a summary is given in the table below.

ENVIRONMENTS OF DEPOSITION OF SEDIMENTARY ROCKS

Environment of Deposition			*Common Sedimentary Rocks*
SEA			
Shallow seas (continental shelf)	Littoral (beaches, sandbanks, tidal flats)		Conglomerate, sandstone, shale
	Neritic	Shelf seas in stable areas	Orthoquartzite, current-bedded sandstone, shale, organic and chemical limestones
		Restricted deep basins	Black shale, source rocks of bitumens
Deep seas		Geosynclinal seas in mobile belts	As for shelf seas with in addition greywackes and other turbidites
		Deep seas in stable areas	Shale, deep-water limestone, sometimes greywacke
Abyssal seas			Calcareous ooze, siliceous ooze, Red Clay
MIXED ENVIRONMENTS			
Deltas			Mainly sandstone, shale
Estuaries, lagoons			Shale
LAND			
Flood-plain			Conglomerate, sandstone, shale
Lakes	with outlet to sea		Sandstone, shale, fresh-water limestone
	in basins of interior drainage		Sandstone, shale, evaporates
Deserts			Sandstone, conglomerate, breccia
Piedmont (intermontane basins, alluvial fans)			Conglomerate, breccia, arkose, sandstone
Areas of glaciation			Tillite

Chapter ten

9. Reconstruction of Palaeogeography: Facies Changes

The area over which any one type of sediment can be deposited at a given time is limited by the geographical extent of the environment in which it can be formed. Sooner or later, as the sediment is traced laterally, it must either come to an end or *change its facies* to conform to a change in the environment. If we mark out on a map the distribution of the rocks of different sedimentary facies deposited at any one time, we can deduce the distribution of the corresponding environments. From these, in turn, we may be able to reconstruct the positions of land and sea, that is, the *palaeogeography* of the earth's surface. Rocks of marine facies mark the sites of the seas, littoral deposits may define shore-lines, and fluvial or desert deposits the land-areas—we should remember also that, since the lands are more often regions of erosion than of deposition, the position of land-areas may be indicated simply by an *absence* of deposits of a particular age. We can illustrate the kind of reasoning involved in the construction of a palaeogeographical map by an example from the British Isles.

The rocks deposited in the British area, during the early part of the period that the geologist calls the Carboniferous, had a broad distribution as shown in Fig. 10.7. We can comment on this map from south to north and so justify the palaeogeographical map also shown in the figure. In the south, where Devonshire now is, rocks of this age are shales and mudstones with marine fossils—deepish-water marine deposits. Farther north, in South Wales and the Midlands, the facies changes, the rocks being mainly limestones which, though thick, are of shallow-water type with lagoonal muds and shell-banks. The limestones, and indeed all rocks of this age, are absent over a central region and a land mass must have stretched from the present Irish Sea across Central Wales and the Southern Midlands. This land—geologists have named it 'St. George's Land'—must have been low, with no great streams draining into the shallow seas around it, for the limestones which fringe it contain very little detrital material such as would be brought to the sea by actively eroding streams.

Farther north, in Yorkshire, shales and sandstones are intercalated with limestones, and farther north still, in Northumberland and the Midland Valley of Scotland, seams of coal enter the succession and limestones become thinner and fewer. It is clear that the clean seas around St. George's Land gave place northward to muddy estuaries and to deltaic flats and swampy forests bordering a northern land.

10. Changes with Time: Correlation and Succession

The reconstruction of the geography of early Carboniferous times given above can only be relied on in outline, because the series of rocks on whose facies it is based represents the deposits of a very long time, probably some tens of millions of years; during this time, there must certainly have been fluctuations in the extent and depth of the sea, the height of the land and so on; but these cannot be detected unless we can split the early Carboniferous rock-series into smaller units deposited over shorter and shorter lengths of time. Thus an essential preliminary for the study of facies changes is to *establish the contemporaneity* of the deposits concerned.

Correlation. Two beds or groups of beds are *correlated* when it is established that they were formed at the same time. This can only be done properly when the beds can be identified by the presence of characteristic fossils according to William Smith's Second Law mentioned in Chapter 1 (p. 2) and it will be as well to postpone further discussion until the fossils have been dealt with. For the present, we will take it for granted that correlation of strata is possible.

Succession. The *succession* or *sequence* of beds arranged in order from oldest to youngest provides the *stratigraphical succession* which is the basis for interpretation of the history of deposition. If a single group of beds is likened to a 'still' from a film, then the stratigraphical succession is like

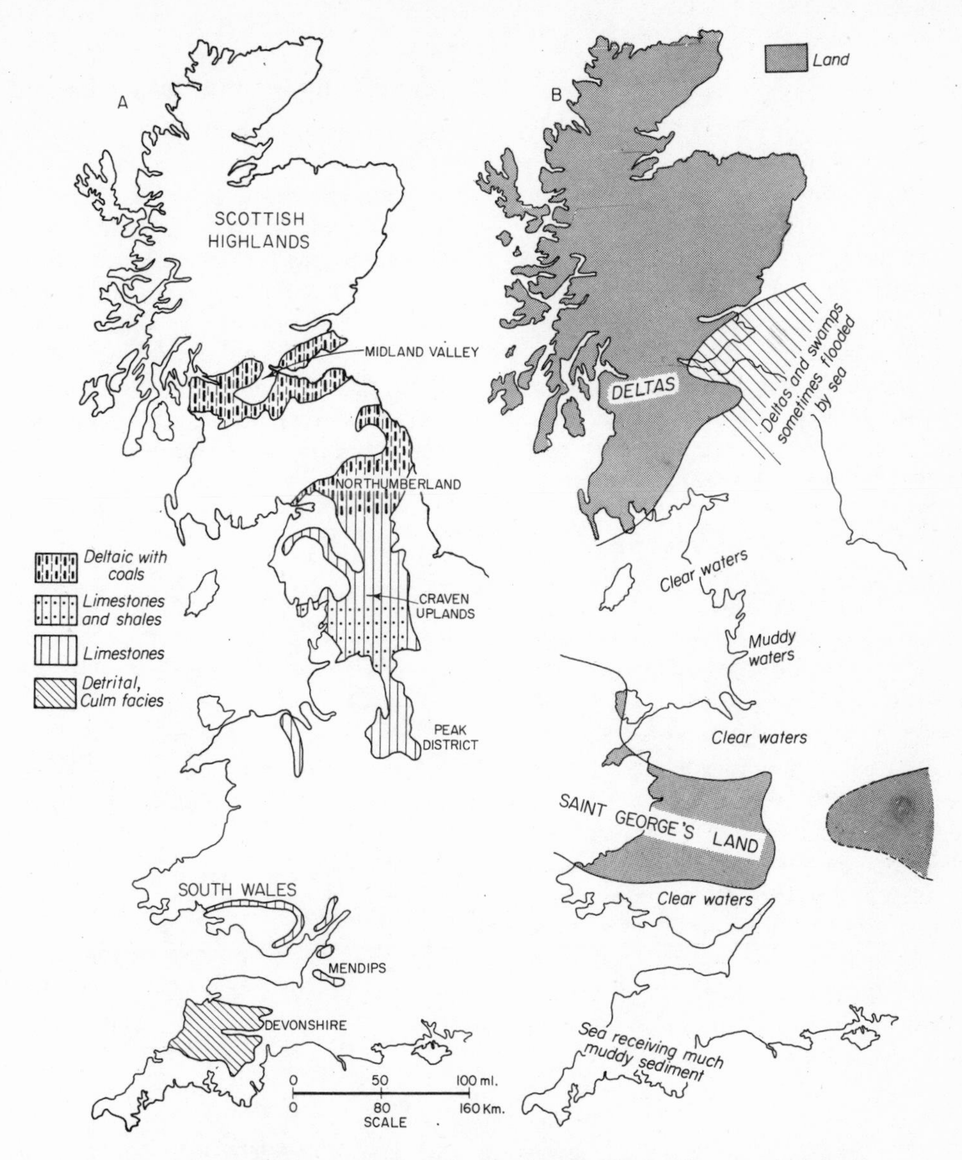

Fig. 10.7 *Facies variations in the Lower Carboniferous of Britain (for further details see Chapter 17)*

A. Facies map B. Palaeogeographical map based on A

the film itself—a succession of 'stills' following one another in chronological order.

It will be seen from William Smith's Law of Superposition that the record of deposition provided by an undisturbed succession must be read *from the bottom upwards*. For this reason, successions are conventionally written down with oldest members at the bottom; and the terms 'Lower' and 'Upper' are used to distinguish between older and younger portions of any group.

Once the stratigraphical succession is established, it provides a history of *changes in the environment of deposition*. A continuous sequence changing upwards from sandstone to shelly limestone, for example, might indicate that the supply of sandy detritus brought into a shallow

Chapter ten

sea gradually became less, so that in the end only organic debris accumulated there. Again, the appearance at a certain level in a sequence of sedimentary rocks of beds of tuff marks an outburst of volcanic activity.

Transgression and regression. Two kinds of change in a succession are of special significance —they are those associated with the advance of the sea on to the land (*transgression*) and with the retreat of the sea (*regression*). Both events have been repeated many times in earth-history and leave distinctive records in the stratigraphical succession (Fig. 10.8).

During a period of *transgression* the sea at any selected point becomes deeper. The first deposit laid down will usually be of littoral facies, formed along the advancing sea-shore. It will be followed by sediments of progressively deeper-water facies, so that a beach conglomerate might be followed by current-bedded sandstone and this by shales or possibly limestones. On the other hand, a period of *regression* is marked by a shallowing of the sea and the progressive emergence of a land-surface. It could accordingly be recorded by a succession from deep-water shales to shallow-water sandstones, then to sediments of lagoonal or estuarine facies and finally, perhaps, to terrestrial deposits of eolian or fluviatile origin. Marine fossils in the lower beds may give way to those of brackish waters and finally to land-dwelling forms.

11. Breaks in the Succession: Unconformity

The succession of sedimentary rocks is nowhere complete in the sense that every moment of time is represented by a layer of rock. There are always pauses between acts of deposition—these pauses may represent the intervals between one high tide and the next, or between floods spreading sediment over an alluvial plain, storms stirring up turbidity currents and so on. Such short breaks, of course, do not affect the continuity of the record as the geologist sees it. Longer breaks in deposition produce *non-*

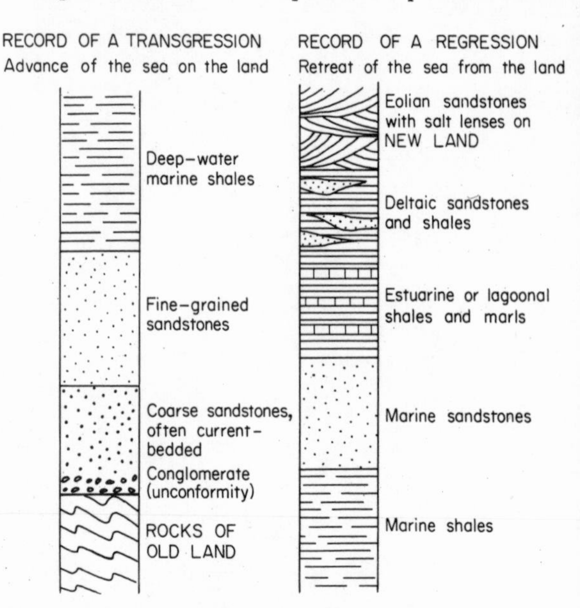

Fig. 10.8

Transgression and regression: the columns illustrate successions which could be laid down during periods of transgression and regression

Fig. 10.9

Unconformity: an Old Red Sandstone conglomerate, of Devonian age, rests with angular unconformity on folded and metamorphosed rocks of the Dalradian Series (late Pre-Cambrian to Cambrian)

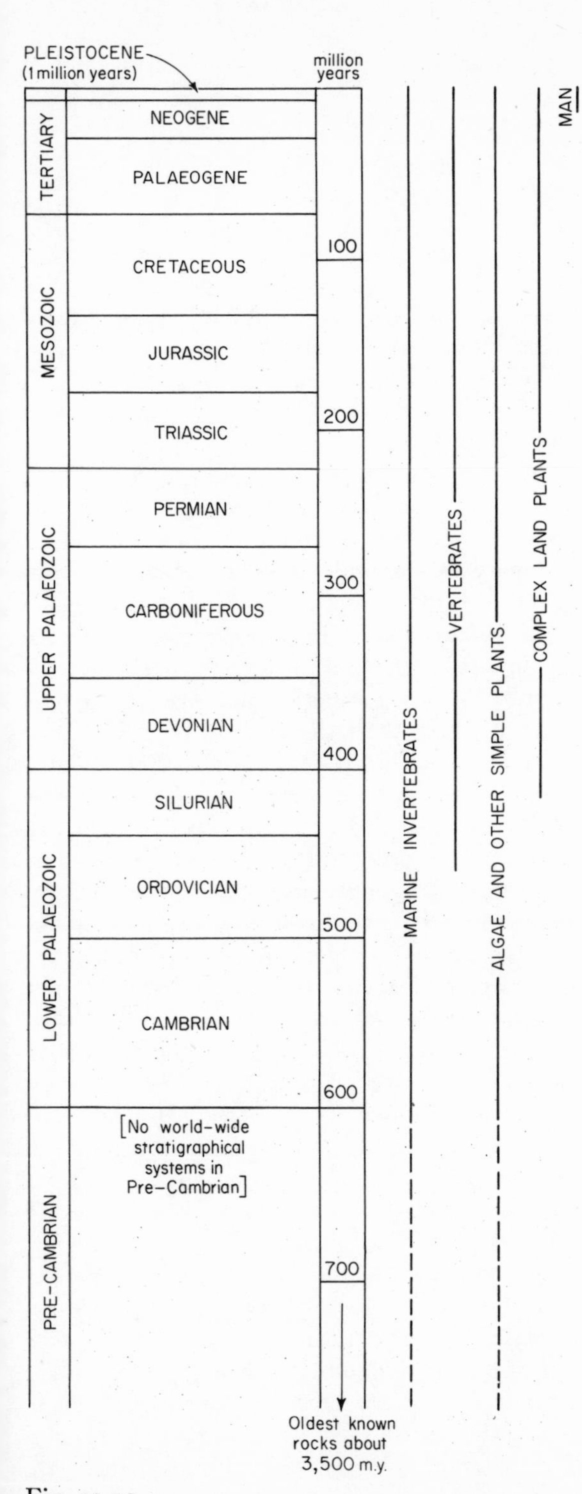

Fig. 10.10
The stratigraphical column (*compare* Fig. 15.2)

sequences in the succession—a part of the possible sequence just is not there. This kind of break can usually only be recognised in fossiliferous rocks where strata containing fossils characteristic of a particular period are found to be absent.

More important interruption of deposition takes place when the area of sedimentation is *disturbed by earth-movements*. The strata are tilted or folded and deposition is usually brought to an end because the disturbed rocks are pushed up above sea-level. Erosion of the uplifted area follows and when deposition is resumed at a later period the new sediments are laid down discordantly on the eroded edges of the tilted or folded sequence. The surface which separates older and younger sequences arranged in this way is called an *unconformity* (Fig. 10.9); it clearly records a violent interruption of sedimentation and is connected with earth-movements taking place below the surface.

12. The Stratigraphical Column

We have seen how a stratigraphical succession can be recognised in any area by determining the order of deposition from oldest to youngest. By fitting together the successions of many different areas, geologists have established the succession of strata throughout the whole 600 million years of geological time since fossils first became plentiful. This complete succession of the fossiliferous rocks is called the *stratigraphical column*. It is shown in Fig. 10.10, where the names given to successive portions of the succession are listed from oldest at the bottom to youngest at the top. Each named portion constitutes a geological *system* deposited during a geological *period*. It is important to remember that this column only covers about one-fifth or one-sixth of the whole record of geological time. The rocks older than the Cambrian, collectively called *Pre-Cambrian*, do not contain fossils which can be used in correlation and are therefore more difficult to classify according to age. We shall discuss the history of deposition recorded in the stratigraphical column and in the sequence of Pre-Cambrian rocks in later chapters.

11

FOSSILS

This chapter deals briefly with the main groups of animals and plants which are commonly found as fossils and gives some account of the ways in which they are of interest to geologists. It is seen that fossils not only provide records of the history of life but also supply evidence concerning the age and environment of deposition of the rocks in which they occur.

1. Fossils as Records of Past Life

The living animals and plants of today are descended from older organisms, these from ones which existed still longer ago and so on through the whole history of life on earth. Samples of past generations are provided by *fossils*, the remains of ancient organisms entombed in sedimentary rocks. Fossils are of great interest both to the biologist and the geologist. They supply the only direct evidence concerning the history of living things and from this evidence we can see that organisms of the past were very different from those of today. By comparing the fossil forms characteristic of different geological periods we can gain some idea of the effects of *evolution*, the process by which ancestral types of animals and plants have been modified and diversified with the passage of time.

For the geologist, fossils are also of particular value in the study of geological history. Because the process of evolution has produced progressive changes in living organisms, fossils may be used as *indices of age* of the rocks containing them. The use of fossils in this way makes possible the *correlation* of strata of the same age and the placing of strata in their correct positions in the stratigraphical column. Differences in fossil content may make it possible to divide a sequence of strata into small sections or *zones* each characterised by a particular association of fossils: each zone can be named after one of the characteristic forms which is then called the *zone-fossil*. We shall meet with many examples of the use of fossils as indicators of geological age in the chapters on historical geology (Chapters 15–17).

Fossils also provide evidence of the *environmental conditions* under which deposition took place, for the fossils of a sedimentary rock are usually (though not always) derived from organisms whose natural habitat was the environment in which the sediment was deposited. For example, a marine clay may contain the remains of deep-sea organisms, an estuarine clay organisms such as worms or mussels which thrive in the intertidal zone and a lake-clay fresh-water organisms, with insects and pollen blown in from the surrounding land. Sediments laid down in hot climates will contain fossils different from those of polar regions, those of swamps will be different from those of deserts and so on. The study of present-day animals and plants in relation to their environments is called *ecology*; and *palaeoecology* is the corresponding study of fossils in relation to the environments of the past. In interpreting the evidence of fossil faunas and floras, we must make use of our knowledge of the processes of sedimentation. Fossil remains may be moved about by waves and currents and may therefore be abraded, sorted and transported like other particles of similar size. For this reason one cannot assume unquestioningly that *all* the fossils contained in a rock represent organisms which lived in the environment of deposition. The bones of fresh-water fish, for example, may be swept out to sea by rivers in flood and mingled with marine fossils; marine shell fragments may be blown back from a beach and incorporated in eolian sand-dunes and so on. Furthermore, fossils weathered out of older strata may be redeposited as *derived fossils* in a new sediment. A *life assemblage* is an association of fossils representing organisms which could all live in the same environment. The fossils in such an assemblage, having suffered little transport, may often be almost undamaged and some may remain in their positions of growth—examples are seen in reef-limestones and in the seat-earths of coal-seams. A *death assemblage* is a chance association of fossils brought together by the surface agents of transport. Some at least of the fossils are likely to be incomplete or damaged. We can classify the more important environments of living communities from a geological point of view as shown in Fig. 11.1.

2. Fossilisation

The soft parts of dead animals and plants are quickly attacked by bacteria or other organisms, and are soon destroyed. The more durable hard parts—such as shell and bone, horny and woody materials—are then exposed to weathering and erosion and these too soon disappear unless they are covered up by a protective layer of sediment. Freak circumstances occasionally preserve organisms complete with their soft parts, as is illustrated by flies in amber or mastodons entombed in the frozen mud of Siberia. But most fossils represent only the hard parts of the original organism. As burial is necessary for preservation it is natural that marine organisms are much more often fossilised than those of terrestrial environments, since the sea is mainly a region of deposition, and the land of erosion. We must remember that fossil remains do not provide a complete, or even a representative, sample of the life of past periods—they include relatively few soft-bodied organisms and relatively few terrestrial organisms.

The hard parts which provide the common fossils are made of a variety of substances. Some are organic compounds, such as the horny

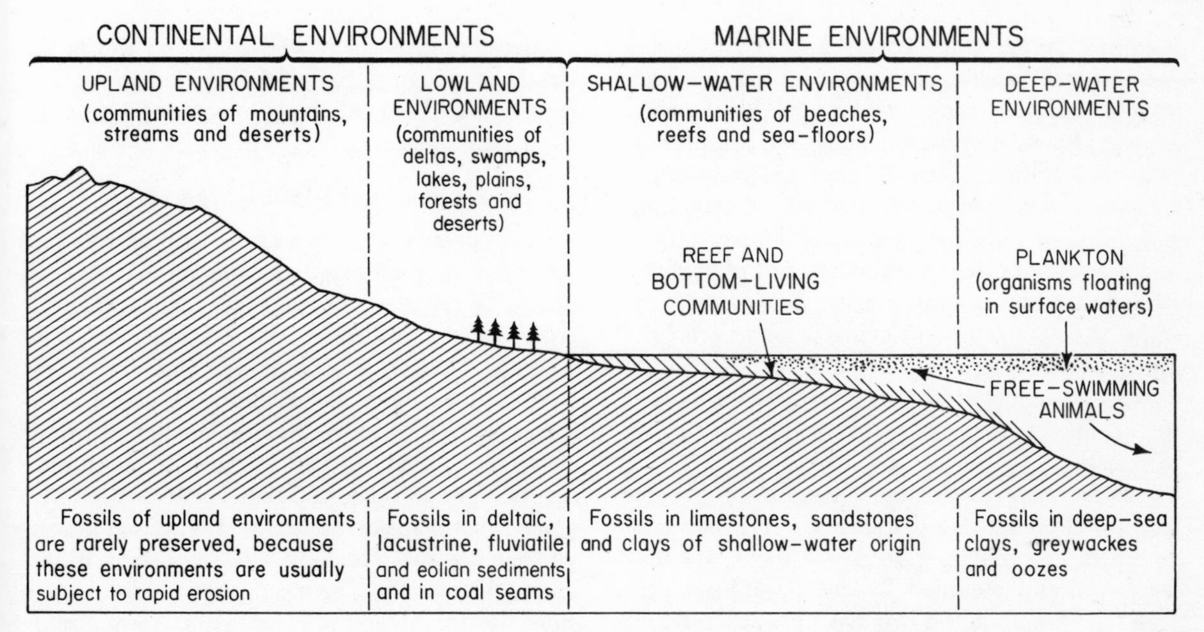

Fig. 11.1 *The principal environments of life*

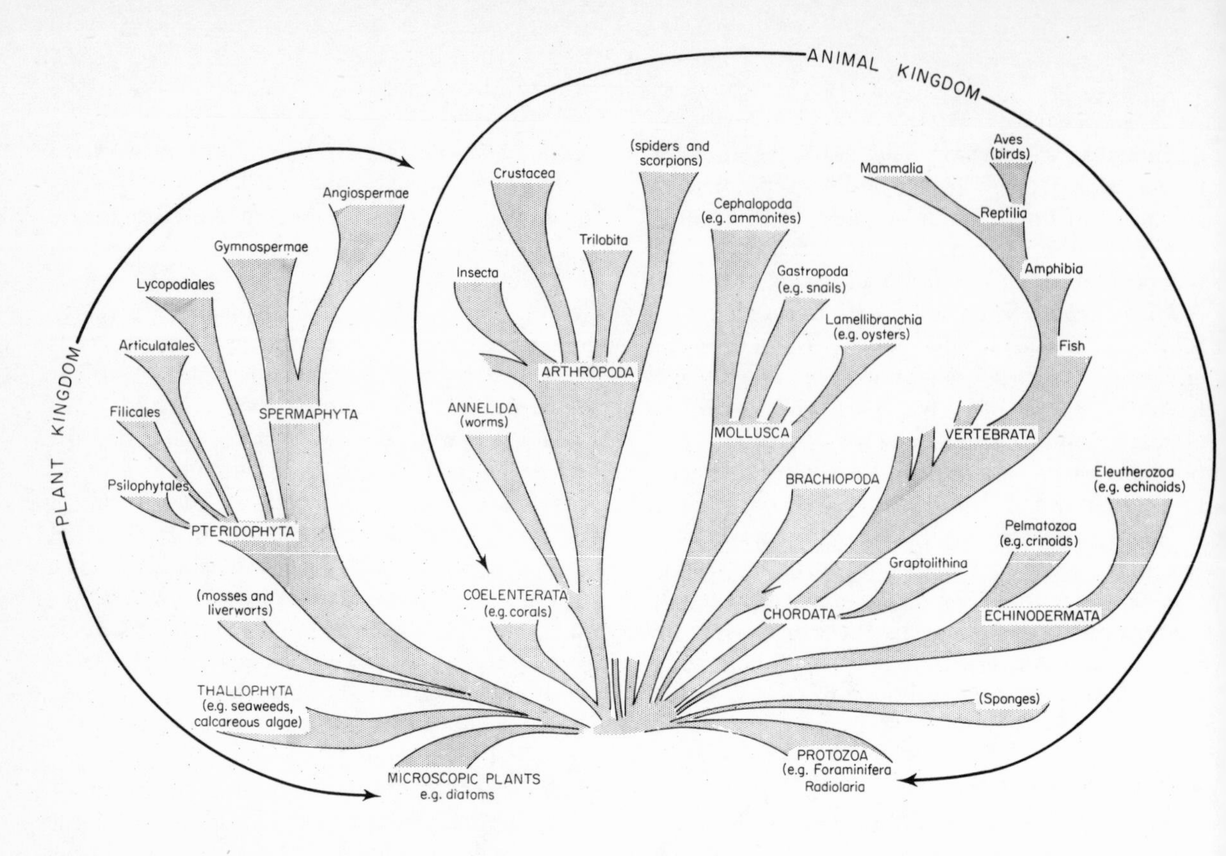

Fig. 11.2 *Relationships of the main groups of animals and plants*

material, *chitin*, which covers insects and is the basis of many shells, or the waxy substance *cutin* which encloses many plant-spores and pollen grains. Others are made of inorganic compounds extracted from the sea or from groundwater—*calcium carbonate* makes the majority of shells and *silica* is used by a few animals and plants. *Bone* is a mixture of calcium carbonate and phosphate. Still other fossils are simply *impressions*, the imprint of the organism being preserved on the hardened bedding-plane. Finally, *trace-fossils* are structures such as footprints or burrows produced in sediments by the activity of the living organism.

Fossils undergo further changes long after burial. They may be flattened by the weight of sediment accumulating on top of them and may be chemically modified by circulating groundwater. *Solution* of the original material leaves a hollow space or *mould* with the form of the fossil—later deposition in this space gives a *cast* of the fossil. Exchange of material may lead to *replacement* of the fossil substance—by silica, pyrites or other materials.

3. The Animal and Plant Kingdoms

Most present-day organisms can be grouped into the two great *kingdoms*, the animals and the plants. Within each kingdom, we can immediately recognise a number of major groups or *phyla* (singular *phylum*) in each of which the organisms have a fundamentally different plan of construction. Within each phylum, smaller groups or *classes* show characteristic variants of the fundamental plan and can be further sub-divided into *orders*, *families*, *genera* and finally *species* distinguished by progressively less important differences. The smallest group or *species* represents a single kind of animal or plant recognisably different from all others.

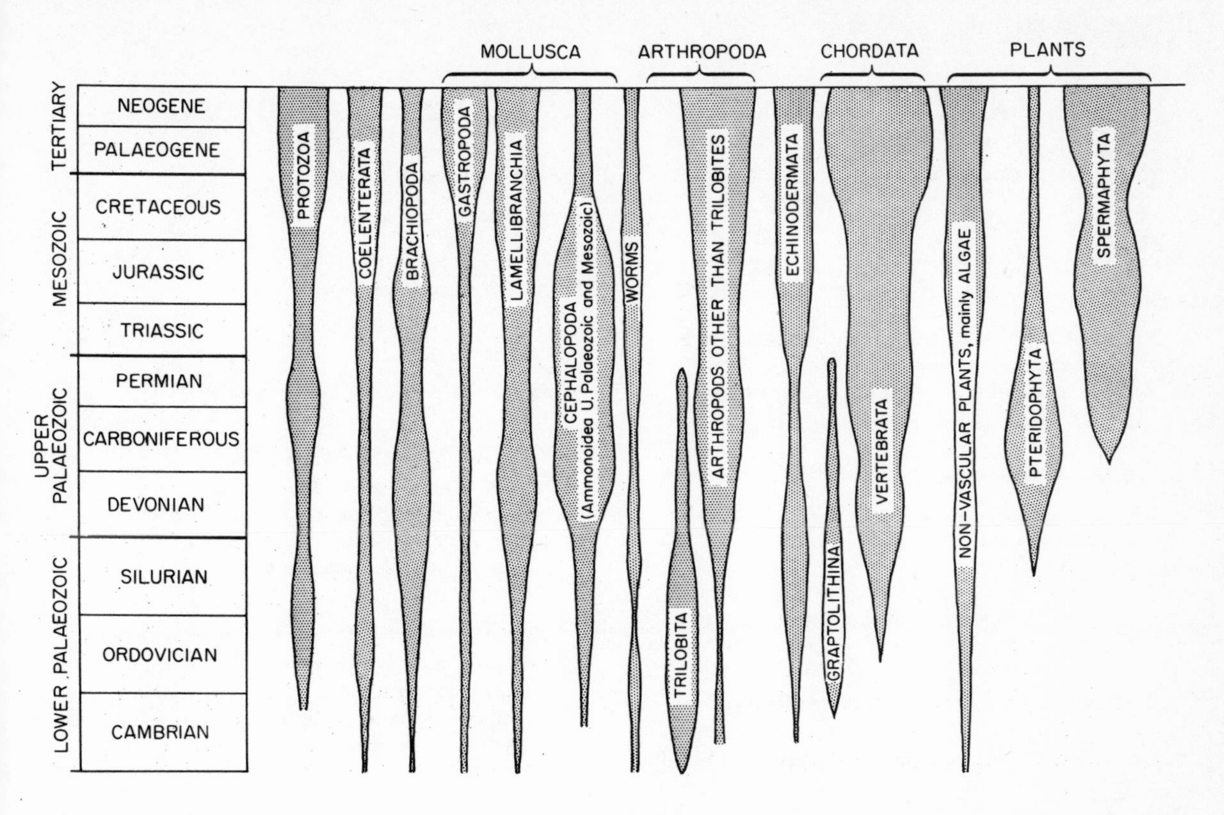

Fig. 11.3 *Time-ranges of the main groups of animals and plants*

We can illustrate this classification by reference to the domestic dog. The name of this species is *Canis familiaris*; with other dog-like creatures such as the fox, *Canis vulpes* and the wolf, *Canis lupus*, it belongs to the genus *Canis*, a branch of the family *Canidae*. The Canidae are grouped with the Felidae (cats) and certain other families in the order of flesh-eaters or *Carnivora* of the class of warm-blooded *Mammalia* in the phylum *Chordata* of the animal kingdom. To some extent (though one must not press the point too far) the system of classification records the history of the species. Thus all members of the genus *Canis* are very much alike and have probably evolved, at no very distant time, from a single ancestral stock. The dog and the cat families, though now noticeably different, had common ancestors at some more remote time, and so on through the larger groups. We shall meet in later pages examples of the evolution of several modern groups from single stocks by the splitting of *lineages* or lines of descent.

Outlines of the classification of the animal and plant kingdoms are given in the Tables on page 124; in Fig. 11.2 an attempt is made to indicate the possible relationships of the main groups, and in Fig. 11.3 the ranges of these groups through geological time are shown. The more important divisions are described from a geological point of view in the remainder of this chapter. Groups which are of little importance to the geologist are dealt with in outline in Fig. 11.4. The student should try to examine at least one example of these minor groups and as many as possible of the more important groups discussed below.[1]

[1] Common fossils from British strata are illustrated in three hand-books published by the British Museum (Natural History) at the prices shown. These are *British Palaeozoic Fossils* (1964) 12/6, *British Mesozoic Fossils* (1962) 12/6, and *British Cainozoic Fossils* (1963) 6/-.

Chapter eleven

THE ANIMAL KINGDOM

(Groups of little importance to geologists are omitted)

I. Non-cellular animals

PROTOZOA

Foraminifera: small non-cellular animals; marine or freshwater.
Radiolaria: small non-cellular animals; marine.

II. Multicellular animals

PORIFERA: sponges.

COELENTERATA

Anthozoa, corals, important marine fossils and rock-builders.

BRACHIOPODA: 'lamp-shells': important fossils.

BRYOZOA (POLYZOA): 'sea-mats'. Complex colonial encrusting organisms.

ANNELIDA: segmented worms. Worm tracks and burrows.

ARTHROPODA

Trilobita, marine, extinct: important fossils.
Crustacea, e.g. crabs, lobsters: not common fossils.

MOLLUSCA

Gastropoda, e.g. whelks, snails.
Lamellibranchia, e.g. mussels, cockles.
Cephalopoda, e.g. nautilus, octopus, *ammonites* (extinct) and *belemnites* (extinct).

ECHINODERMATA

Crinoidea, e.g. sea-lilies.
Echinoidea, e.g. sea-urchins.

CHORDATA

Graptolithina: extinct: position doubtful, marine.
Vertebrata: back-boned animals.
Fish-like forms and true fishes.
Amphibia, e.g. frogs.
Reptilia, reptiles, e.g. snakes.
Aves, birds.
Mammalia, e.g. horse, man.

THE PLANT KINGDOM

I. Non-cellular plants

Diatomacea: aquatic.

II. Multicellular plants

THALLOPHYTA

Algae: e.g. seaweeds, reef-building types.

PTERIDOPHYTA (spore-bearing land plants)

Psilophytales: Rhynie Chert plants.
Lycopodiales: club-mosses: almost extinct.
Articulatales: horse-tails: almost extinct.
Filicales: ferns: abundant today.

SPERMAPHYTA (seed-bearing plants)

Gymnospermae (seed partly enclosed in ovary), e.g. cycads, conifers.
Angiospermae (seed completely enclosed in ovary): the flowering plants.

SPONGES

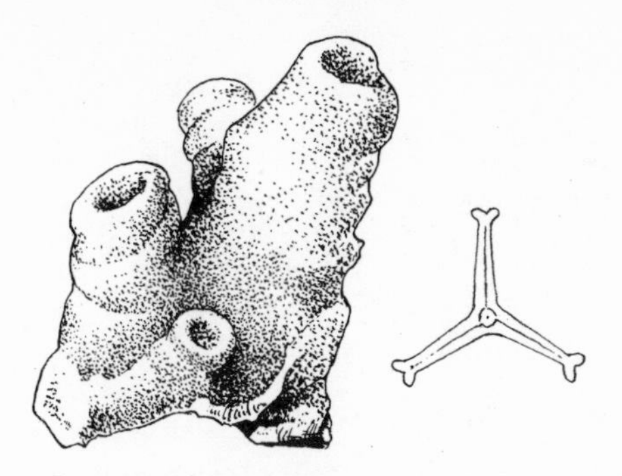

SPONGES are primitive animals living in salt or fresh water. The lobed or bun-shaped body, fixed at one end, is often stiffened by calcareous or siliceous spicules which may be preserved as fossils. The spicule illustrated is greatly enlarged.

BRYOZOA or POLYZOA

POLYZOA or BRYOZOA are small colonial animals forming bushy growths or encrustations on rocks and plants in water. The walls of the colonies are often calcified.

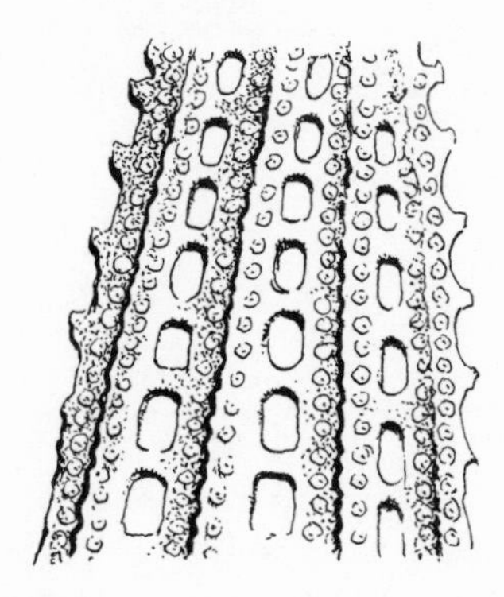

ANNELIDA or SEGMENTED WORMS

ANNELIDA or SEGMENTED WORMS include free-living, burrowing or swimming worms which often produce characteristic markings in sediments. Some species, like that illustrated, build protective tubes of calcium carbonate or cemented sand-grains.

Fig. 11.4 *Fossils of some minor groups*

4. THE ANIMAL KINGDOM: INVERTEBRATA

Protozoa: Foraminifera and Radiolaria

The Protozoa differ from all other animals in that their living substance or protoplasm is not divided into cells. Although the group includes an enormous number of microscopic organisms (among them the familiar amoeba), only two orders provide important fossils. These are (1) the *Foraminifera* which build minute shells or tests, usually made of calcium carbonate and (2) the *Radiolaria* which have delicate and complicated skeletons of silica (Fig. 11.5). The unbelievably numerous remains of foraminifera and radiolaria rain down from the surface-waters to the sea-bottom to build up the oozes of the abyssal oceans (p. 88). In sediments of shallow seas, foraminifera are also widespread and contribute largely to certain limestones such as the Tertiary *Nummulitic Limestone*.

The *Foraminifera* (Ordovician–Recent) are, as a rule, not much bigger than a pinhead, although the largest types may be an inch or more across. They are often very abundant in Mesozoic and Tertiary strata and are used as zone-fossils for correlation in oilfields because their small size makes it possible to extract them from borehole samples. The *test*, which in life houses the main part of the protoplasm, consists either of a single capsule with one large opening or, more frequently, of a number of *chambers* opening into one another and arranged according to some simple pattern as shown in Fig. 11.5. Different genera are distinguished by the shape and arrangement of the chambers and the composition of their walls which are most frequently made of calcium carbonate.

Coelenterata: Corals

The coelenterates (Pre-Cambrian–Recent) are a group of simply constructed aquatic animals including the jellyfish, sea anemones and corals. Most members of the phylum have gelatinous bodies and are seldom fossilised; but a few groups,

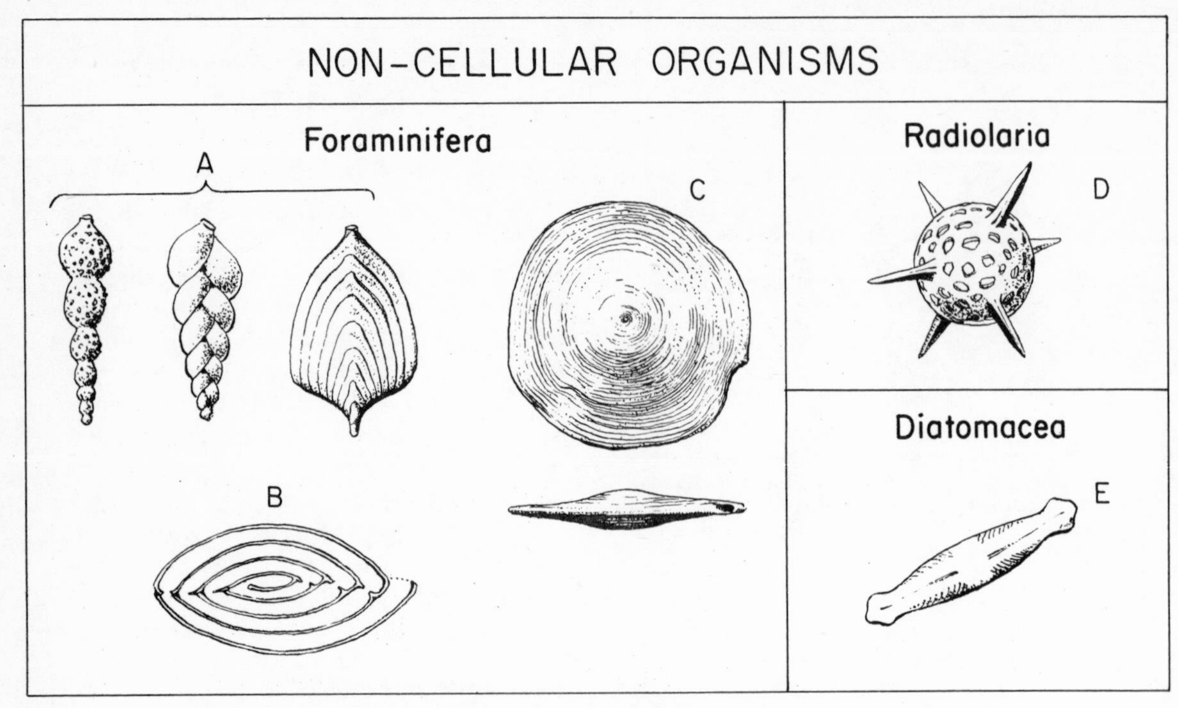

Fig. 11.5 A. Three foraminifera with differing arrangements of the chambers (all greatly enlarged)
B. A fusilinid (late Palaeozoic) in cross-section (×2)
C. A nummulite (early Tertiary), vertical and lateral views (×2)
D. A radiolarian (greatly enlarged)
E. A diatom (greatly enlarged)

living fixed to the sea-floor, protect the body with a covering of calcium carbonate. These groups, which can be loosely called 'corals', not only provide numerous fossil forms, but also act as rock-builders, making *coral reefs*. Corals, or at least the reef-building kinds, live mainly in clear, warm shallow waters and their fossil remains are therefore most common in shallow-water limestones.

Two classes of coelenterates include coral-like organisms—(i) The *Hydrozoa* in which the gelatinous body is built round a large undivided cavity, and (ii) the *Anthozoa* in which the central cavity is partially divided by radial partitions. The *true corals* are anthozoans, but we will also mention briefly the extinct *stromatoporoids* which are thought to have been hydrozoans.

The *anthozoan corals* (Ordovician–Recent) have the peculiar habit of being able to live not only as separate individuals (*solitary corals*) but also as colonies in which many individuals are united by a common branching stalk. It is these *colonial corals* which are the main reef-builders for, although the size of the individual animal or *polyp* is seldom more than a few inches, colonies of many polyps make bushy or compact stony masses several feet in diameter (Fig. 11.7).

The skeleton of a solitary coral, or of one polyp in a colony, consists of a deep cup or *theca* fixed at one end to the rock or to the stalk of the colony and open at the other (Fig. 11.6). Transverse partitions (*tabulae*) may cut off the lower parts of the theca. Its central space may be partly divided by radial *septa* projecting inward and its wall may be strengthened by small tangential plates or *dissepiments*. Three groups are

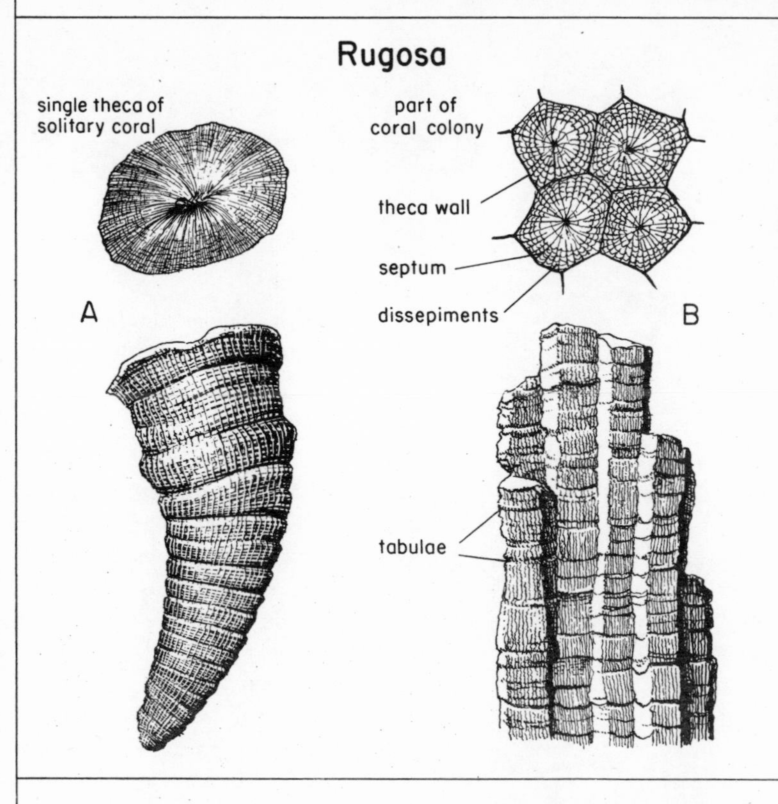

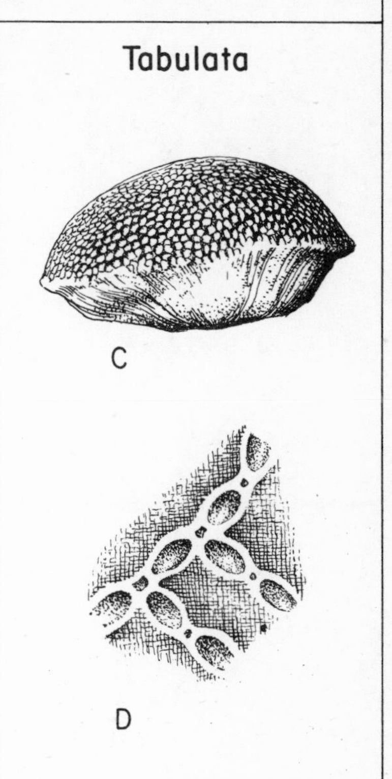

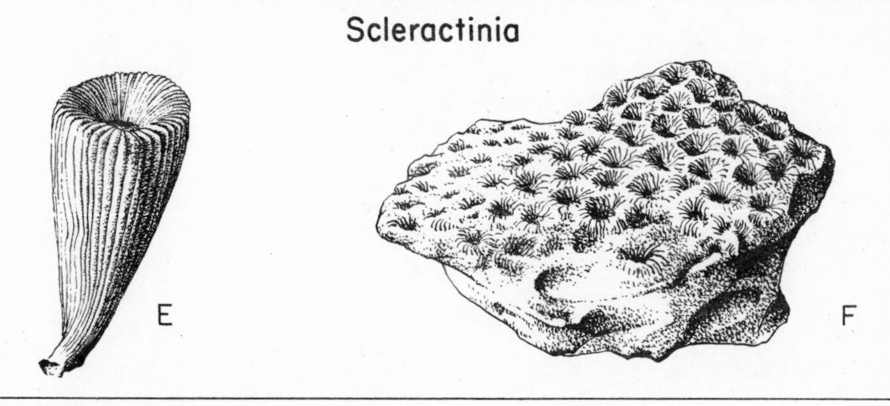

Fig. 11.6 A. Lateral and vertical views of *Palaeosmilia* ($\times\frac{1}{2}$)
B. Part of a colony of *Lithostrotion* ($\times\frac{2}{3}$) with a vertical view of four thecae
C. *Favosites* colony ($\times\frac{3}{4}$)
D. Part of a *Halysites* colony to show arrangement of thecae ($\times 4$)
E. *Parasmilia* ($\times 1\frac{1}{2}$) F. *Isastraea* colony ($\times\frac{3}{4}$)

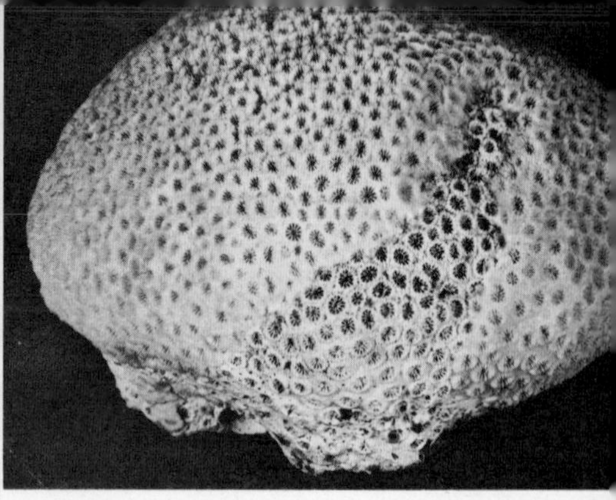

Fig. 11.7
Above *Coral colony from the Persian Gulf* ($\times \frac{1}{5}$)
Left *Corals growing in the Great Barrier Reef, Australia*

distinguished by differences in the structure of the theca as shown on page 127; further subdivisions within these groups are made according to its shape and, in colonial forms, to the arrangement of polyps in the colony.

(*a*) The *Tabulata* (mainly Palaeozoic) are invariably colonial and have simple thecae in which tabulae are the only internal partitions. The thecae are very long and tubular and may be joined together in various patterns—in branching lines as in *Halysites* or in compact masses as in *Favosites* (Fig. 11.6C and D).

(*b*) The extinct *Rugosa* (Palaeozoic) include both solitary and colonial forms. The theca is usually shorter and stronger than in tabulates, and has a characteristic ribbed surface. It is divided internally by septa in such a way that there is a noticeable gap at one point around the circumference—the presence of this gap gives a bilateral symmetry to the theca. Tabulae and dissepiments are also often present. Solitary rugose corals can be exemplified by *Zaphrentis*, common in the Carboniferous Limestone, colonial types by *Lithostrotion*, occurring in the same formation (Fig. 11.6B).

(*c*) The *Scleractinia* (Mesozoic–Recent) are the common corals of the present day. They resemble the rugose corals in the occurrence of septa, tabulae and dissepiments, but differ from them in that the septa are regularly spaced around the circumference, so that the symmetry is radial rather than bilateral. *Parasmilia* provides an example of a solitary form, *Isastraea* of a colonial form (Fig. 11.6E and F). As in the other two groups, the colonial corals include both loosely branching and tight compact types (Fig. 11.7).

The extinct hydrozoan *Stromatoporoidea* (Palaeozoic) were exclusively colonial coelenterates and the fossils derived from them consist simply of concentrically arranged calcareous laminae making up stony lumps and nodules a foot or more across. There are no individual thecae and it is thought that the laminae represent successive platforms on which the polyps of the colony were supported.

Brachiopoda

The brachiopods are marine shelled animals which live as a rule fixed to the sea-floor by means of a short stalk. These animals, popularly known as 'lamp-shells', were very numerous in Palaeozoic and Mesozoic times but are comparatively scarce at the present day. They are found as fossils in most types of shallow-water marine sediments.

The body of the brachiopod is enclosed in a horny or calcareous shell, generally an inch or so across, which consists of two unequal *valves* (Fig. 11.8). The *ventral valve* is usually the larger of the two and is pierced in most groups by an aperture

PALAEOZOIC BRACHIOPODS

Inarticulata

Lingula
A

Orbiculoidea
B

Articulata

A productid
C

An orthid
E

A spiriferid
F

Leptaena
D

MESOZOIC BRACHIOPODS

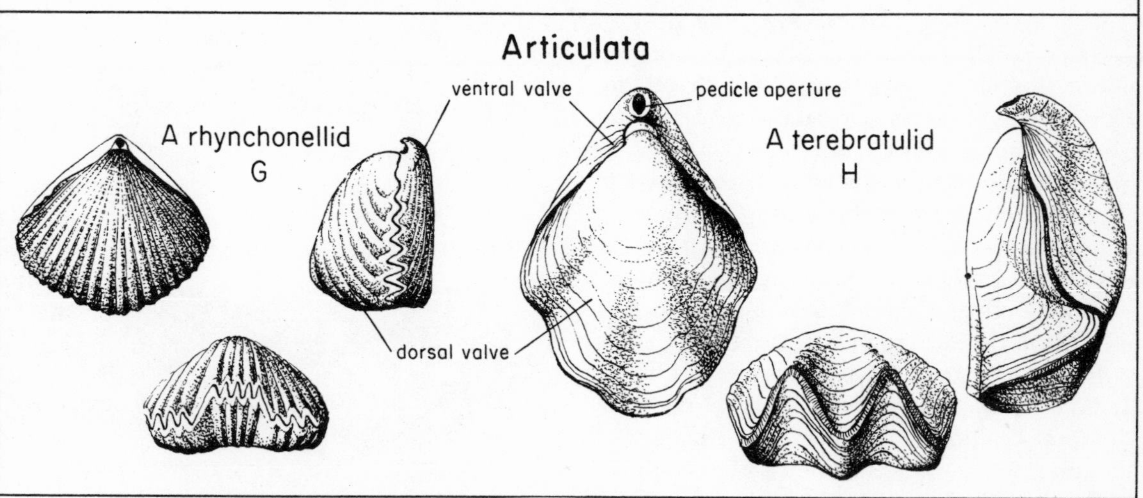

Fig. 11.8 A. *Lingula*, no pedicle aperture
B. *Orbiculoidea*, pedicle aperture near centre of ventral valve
C, D, E. Examples of the first group of Articulata
F, G, H. Examples of the second group of Articulata
(All about natural size or somewhat reduced)

through which emerges the tough stalk or *pedicle*. The *dorsal valve* in some groups carries on its inner surface complicated projections (the *brachial supports*) which strengthen the 'arms' used in propelling food towards the mouth. The valves are opened and closed by muscles and are usually also hinged or *articulated* along a toothed *hinge-line* near the pedicle aperture. The *hinge-teeth* project from the ventral valve and fit into *sockets* in the dorsal valve.

The brachiopods are classified according to the characters of the hinge and the brachial supports. The two major divisions are (i) *Inarticulata*, and (ii) *Articulata*.

(i) Inarticulata: primitive forms with no hinge-structures or brachial supports. In the Inarticulata, the valves are held together simply by muscles and are horny rather than calcareous. Two types of structure can be distinguished. One is exemplified by *Lingula* in which the pedicle emerges between the valves and not through the ventral valve (Fig. 11.8A). *Lingula* is one of the longest-lived of all known genera and has survived with little change for some 500 million years from the Ordovician to the present day. In the second type (e.g. *Orbiculoidea*), the pedicle projects through a hole near the middle of the ventral valve, or the valve is cemented directly to a rock. Both types of Inarticulata were already in existence in the Lower Palaeozoic.

(ii) Articulata: more advanced forms with calcareous valves articulating along a toothed hinge-line. In the Articulata, there is always a pedicle aperture near the hinge-line in the ventral valve. Two subdivisions are distinguished by differences in the structure of the shell around the aperture. The first sub-division, most common in Palaeozoic rocks, includes forms such as *Orthis*, *Leptaena* and *Productus* (Fig. 11.8). The second includes the bulk of the Mesozoic and Tertiary brachiopods as well as some of Palaeozoic age. It is characterised by the occurrence on the dorsal valve of *brachial supports*, spiral in *Spirifer*, hook-like in *Rhynchonella* and looped in *Terebratula*.

Arthropoda

The phylum Arthropoda includes the crabs, insects, spiders and centipedes which all have two fundamental features in common—(i) the body is built up of a number of similar *segments* arranged one behind the other, and (ii) it bears several pairs of jointed limbs or *appendages*. Both body and appendages are enclosed in a chitinous or calcified skeleton. The majority of arthropods are free-moving animals and many are equipped with elaborately constructed eyes and other sense organs.

Trilobita. Although most classes of arthropods provide occasional fossils, only one group is of importance to the general geologist. This is the extinct group of the *trilobites*, marine animals which appear to have been bottom-dwellers, crawling, swimming or lying half-buried in sand (see Fig. 11.9). Trilobites were abundant throughout the Lower Palaeozoic, a few survived through the Upper Palaeozoic but all died out by the end of that era. They are widespread in shallow-water sediments and provide useful zone-fossils in Cambrian strata.

Trilobites have a flattened body, roughly oval in outline, usually between an inch and a foot in length, and bear at first glance some resemblance to an enlarged woodlouse. The upper surface is covered by strong chitinous armour in which the

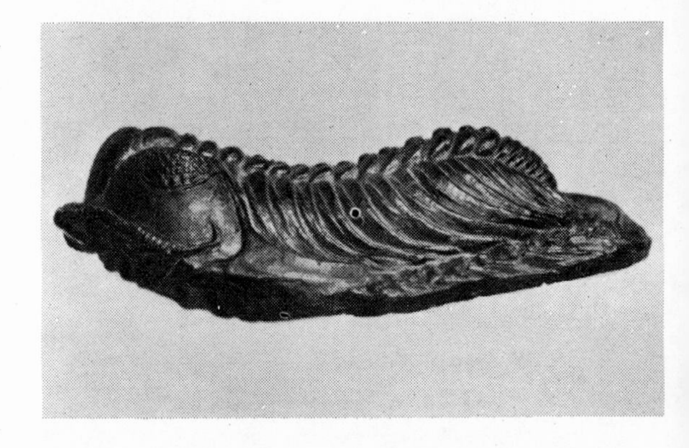

Fig. 11.9 *Lateral view of a trilobite showing its position at rest on the sea-floor*

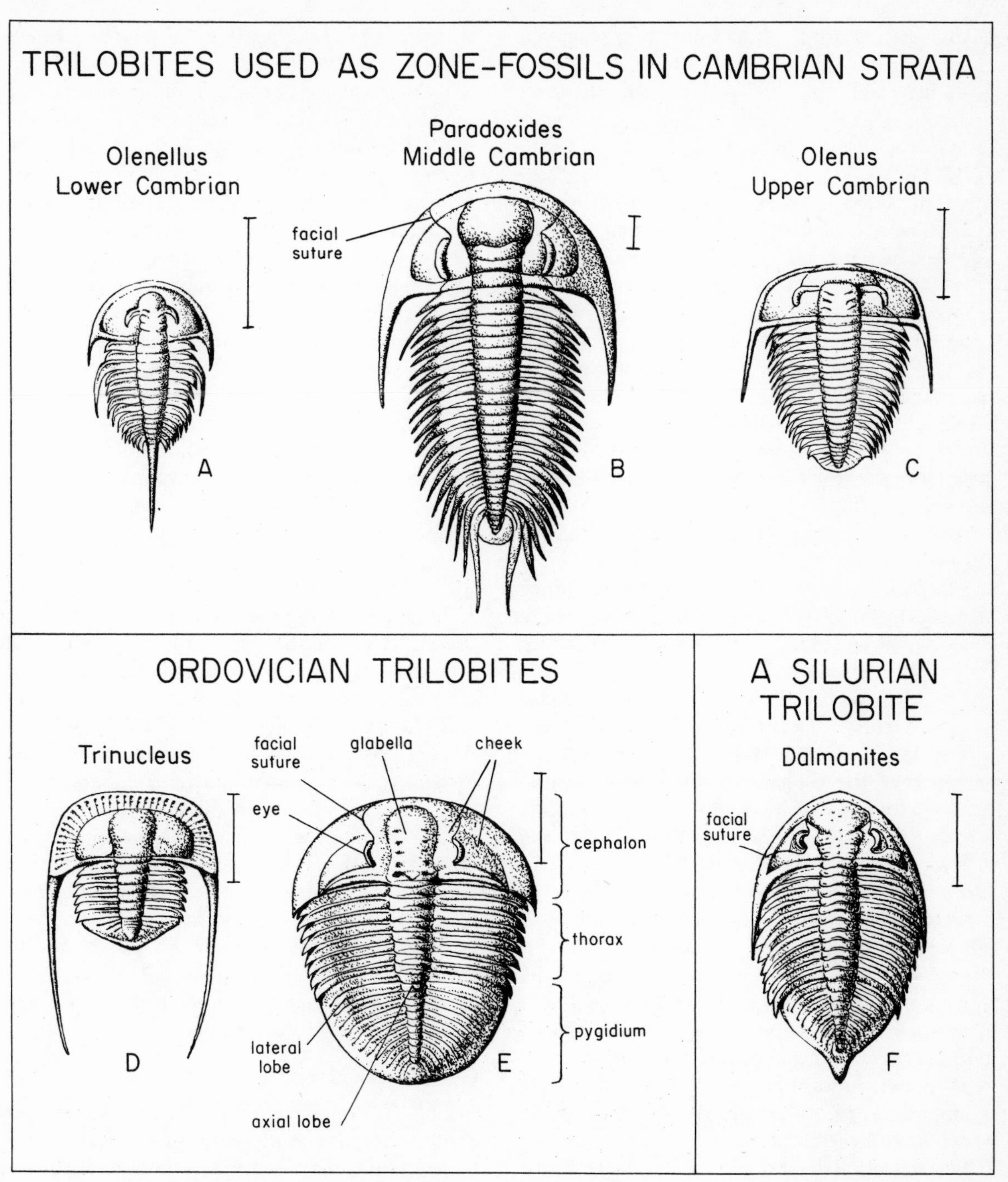

Fig. 11.10 *Trilobites, characteristic fossils of the Lower Palaeozoic*

The annotated specimen (E) is related to *Ogygia*. A and D are hypoparian; B, C and E are opisthoparian and F is proparian

characteristic arthropod division into segments is clearly apparent (Fig. 11.10). The lower surface (seldom visible in fossils) bears the feathery appendages—a pair to each segment—which were probably used both for swimming and as gills.

At the front of the animal, the first few segments are fused together to make a *head* carrying eyes and a pair of antennae. The skeleton covering the head segments is united into a single shield or *cephalon*. Behind it lies the *thorax* consisting of a number of separate segments (from two to twenty-nine according to genus) and at the rear the tail or *pygidium* in which several segments (again the number varies) are fused together. The whole dorsal skeleton is divided lengthwise into three portions, an arrangement which gives the group its name: the central or *axial lobe* is humped and contains the main part of the body while the *lateral lobes* on either side are mere flaps protecting the appendages.

The *cephalon* shows a number of features which are used in the classification of the trilobites. In early Cambrian forms the raised axial lobe, called the *glabella*, is usually crossed by grooves which indicate the original divisions between the fused head-segments. In most later and more advanced forms, the glabella is undivided. The lateral lobes or *cheeks* of the cephalon are in many families crossed by a line which is perhaps the place where the skeleton split during moulting. This *facial suture*, when present, marks off an inner *fixed cheek* from an outer *free cheek*—the presence or absence of the facial suture and the course it follows across the cheek provide a basis for classification as explained below. The *eyes*, usually situated on the free cheek, are made of a number of lenses arranged side by side as in the eyes of modern insects; since trilobites were in existence in Cambrian times it is probable that they were the first living creatures capable of looking at the world around them.

The earliest trilobites, found in Lower Cambrian rocks, are mostly rather primitive. The glabella still shows grooves marking the divisions between the head-segments. There is no facial suture on the upper surface of the cephalon, which is said to be *hypoparian* (see Fig. 11.10). The eyes are large and are attached to the glabella by ridges. The thoracic segments are numerous but the pygidium is small. One genus of this primitive kind, *Olenellus*, is used as a zone-fossil of the Lower Cambrian (Fig. 11.10A). The trilobites of Middle Cambrian and younger strata do not as a rule retain the primitive characters mentioned above, but show instead a variety of specialised features. *Opisthoparian* trilobites in which the facial suture runs to the hinder margin of the cephalon include *Paradoxides*, the zone-fossil of the Middle Cambrian, and *Olenus*, the zone-fossil of the Upper Cambrian. *Proparian* types in which the suture runs to the outer margin of the cephalon include *Dalmanites*. Some specialised forms such as *Trinucleus* are hypoparian and it is believed that these may be descended from ancestors which possessed a facial suture.

Mollusca

The molluscs are the common 'shellfish' of the present day. We shall mention only three of the numerous classes, the *Gastropoda* (snails, whelks, etc.), the *Lamellibranchia* (mussels, clams, etc.), and the *Cephalopoda* (octopus, cuttlefish, etc.). All are characterised by the presence of a calcareous (or occasionally horny) shell which is in two pieces or *bivalved* in the lamellibranchs, but *univalved* and generally coiled into a spiral in the other two classes. The cephalopods are entirely marine, the lamellibranchs marine and freshwater, the gastropods marine, fresh-water and terrestrial. All provide important fossils; and the extinct group of cephalopods known as *Ammonoidea* supplies many zone-fossils used for Upper Palaeozoic and Mesozoic rocks.

Gastropoda

The *Gastropods* are, with a few exceptions, easily recognisable because the univalve shell housing the body is twisted into a distinctive rising or *helicoid spiral*, familiar in the common snail. The shell (Fig. 11.11), made all in one piece and generally not more than a few inches across, is really a narrow cone wound tightly on itself. In

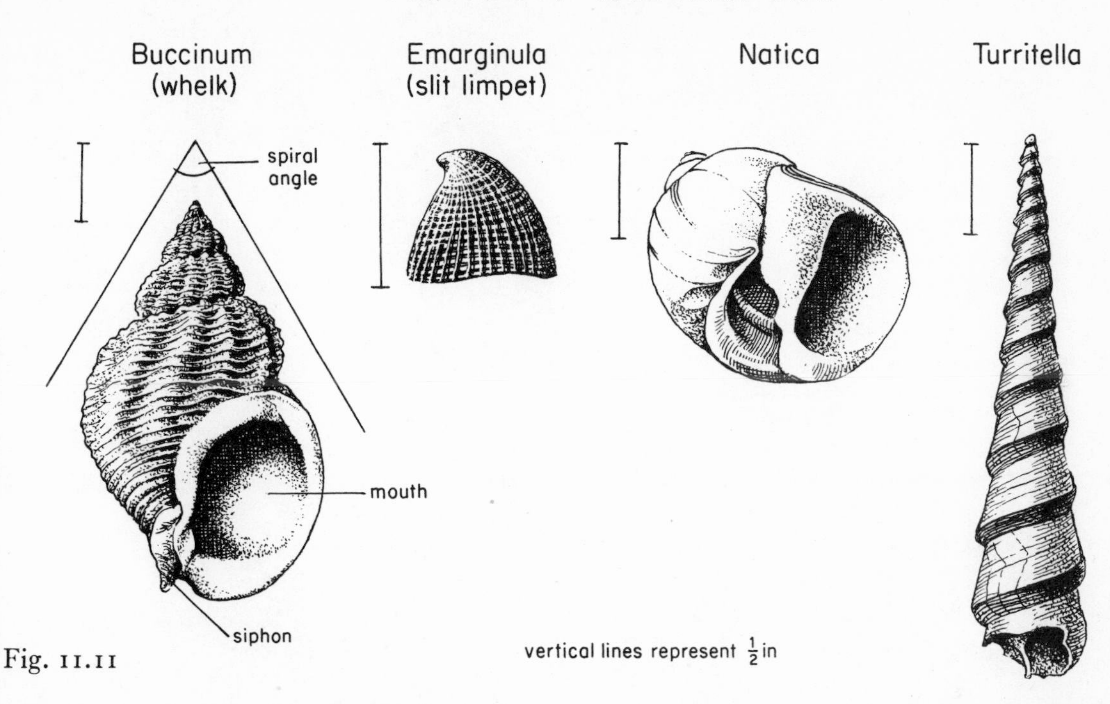

Fig. 11.11

most genera, the opening or mouth of the cone lies on the right when the shell is held point upward; and the spiral is then said to be *dextral*. A few are coiled the other way round and are said to be *sinistral*. The *spiral angle* is the angle between the tangents to successive whorls. A small spiral angle gives a long pointed shell like that of *Turritella*; a large angle gives a squat shell like that of *Natica* and an angle of about 180° a plane spiral as in *Euomphalus*. The shell aperture or *mouth* through which the body can project in life has a characteristic shape in many genera—its rim may be thickened or ribbed, or it may be drawn out into a spout or *siphon*. The surface of the shell is often ornamented with ridges, knobs and colour-patterns which help to distinguish particular genera.

Lamellibranchia

The lamellibranchs are mostly bottom-dwelling animals seldom more than a few inches in length. Some are fixed types like the oyster, others burrow in mud or sand and many plough their way laboriously along by means of a muscular 'foot' which can be protruded from the shell. They occur in many shallow-water marine sediments and in fresh-water sediments such as those of the Coal Measures where they provide useful zone-fossils.

The shell consists of two *valves* which are placed laterally and which are usually mirror-images of one another (Fig. 11.12); there is no aperture like the pedicle aperture of the brachiopod shell (p. 129). The valves are opened and closed by muscles and, as a rule, articulate along a toothed hinge-line. The structure of the hinge and arrangement of the teeth provide useful diagnostic characters and four kinds of *dentition* may be distinguished, as illustrated in Fig. 11.12.

(*a*) *Taxodont dentition* in which the hinge-teeth are set in a row, perpendicular to the hinge-line, in one valve and fitting into corresponding sockets in the other; this primitive arrangement is seen in the genus

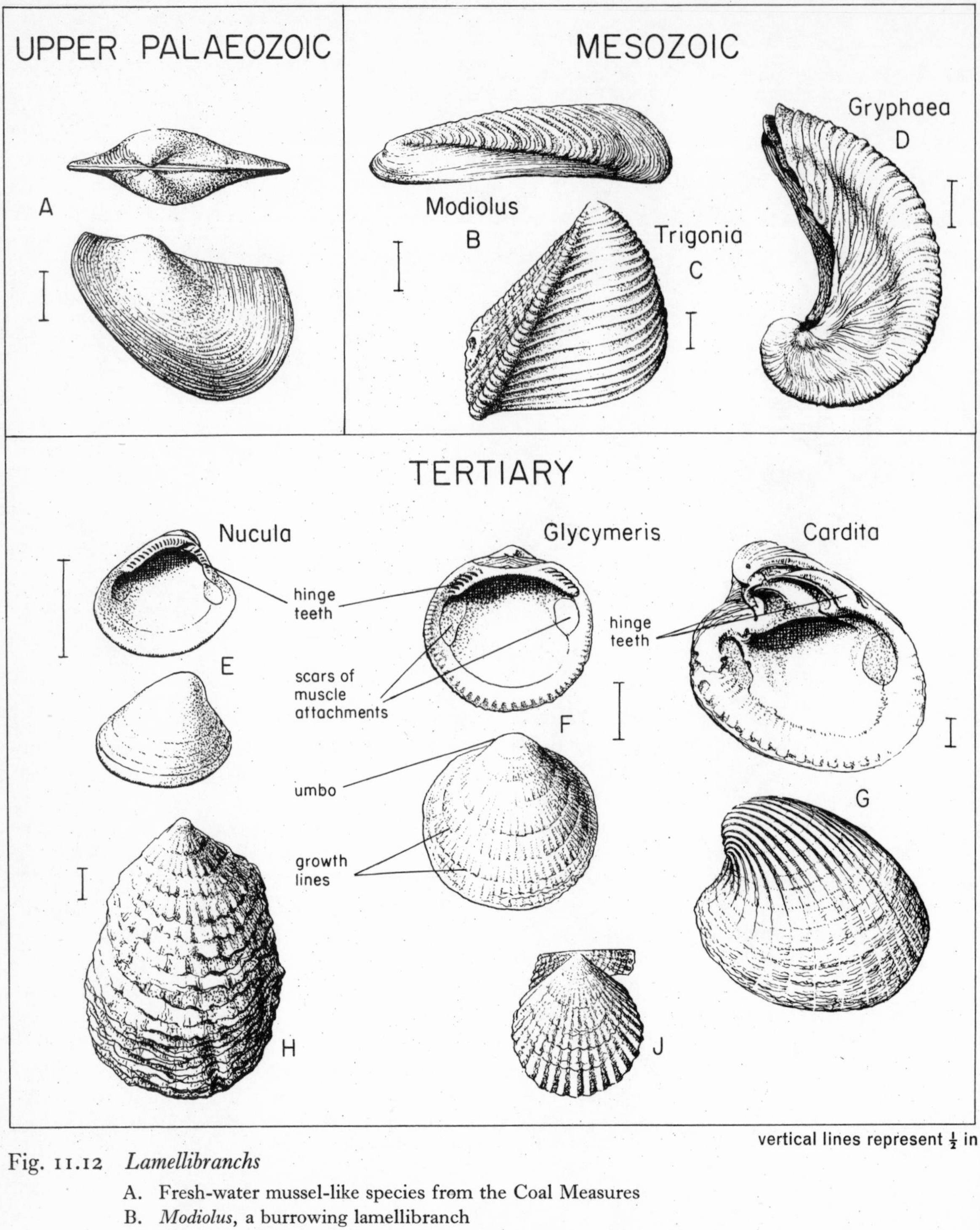

Fig. 11.12 *Lamellibranchs*

A. Fresh-water mussel-like species from the Coal Measures
B. *Modiolus*, a burrowing lamellibranch
C. *Trigonia*, a genus which often has a strongly ornamented shell
D. *Gryphaea*, an oyster-like genus
E, F, G. Three genera illustrating different types of dentition
H. A species of oyster
J. A species of scallop

Nucula (Fig. 11.12E) which has been in existence since Silurian times.
(*b*) *Actinodont dentition* which differs from the taxodont type only in the fact that the teeth are arranged radially as in *Glycymeris* (Fig. 11.12F).
(*c*) *Heterodont dentition* in which the teeth are fewer and more complicated in shape and arranged as in *Cardita* (Fig. 11.12G).
(*d*) *Dysodont dentition* in which hinge teeth are feebly developed or absent as in the fixed mussels and oysters (e.g. *Ostrea*).

The shape of the shell tends to vary with the mode of life. Swimming forms such as the scallop have nearly symmetrical valves, whereas burrowing forms are often greatly elongated (e.g. *Modiolus*); in fixed forms there is sometimes a marked difference between the two valves as in the oyster, *Ostrea*, where the upper valve forms a kind of lid on the larger valve. The outer surfaces of the valve are marked by *growth-lines* concentric about a starting-point or *umbo* near the hinge and are often ornamented by ribs, knobs or colour-patterns. The inner surfaces show a number of markings which represent the points at which the muscles were attached in life. These are shown in Fig. 11.12F.

Cephalopoda

The *Cephalopods* were of great importance in Upper Palaeozoic and Mesozoic times, though they are represented at the present day only by a few types such as the cuttlefish, squid and octopus. They differ from most molluscs in that the majority were, and are, capable of swimming freely in the open sea; and their fossil remains are therefore found in both deep-water and shallow-water sediments. Three sub-classes are recognised:

(i) *Nautiloidea*, relatively important in the Palaeozoic era, rare in younger rocks, and now represented only by the genus *Nautilus*.
(ii) *Ammonoidea*, abundant in Upper Palaeozoic and Mesozoic strata, now extinct.
(iii) *Dibranchiata*, the extinct belemnites and the present-day cuttlefish, squid and octopus.

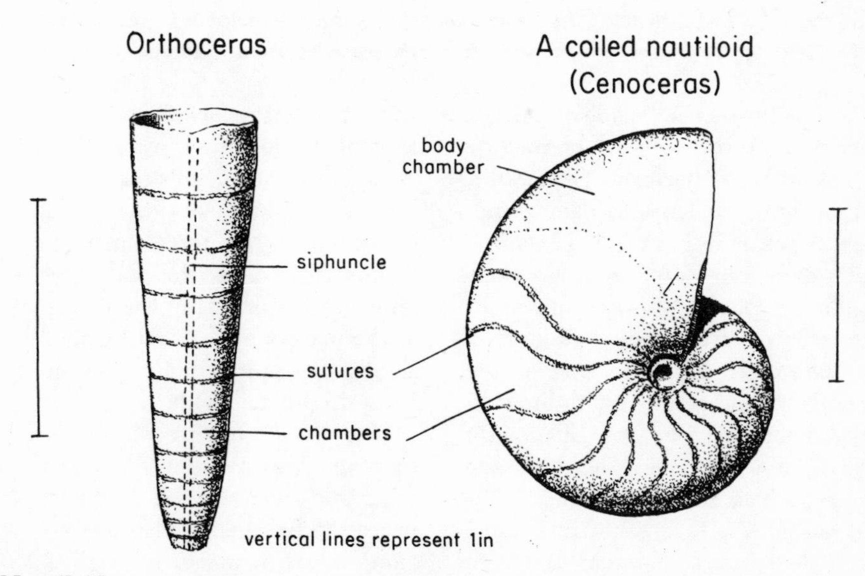

Fig. 11.13 *Nautiloids*
Orthoceras is Palaeozoic, *Cenoceras* is Mesozoic

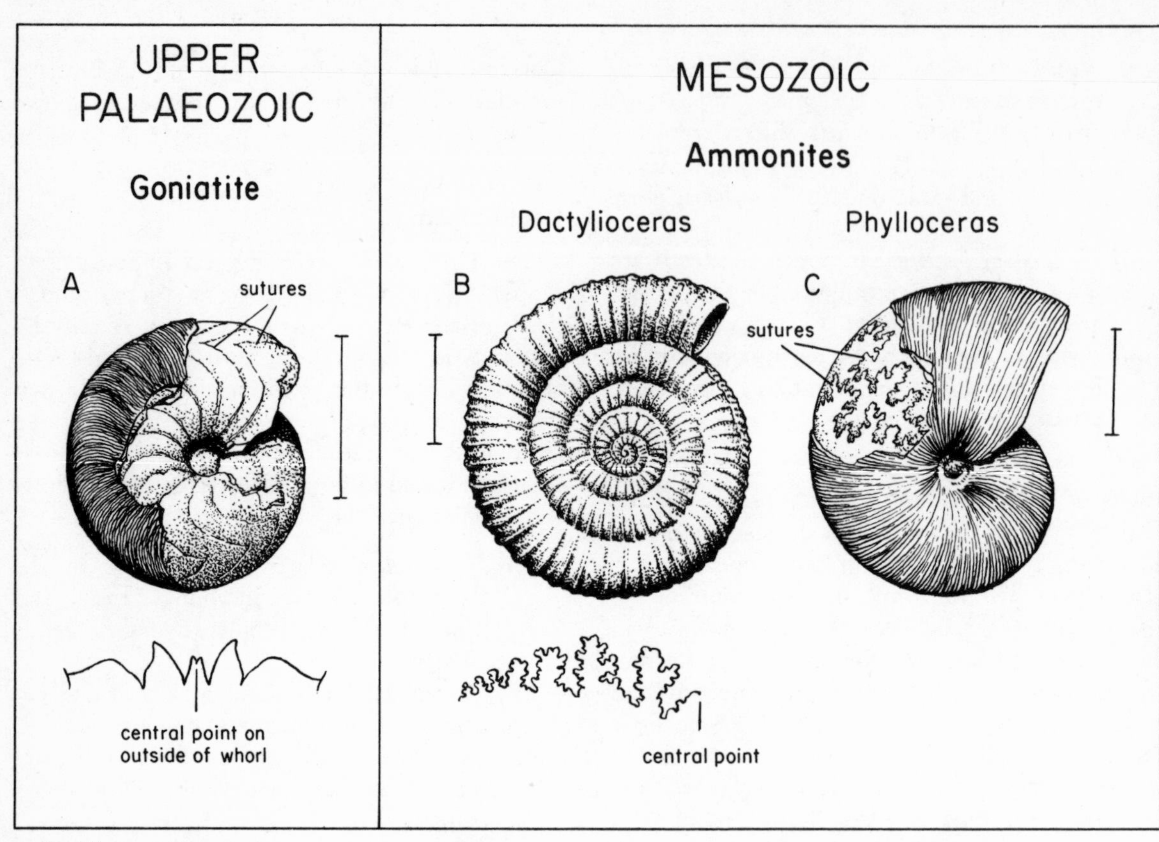

Fig. 11.14 *Ammonoids. The characteristic sutures of goniatites and ammonites are shown diagrammatically below and can be seen where the shell is partly worn away in* A *and* C

(i) The *Nautiloidea* have a shell made of a single valve in the form of a narrow cone which may be either straight, as in the Palaeozoic *Orthoceras*, slightly curved, or tightly coiled in a plane spiral, as in the modern *Nautilus* (Fig. 11.13). The space within the shell is divided by transverse partitions or *septa* (singular, *septum*) into a number of *chambers*; these septa meet the shell-wall along a smooth line or *suture*. In life, the body of the animal occupies the chamber nearest the mouth of the shell; the other chambers are filled with gas and a narrow tube containing living tissue (the *siphuncle*) passes back through them.

(ii) The *Ammonoidea* (Devonian–Cretaceous) had a chambered shell basically similar to that of the nautiloids and were probably developed from nautiloid ancestors. The shell (usually between two and eight inches across) is generally coiled in a plane spiral, but is very occasionally straight, only partly coiled, or coiled in a helicoid spiral. Each succeeding whorl may touch the outer edge of the one within, to produce an *evolute* spiral; or it may entirely envelop the inner whorls to give an *involute* spiral (Fig. 11.14). The *siphuncle* lies at the margin of the shell, generally on the outer side of the spiral but in one group (the *clymeniids*) on the inner side.

The main feature distinguishing the ammonoids from the nautiloids is the shape of the septa and their sutures. The part of each septum nearest to the shell-wall is crinkled, and the suture along which it meets the wall is therefore not a smooth line as in nautiloids but a line which is wavy, zigzagged or intensely crenulated. In the *goniatites* and clymeniids (both Upper Palaeozoic), the suture simply makes a few large zigzags but in the Mesozoic *ammonites* the suture is much more complicated, as shown in Fig. 11.14A and B. Both goniatites and ammonites are extensively used as zone-fossils, different types being distinguished by the pattern of their sutures, the shape and ornamentation of the shell and the shape of the shell mouth.

(iii) The *Belemnites* (Trias–Cretaceous) had a straight, chambered shell and were perhaps descended from a type like the nautiloid *Orthoceras*. The chambered portion or *phragmacone*, however, was so much reduced in size that it could not contain the whole body (Fig. 11.15). Its wall was thickened towards the apex and commonly produced into a solid rod, the *guard*, which is the part most frequently fossilised. An elongated plate, the *pro-ostracum* (possibly analogous to the 'bone' of the present-day cuttlefish) extended forward from the opposite end of the phragmacone. Belemnites are classified largely on the shape of the guard.

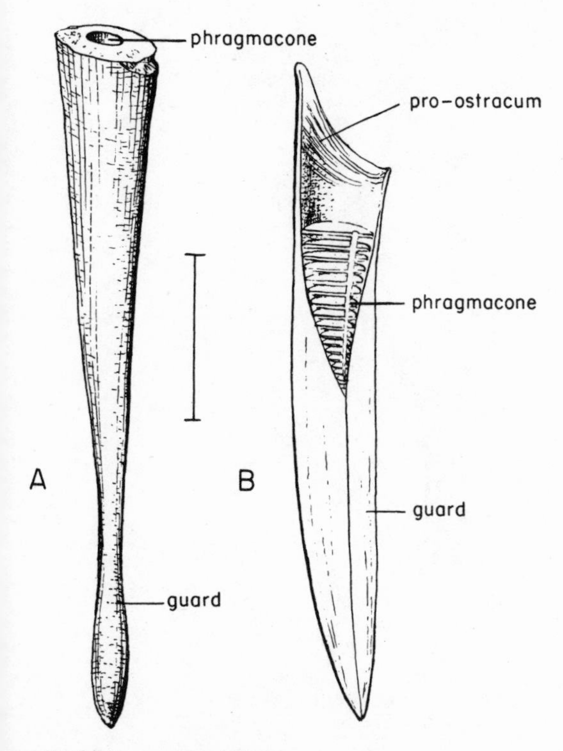

Fig. 11.15 *Belemnites*

A. An incomplete specimen
B. Section of a complete specimen of a different genus

Echinodermata

The echinoderms are bottom-living marine organisms whose fossil remains are common in shallow-water and in moderately deep-water sediments. Their skeleton is characterised by a *five-fold symmetry* unique among animals. It is made up of a large number of plates or *ossicles* which are usually fitted together into a rigid or flexible *test* enclosing the body. Each ossicle is a single calcite crystal and often exhibits the recognisable calcite cleavage when broken. The group includes the *sea-urchins*, *starfish* and *sea-lilies* as well as some other types of little importance to us. The majority of forms are only a few inches across.

The Echinodermata are classified into two sub-phyla, of each of which we shall take a single class as an example:

(i) Sub-phylum *Pelmatozoa*, forms living fixed to the sea-floor by means of a stalk: example, the class *Crinoidea*, the sea-lilies.
(ii) Sub-phylum *Eleutherozoa*, free-living forms: example, the class *Echinoidea*, the sea-urchins.

Crinoidea. The crinoids were very abundant in the Palaeozoic and early Mesozoic when their innumerable stem-ossicles were often piled up by currents to produce *crinoidal limestones*; they are now rare. The crinoid skeleton (Fig. 11.16) consists of three parts—the *stem*, made of a column of ossicles sometimes many feet in length; the small rounded *theca* containing the chief body organs; and the five *arms* mounted on the theca and arranged to produce the five-rayed symmetry characteristic of the echinoderms. The *theca* is a cup made of regularly arranged ossicles, perched on the stem and covered in by a kind of roof or *tegmen* pierced by the mouth or by channels leading to the mouth. The arms which rise from the rim of the cup may be very long and flexible and are often branched.

Echinoidea. The sea-urchins include a few Palaeozoic fossils and very numerous Mesozoic and Tertiary forms. They have neither stalk nor arms, live crawling freely on rocks or burrowing in sand, and are normally oriented with the *oral surface* containing the mouth face-downwards instead of with the oral surface facing upwards as in the Crinoidea. We can consider the class in two groups, comprising the *regular* and *irregular* echinoids. The *regular echinoids* such as *Echinus*, the modern sea-urchin, have the shape of a slightly flattened sphere and show the customary five-fold symmetry of the echinoderms (Fig. 11.17). The main part of the test, the *corona*, has openings at the two poles. The larger opening is on the oral surface and is covered in life by a membrane called the *peristome* surrounding the mouth. The other, *aboral* opening is filled in by two circles of five ossicles each, surrounding a membranous *periproct*, the ten plates and the periproct together making the *apical system*. The plates of the *corona* itself are arranged so as to define five special vertical zones known as the *ambulacra*; each ambulacrum is made by two columns of plates pierced by pores which in life connect the body with the flexible *tube-feet* used in locomotion. Each *inter-ambulacrum* is filled in by two columns of larger plates without pores.

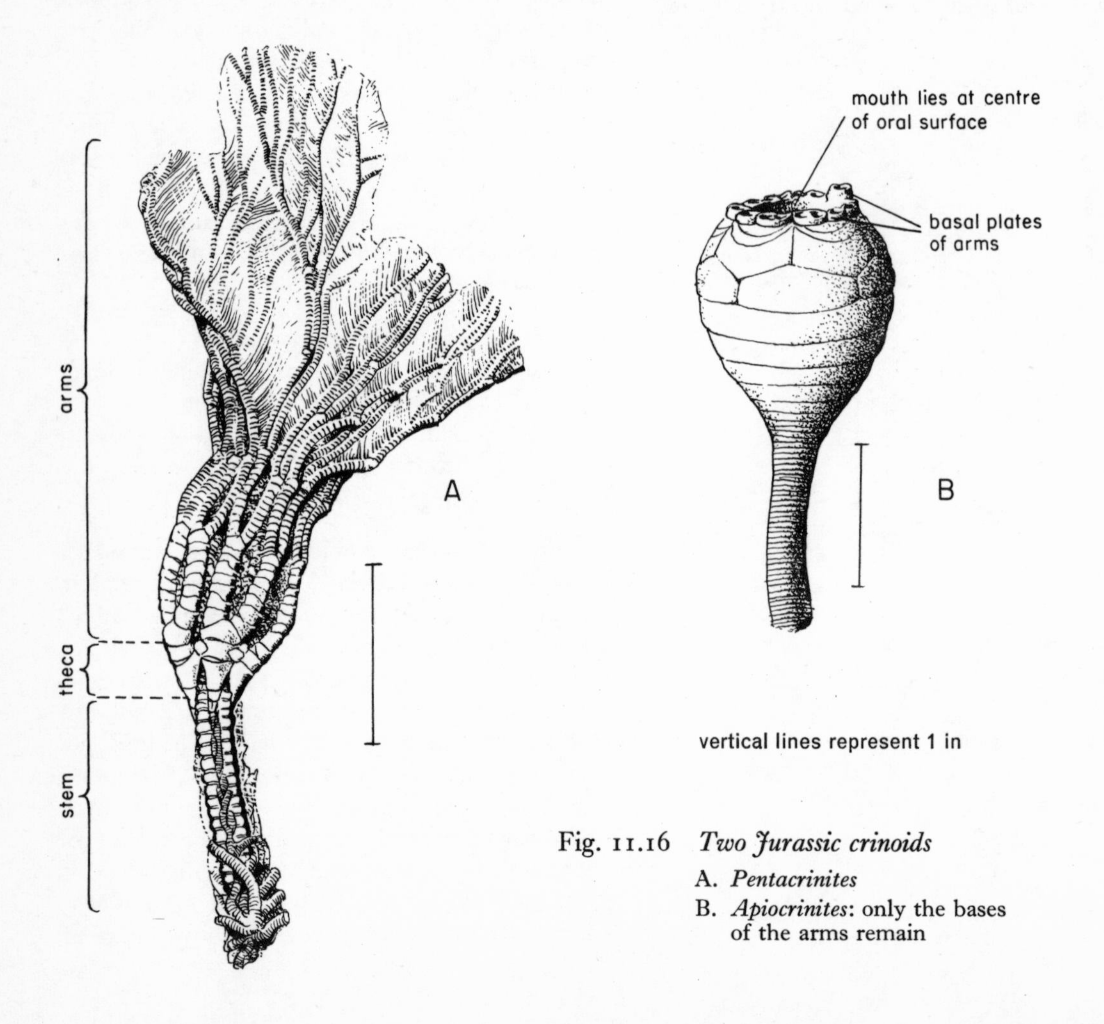

Fig. 11.16 *Two Jurassic crinoids*

A. *Pentacrinites*

B. *Apiocrinites*: only the bases of the arms remain

MESOZOIC ECHINOIDS

A regular echinoid

Irregular echinoids

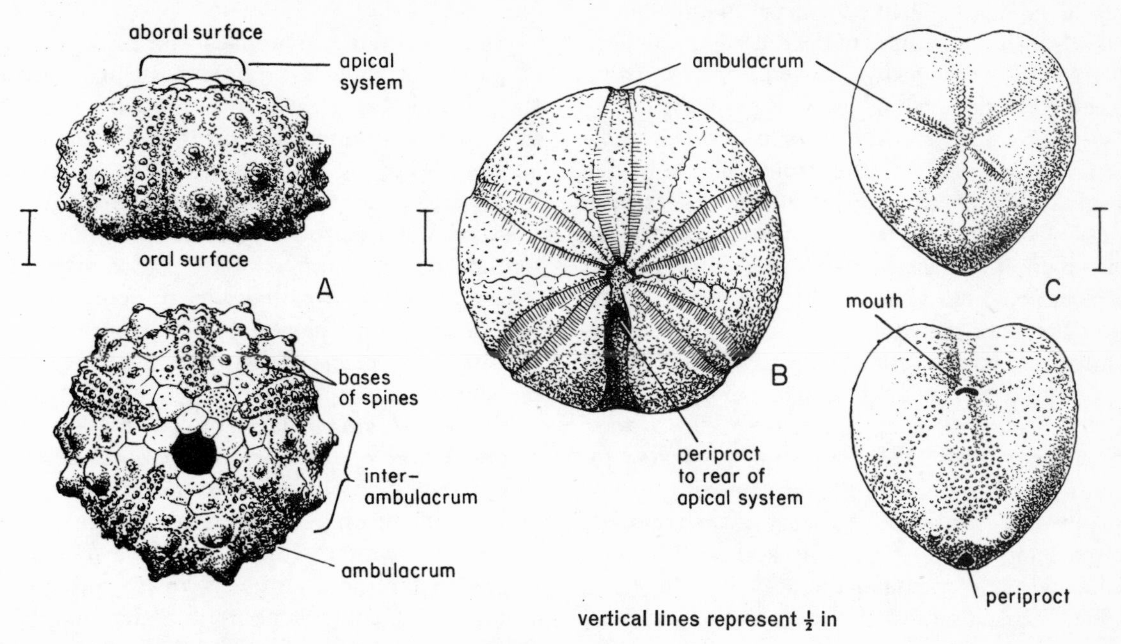

Fig. 11.17 A. *Acrosalenia* (Jurassic) lateral and aboral views
B. *Clypeus* (Jurassic) showing petaloid ambulacra
C. *Micraster* (Cretaceous), an early type of heart-urchin

Many plates bear knobs which articulate in life with large movable *spines* (loose echinoid spines are fairly common fossils).

In the *irregular echinoids* which developed from regular forms during the Mesozoic, the shape of the test has become modified for living on, or burrowing in, sandy beaches. The oral surface is usually flattish and the test as a whole may be conical, disc-like or heart-shaped (the burrowing heart-urchins). Moreover, a *bilateral symmetry* is superimposed on the old five-fold pattern as a result of migration of the periproct from its aboral position into one of the inter-ambulacra; this migration may shift the periproct on to the circumference of the test while at the same time the mouth may move forward in the ambulacrum diametrically opposite. Other modifications involve changes in shape of the ambulacra as illustrated in Fig. 11.17.

Graptolithina

The extinct *graptolites* (Cambrian–Carboniferous) were small *colonial* marine organisms which were very abundant in the Lower Palaeozoic. Many were *pelagic*, drifting passively in the surface-layers of the sea, buoyed up by floats or attached to floating seaweed; their remains are therefore widely distributed in deep-water shales as well as in shallow-water sediments and provide useful zone-fossils for strata of the Ordovician system. It is thought that the graptolites are a branch of the phylum *Chordata* to which the vertebrates described in the next section also belong.

Graptolite colonies are delicate chitinous structures, usually only an inch or so long and little thicker than a pencil-marking (graptolite from Greek *graphein*, to write). Each colony consists of one, two, four or more *stipes* or stalks all springing from one starting-point and each bearing a row of minute *thecae* which house the individual members of the colony. The structure is usually flattened during fossilisation into a mere film in which the outlines of the thecae, triangular, squarish or occasionally hooked, project from the stipe like the teeth of a saw (Fig. 11.18).

Two principal sub-divisions of the Graptolithina are recognised, the *Dendroidea* and the *Graptoloidea*. *Dendroid* graptolite colonies are bushy, with many branches bearing thecae of more than one shape, and can be exemplified by *Dictyonema* (Fig. 11.18). The *graptoloids* have fewer branches, generally one or two, less commonly four or more, and the thecae of one colony are generally all alike. The starting-point of each colony is a small conical structure, the *sicula*, which housed the first individual and floated mouth-downward with a thread or *nema* projecting upwards from it. The colony was developed by the budding-off of new thecae first from the sicula and then from later thecae.

The first graptolites, appearing in the youngest Cambrian rocks, were bushy dendroids. These were joined at the beginning of the Ordovician by the first graptoloids with two, four or more stipes. The many-branched forms died out rapidly and through most of the Ordovician the dominant graptolites were two-branched or *biserial* forms; in the Silurian and later periods they were usually *uniserial*, consisting fundamentally of a single stipe. A second change which took place with time concerned the attitude of the stipes. In most of the early biserial and many-branched graptoloids, the stipes hung downwards from the sicula and are said to be *pendent* (the 'tuning-fork' graptolites). In later forms, the stipes diverged at wider and wider angles until they were *horizontal* and then *reclined*, finally coming together again, with thecae facing outwards to produce *scandent* forms. These evolutionary changes provide a basis for the use of graptolites as zone-fossils as illustrated in Fig. 11.18. After the Silurian period, the graptolites became of little importance and they are not used as zone-fossils in Upper Palaeozoic rocks. The group finally became extinct at the end of the Palaeozoic.

5. THE ANIMAL KINGDOM: VERTEBRATA

The vertebrates, or animals with *backbones*, are of particular interest to human beings because they include not only man himself but also the other mammals and the birds, reptiles, amphibia and fish. They are not, as a rule, very common as fossils and can seldom be used as zone-fossils; this fact is not surprising, since the group contains a high proportion of terrestrial organisms whose chances of fossilisation are poor. In spite of this handicap, however, the history of their evolution is recorded with remarkable completeness (Fig. 11.19).

The *Vertebrata* (Ordovician–Recent) constitute a sub-phylum of the phylum *Chordata* to which the extinct graptolites and some other obscure groups also belong. The vertebrates are characterised by the internal *skeleton of bone or cartilage* which, in the more advanced forms, provides a strong yet flexible support for the body and limbs, the bones articulating by means of joints. This arrangement made possible the development of large, powerful animals capable of living not only in marine environments but also on land. Many changes in the structure of the skeleton during the course of evolution were clearly concerned with the process of *adaptation to environment*—that is, they helped the animal to live more efficiently in its particular habitat or to take up life in a new habitat.

The earliest vertebrates, appearing in Ordovician times, were fish-like aquatic creatures covered with bony plates which united over the head in a protective shield (Fig. 11.19). They had no paired fins such as the true fishes have, and were jaw-less, drawing in small food particles by suction—because of this last character, they are called the *Agnatha* (Greek, without jaws): the modern lamprey is a survivor of these primitive creatures. In Silurian times a second class of fish-like aquatic forms appeared—the *Placodermi* (Silurian–Permian). These had movable jaws capable of grasping or catching more varied foods, and were protected by bony plates; some

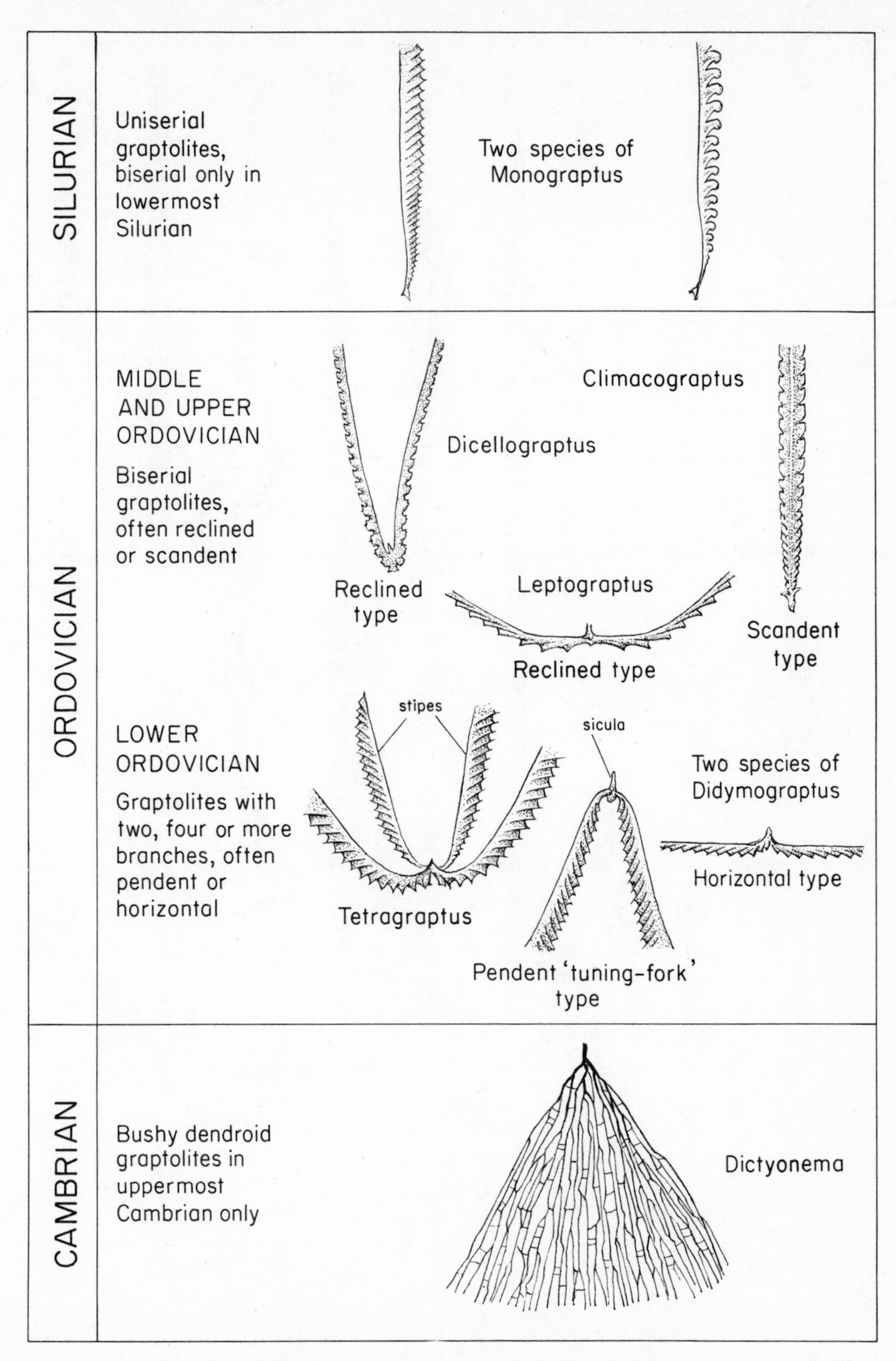

Fig. 11.18 *Graptoloids arranged to illustrate their use as zone-fossils in Lower Palaeozoic strata (somewhat enlarged)*

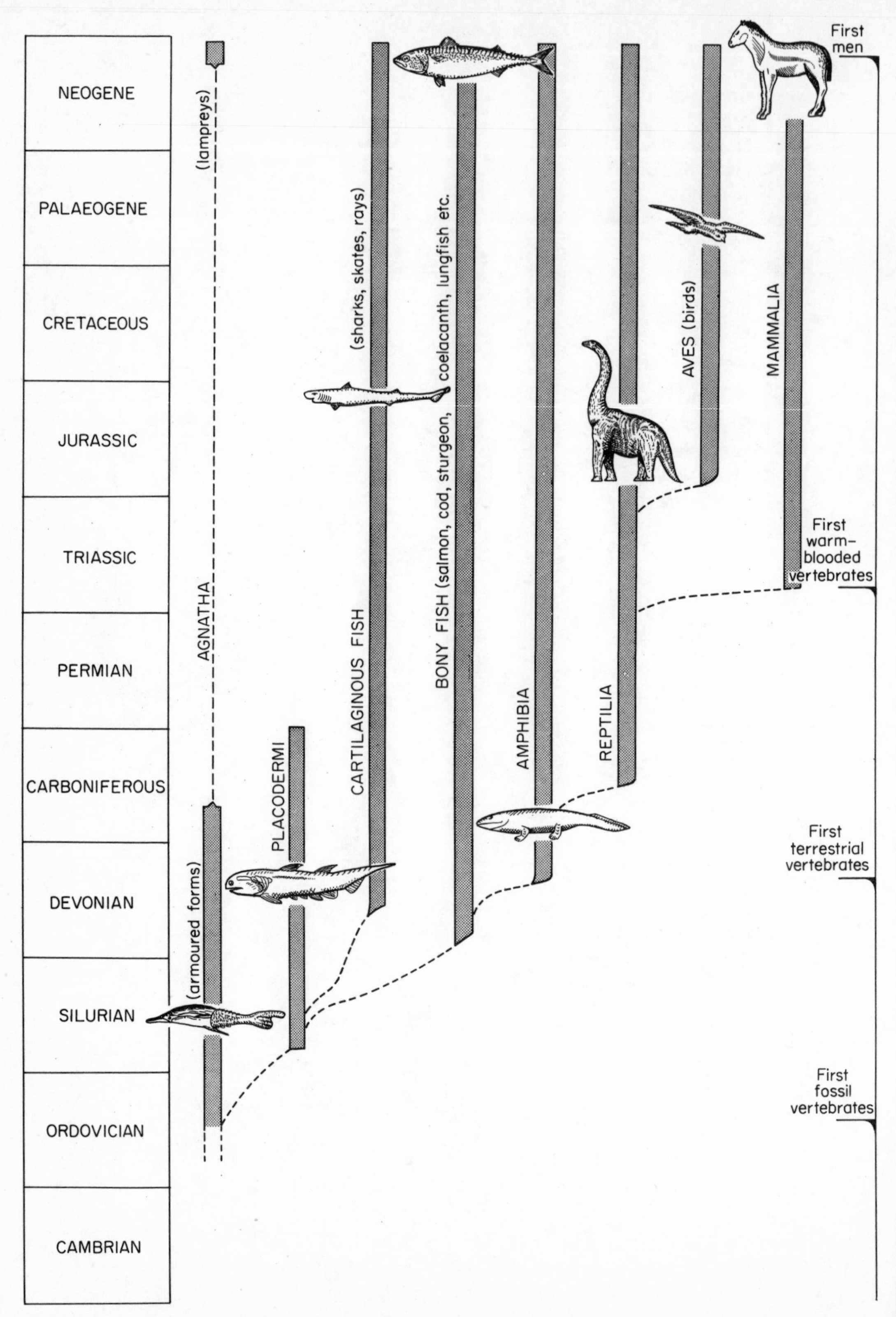

Fig. 11.19 *The evolution of the vertebrates*

forms possessed rudimentary paired flaps or fins.

The *true fish*, of which there are several groups, evolved in the Devonian from Placoderm ancestors. They had efficient jaw mechanisms and two pairs of movable fins. The bony external armour inherited from the ancestral stock gradually became lighter and flimsier and was finally reduced to thin *scales* or small *denticles*. The internal skeleton—the *backbone*, *skull* and *limb-supports*—on the other hand, tended to become more complex. From the early groups of fish are descended not only the present-day fish but also all the terrestrial vertebrates. One Devonian group of fresh-water fish which lived habitually in waters subject to drought had developed a primitive lung to serve for breathing when stranded in stagnant water or mud; these creatures had also modified their fins into fairly rigid stilt-like organs by means of which they could flounder along on land. With the adaptation of these fish to their peculiar environment, the stage was set for the appearance of terrestrial vertebrates (Fig. 11.20).

The *Amphibia* (Devonian–Recent), the first terrestrial forms to appear, lived partly on land and partly in water. Their larvae (comparable with the tadpole of the modern frog) were fish-like and breathed by means of gills; but the adults had lungs and could move about, often rather clumsily, on legs derived from the stilt-like fins of their ancestors (Fig. 11.20B). These limbs terminated in *feet* made on a five-fingered or *pentadactyl* pattern which was inherited by all later vertebrate groups. The early amphibia of the Upper Palaeozoic era were virtually the only land animals of their time and were larger and more varied than amphibia of any later period. They began to decline when more efficient groups appeared and they are insignificant at the present day.

The *reptiles* appeared in late Carboniferous times as an offshoot of the main group of amphibia. They were better adapted to terrestrial life—having developed a type of egg which could hatch out on land they had no need to return to water to breed—and soon split into a number of groups adapted to different modes of life (Fig. 17.9).

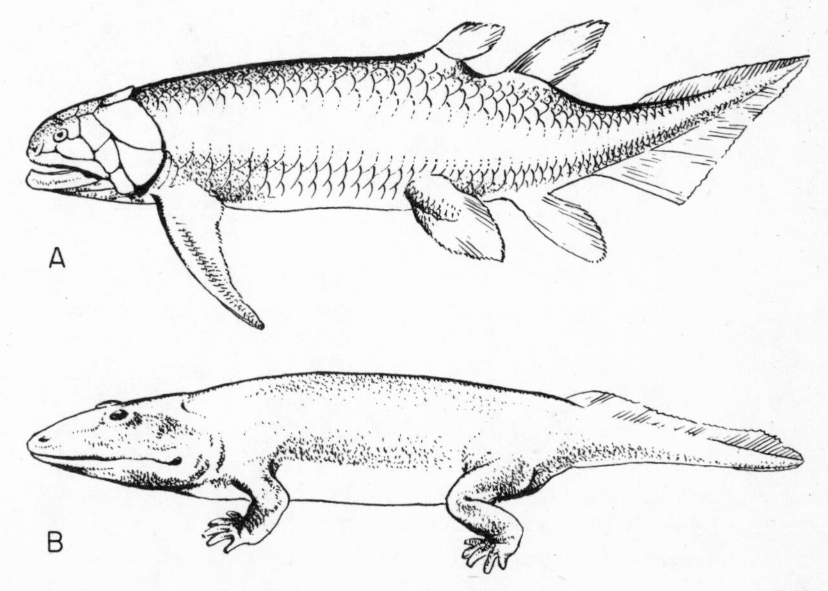

Fig. 11.20
The conquest of the land

A. A Devonian fish of the group which provided ancestors for the terrestrial vertebrates
B. An early amphibian

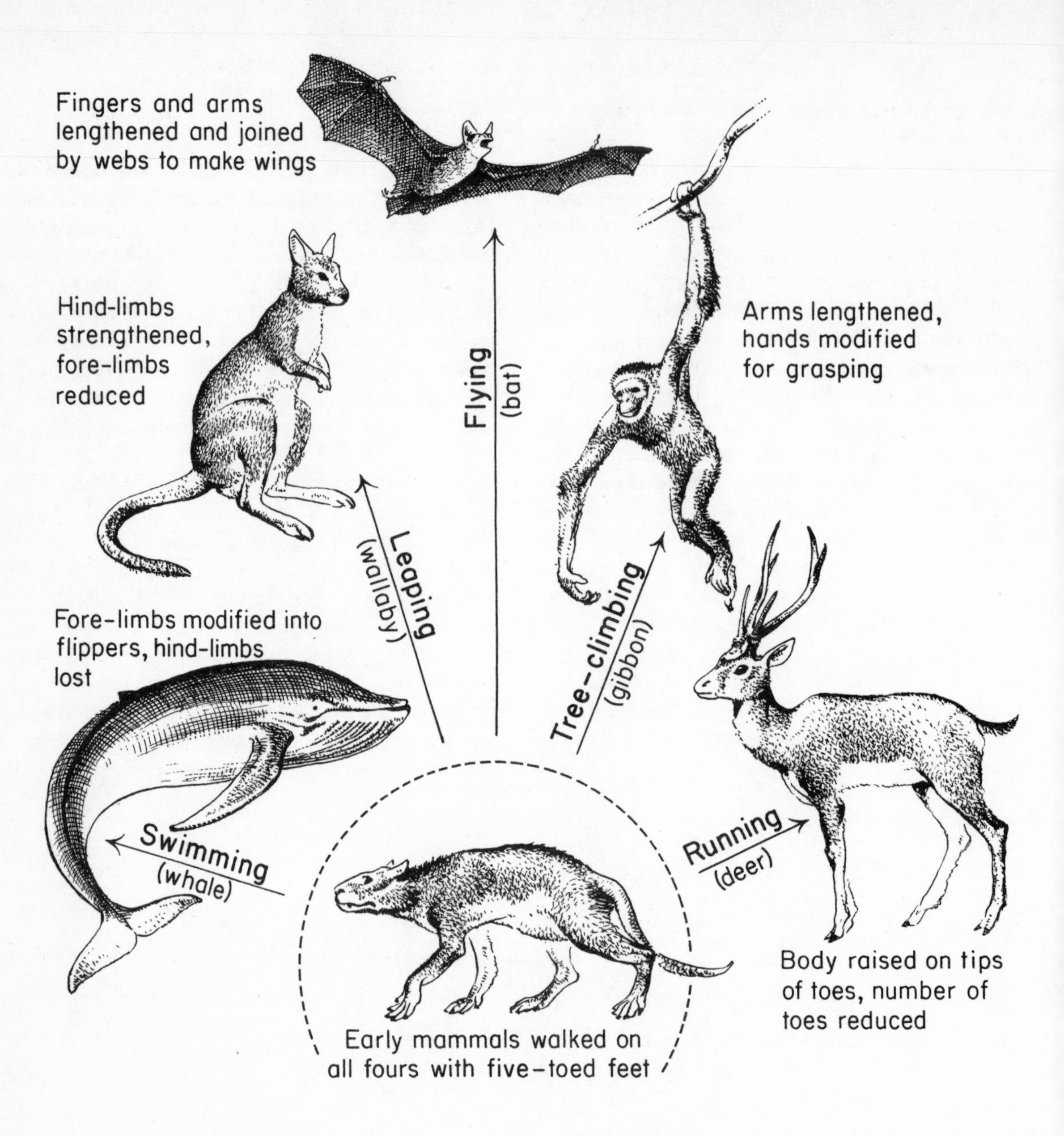

Fig. 11.21 *Adaptation among mammals. The diagram illustrates some of the ways in which mammals have become specialised for different methods of locomotion*

Throughout the Mesozoic era they were the dominant land animals and included also aquatic and aerial forms; but they declined at the end of the era and are now little more numerous than the amphibia. Many of the Mesozoic reptiles were of enormous bulk (30–40 tons), some of the *dinosaurs* and the marine *plesiosaurs* and *ichthyosaurs* being examples. From early swift-running reptiles were derived the flying reptiles or *pterodactyls* and the *birds*.

Early in the history of the reptiles there appeared a line of *mammal-like reptiles*. At some time in the development of this lineage, the distinctive mammalian traits of warm-bloodedness and the production and suckling of live young were acquired, and the class *Mammalia* (Triassic–Recent) thus came into existence. The mammals remained of little importance until the end of the Mesozoic era when, with the decline of the reptiles, they produced many new groups adapted to almost every possible habitat. The groups which evolved during the Tertiary included, to name only a few, the numerous divisions of grazing animals, the *ungulates*, the flesh-eating predators, the *carnivores*, the gnawing *rodents*, the marine *whales* and *dolphins* and the flying *bats*. We illustrate in Fig. 11.21 one aspect of adaptation among mammals—the modifications of the limbs which enable certain groups to move about in ways different from those of their ancestors. One line of special importance led from small unspecialised *insectivores* (the shrews and hedgehogs are modern members of the same group) to the order of *primates* including the monkeys, apes and man (Fig. 17.17). Some further details of fossil man will be found in Chapter 17.

6. THE PLANT KINGDOM

Non-cellular plants. The *non-cellular plants* are microscopic organisms which perhaps have more in common with the Protozoa than with the rest of the plant kingdom. The only types which we need mention are the *diatoms*, which float passively among the plankton in sea and lake waters; their minute siliceous *tests* (Fig. 11.5), rain down on to the sea or lake bottom to join the sediment accumulating there. *Diatomite* is a sediment made largely of such tests.

Fig. 11.22 *Fossil trees in the position of growth:* Lepidodendron *stumps in Carboniferous strata, Glasgow*

Chapter eleven

Multicellular plants. Among the *multicellular plants*, two very unequal sub-divisions can be recognised, first the relatively simple *non-vascular plants* which are mostly aquatic, and second, the more complex *vascular plants* which are predominantly terrestrial. The *non-vascular plants* (Thallophyta) include the seaweeds or *algae* some of which are found even in Pre-Cambrian rocks. Most of these plants are made entirely of soft tissues which do not leave identifiable fossil remains (though they may contribute to *sapropel*, see page 115). *Calcareous algae*, however, build up coatings of calcium carbonate into nodular masses a foot or so across which show a concentric layering when broken. Such masses contribute to many *reef-limestones*.

The *vascular plants* have more complicated internal structures in which specialised tissues are developed to transmit water and foodstuffs. They include two principal divisions, the *Pteridophyta* or *spore-bearers* and the *Spermaphyta* or *seed-bearers*. The *Pteridophyta* (Silurian–Recent) were the first to appear and provided most of the Palaeozoic land-floras, including many trees, climbers and bushes in the luxuriant Coal Measures forests (Figs. 11.22, 17.4). Since Palaeozoic times, the pteridophytes have been comparatively insignificant, being represented mainly by small club-mosses, horsetails and ferns. The Palaeozoic forms belonged, for the most part, to the same three groups, but were far more varied and abundant. The *Lycopodiales* (Devonian–Recent)—the group now represented by the *club-mosses*—included forest trees up to a hundred feet tall. A common genus was *Lepidodendron*, in which the tall trunk was patterned by the lozenge-shaped scars of old leaf-bases; roots of the same tree are known as *Stigmaria* and leaves as *Lepidophyllum*. The *Articulatales* (Devonian–Recent) now represented by the *horsetails*, are characterised by hollow, jointed stems often showing prominent ribs (Fig. 17.4). *Calamites*, of this group, formed large bamboo-like clumps in Coal Measure forests. Finally, the ferns or *Filicales* (Devonian–Recent), characterised by large frond-like leaves carrying spore-bearing capsules on their lower sides, formed much of the undergrowth of these forests.

The *Spermaphyta* (Carboniferous–Recent) were developed from pteridophyte ancestors and by the Mesozoic era had become the dominant plants. One of the earliest spermaphyte groups was the *Pteridospermae* (Carboniferous–Jurassic), consisting of fern-like plants which grew abundantly in the undergrowth of Coal Measure forests. One genus, *Glossopteris*, was widely distributed in Permo-Carboniferous rocks of the southern hemisphere. Other early groups included the *cycads* and *conifers* both of which survive to the present day. The first *flowering plants* or *angiosperms* appeared in the Jurassic and by early Tertiary times this single group dominated all other land-plants. The evolution of the group, and especially the appearance of the *grasses*, had a profound effect on the food-supplies of land-animals and took place side by side with the evolution of the Mammalia and the increasing complexity of the insects by which many flowering plants are pollinated.

12

METAMORPHISM AND THE METAMORPHIC ROCKS

Rocks which are deeply buried in mobile belts may be subjected to high temperatures and pressures and to conditions of great chemical activity. In these circumstances, metamorphism takes place and the rocks develop new minerals and textures suitable for their new environment. This chapter deals with the metamorphic rocks which may be produced under various conditions and with the geological factors controlling metamorphism.

1. Introduction

The third great class of rocks which we have to deal with is made up of those rocks produced by the transformation or *metamorphism* (Greek, change of form) of any previously existing rock-types. The *metamorphic rocks* have acquired *secondary characters* which are *superimposed on the primary characters* of the parent rocks. Metamorphism takes place at considerable depths in the crust where the temperatures and pressures are higher than those prevailing at the surface. In this environment, the primary minerals and fabrics which suited the environments of formation of the parent rocks are no longer stable, and they give place to new minerals and fabrics more in harmony with the new conditions. This is in accord with the principle of reaction to change of environment referred to in the first chapter of this book.

2. The Environments of Metamorphism

The main factors which control metamorphism are four—the temperature, the confining pressure, the directed pressure and the presence of migrating solutions. The nature of the metamorphic rocks produced depends on these factors and, in addition, on the bulk composition of the parent rock.

(i) *Temperature* is perhaps the most important factor, since the chemical reactions necessary to produce new minerals are speeded up by heat. At temperatures below about 200° C., these reactions are so sluggish that they have little effect, and most metamorphism takes place between, say, 200 and 700° C. In rocks of one particular bulk composition, let us say pelitic rocks, different minerals come into equilibrium at successively higher temperatures and the mineral composition therefore provides a rough guide to the temperature at which metamorphism took place. When the distribution of metamorphic rocks in the crust is examined, it is often found that such temperature-assemblages show a zonal arrangement which is related to the thermal gradient existing at the time of metamorphism.

(ii) The *confining or hydrostatic pressure* in the crust is due to the weight of the overlying rocks and, generally speaking, increases with the thickness of the cover; at depths of 20 kilometres, it is of the order of 6,000 atmospheres. High pressures tend to favour the growth of dense minerals such as garnet (sp. gr. 3·5–4·3) or kyanite (sp. gr. 3·6–3·7) and, moreover, they have the effect of making *ductile* or capable of flow many rocks which are brittle at atmospheric pressures.

(iii) *Directed pressure* or *stress* is due to earth-movements. It tends to distort rocks, as we see in the next two chapters, and leads to the development during metamorphism of various secondary fabrics. In cool rocks, strong stress may result in mechanical breakdown or *cataclasis*. At higher temperatures, it may speed up chemical reactions or crystal growth and may influence the way that

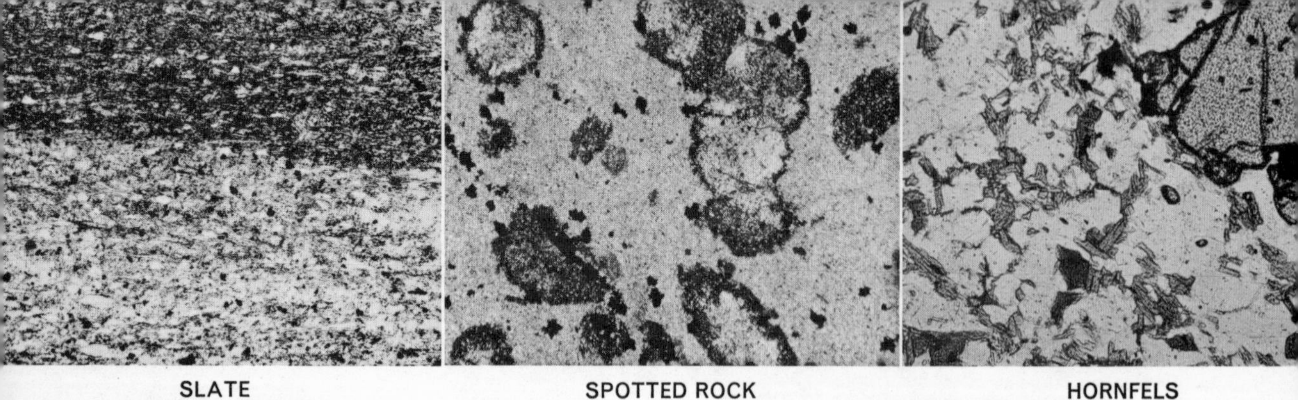

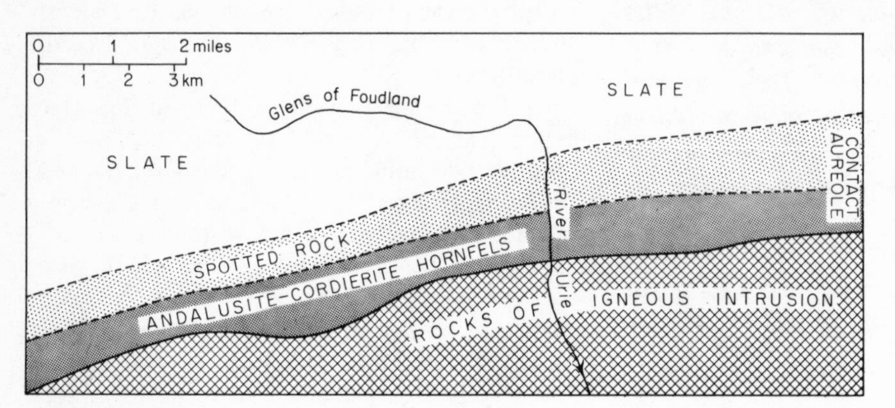

Fig. 12.1
Contact-metamorphism: map of the metamorphic aureole of the Insch gabbro, north-east Scotland; with (above) *a thin slice from each metamorphic zone.*

new minerals are arranged, giving rise to a *preferred orientation.*

(iv) The *migration of elements* during metamorphism takes place mainly via the *pore-fluids* filling the interstices of the rocks. A constant interchange of material between these fluids and the rock-minerals allows new minerals to be built up and old ones to disappear. In addition, substances in the pore-fluids may migrate out of the rock altogether and new substances may be introduced from neighbouring portions of the crust or from bodies of intrusive magma. In this way, changes of chemical composition can be achieved during metamorphism; such processes of chemical change accompanying metamorphism are termed *metasomatism* (Greek, change of body).

3. The Types of Metamorphism

Metamorphism does not take place everywhere in the deeper parts of the crust—this is made clear by the fact that many piles of sedimentary rocks which are known to have been deeply buried at some time in their history have suffered little or no metamorphism. We must therefore conclude that metamorphism at moderate depths takes place only in special circumstances (metamorphism may be more widespread at very great depths, but we are not likely to be able to examine its products). We can recognise three circumstances in which metamorphism is likely to take place and can distinguish three corresponding types of metamorphism—contact-metamorphism, dislocation-metamorphism and regional metamorphism.

(i) *Contact-metamorphism* is caused by the heat of igneous intrusions in the *country-rocks* which they invade. Large intrusions may produce a *contact aureole* of metamorphic rocks up to a mile in width around their borders, while dykes and sills may modify only an inch or so of the country-rocks. The principal control of the metamorphism is *thermal*, but stress and the escape of volatiles from the magma may also play a part.

(ii) *Dislocation-metamorphism* takes place in rocks subjected to strong earth-movements. *Stress* is here the principal control and many of the products are cataclastic rocks.

(iii) *Regional metamorphism* takes place in active segments of the crust, particularly in the *mobile belts* where both strong stresses and unusually high temperatures prevail. In the great fold-belts of the world, regionally metamorphosed rocks occupy broad tracts many hundreds of miles in length. All the metamorphic agents work together during the production of these rocks, since the general mobility of the belts not only results in high temperature and strong stress but also promotes the migration of chemically active fluids. The scale on which regionally metamorphosed rocks are developed is far greater than that on which contact- or dislocation-metamorphism takes place and the products are correspondingly more varied and complex.

4. Contact-metamorphism

During and after the intrusion of a body of magma, heat flows out from it into the country-rocks and thus sets up a temperature-gradient declining away from the contact. The *contact-aureole* produced by metamorphism is therefore usually *zoned*, the most-altered rocks lying nearest to the intrusion. We can illustrate this arrangement with reference to a typical aureole in pelitic country-rocks.

At *Insch* in Aberdeenshire, Scotland, a contact-aureole is developed along the northern margin of a large gabbro intrusion (Fig. 12.1). Outside the limits of the aureole, the pelitic country-rocks are *slates* made up of minute grains of quartz and flakes of sericite and chlorite arranged parallel to one another, so that the rocks part into thin sheets when broken (see later, page 151). When traced towards the gabbro, the slates pass into *spotted slates* in which small rounded nodules of andalusite and cordierite (a silicate of aluminium, magnesium and iron) are seen. Nearer to the contact, the spots develop into better-shaped crystals and biotite appears as flakes filled with inclusions of quartz. Finally, in the innermost parts of the aureole, the slates are completely recrystallised to form rocks known as *hornfelses*. None of the original groundmass is left and the new minerals are arranged to produce a coarser-grained fabric entirely lacking the parallel structure of the parent slates. The main part of the hornfels is made up of biotite, muscovite, quartz and a little feldspar forming a mosaic of shapeless grains; in this mosaic are set large shapeless spots of cordierite, large ragged biotite flakes and well-formed andalusite crystals.

The texture shown by the andalusite-cordierite-hornfelses of the Insch aureole is of a type which is often seen in metamorphic rocks which recrystallised in an environment free from directed stress—the component minerals are generally rounded or irregular and where flaky or elongated minerals are present they show no tendency towards a parallel arrangement. This easily recognised texture (Fig. 12.1) is called the *hornfelsic texture* and is widely developed in the inner zones of contact-aureoles.

Hornfelses in general are tough, massive rocks with no direction of easy splitting. Their component minerals depend on their bulk composition. Pelitic rocks contain micas and often andalusite, sillimanite or cordierite. Basic rocks contain feldspar and amphibole or pyroxene. Impure calcareous rocks contain lime-silicates such as lime garnet, wollastonite or idocrase. Pure limestones give *marbles* made of granular calcite.

Other types of rocks may be developed in contact-aureoles. The intrusion of the magma may set up directed stresses which modify the metamorphic environment and give rise to *contact-schists* similar to the schists described in connection with regional metamorphism. Where the volatile constituents of the intrusive magma escape into the aureole, they may bring about metasomatism. The hot gases streaming through the country-rocks contain such elements as fluorine, boron or sulphur as well as traces of rare metals which may be deposited to produce metallic ores. Common minerals produced by this metasomatic gas-action or *pneumatolysis* are fluorspar (CaF_2), tourmaline (Al-borosilicate) and topaz (Al-fluosilicate). The effects of pneumatolysis are well displayed at the contacts of the Cornish granites.

Chapter twelve

5. Dislocation-metamorphism

Dislocation-metamorphism is, as we have said, a result of the deformation of rocks in zones of strong earth-movements. Its effects are seen particularly where two portions of the crust have moved past one another along thrusts or shear-belts (see Chapter 14). Movement which takes place at high levels in the crust, where the temperature and hydrostatic pressure are both low, simply has the effect of breaking the rocks up; the product is a *breccia* of angular fragments jumbled together in a clayey *gouge*. Such a breccia is not strictly a metamorphic rock at all. More thorough-going deformation results in *cataclasis*, the mechanical destruction of the grains making up the rock. The first signs of cataclasis are provided by the fracturing or bending of grains. Later, the least resistant minerals are broken down to minute fragments and smeared out into streaks. The most resistant minerals survive as relatively large grains or *porphyroclasts*, their outlines streamlined and their surfaces plastered with powdery fragments. The final product is *mylonite* (Greek *mylon*, mill) composed of minute angular fragments bound together by fine dust and enclosing a few remaining porphyroclasts (Fig. 12.2). Many mylonites break readily into thin plates because the mineral debris is streaked out parallel to the plane of movement. Famous examples of mylonite occur along the great thrusts of the North-west Highlands of Scotland described in Chapter 14.

Mylonitisation and cataclasis are largely destructive processes which do not involve the active growth of new minerals. When the rocks subjected to dislocation-metamorphism are hot, however, mineral growth begins to play a more important part in the metamorphic process. New minerals such as chlorite, mica, albitic feldspar, epidote (a silicate of calcium, aluminium and iron) or calcite are developed to form a fine-grained groundmass, in which are enclosed large porphyroclasts and other smaller relics of the original minerals. The new flaky minerals are often arranged with their surfaces parallel to the platy partings produced by streaking-out of the original components, so that the whole rock shows a parallel texture or *schistosity* like that described in the next section.

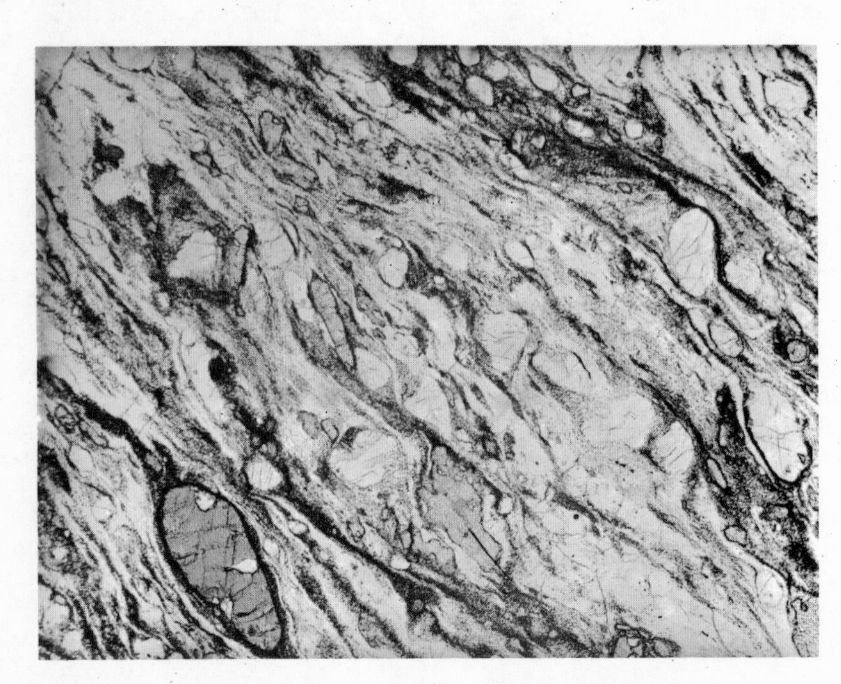

Fig. 12.2
Mylonite, thin slice at high magnification

6. Regional Metamorphism

Regional metamorphism, as its name implies, affects large portions of the crust and as we have noted it is characteristic of the mobile belts. Because mobile belts may take hundreds of millions of years to develop, the rocks within them may suffer regional metamorphism several times over, the minerals and textures of later episodes being superimposed on those of the earlier episodes. The factors controlling regional metamorphism are the high temperatures and pressures, the strong directed stress and the migrating fluids set in motion by the deep-reaching disturbance of the crust. *Migmatites* (p. 53), produced by the mixing of granitic material with country-rocks are often associated with regionally metamorphosed provinces.

The characteristic *textures* of regionally metamorphosed rocks reflect the prevalence of directed stress. The new minerals formed during metamorphism tend to arrange themselves in some particular orientation with respect to the direction of greatest pressure—in slates, for example, the minute flakes of chlorite and sericite usually lie with their largest surfaces perpendicular to the direction of greatest stress. The flakes therefore have a *preferred orientation* in planes perpendicular to this direction. Other preferred orientations are developed in other kinds of rock; flat or flaky minerals tend to lie in planes; elongated minerals may lie parallel to a linear direction and rounded grains may show a preferred orientation of the crystallographic axes. The whole fabric of the regionally metamorphosed rock is thus bound up with the conditions of stress prevailing during the metamorphism.

We can recognise four common types of texture in regionally metamorphosed rocks: the slaty, schistose, granular and gneissose textures. We deal with these in turn below and at the same time describe examples of rocks illustrating each type.

Rocks of slate grade. *Slaty textures* are characterised by the parallel orientation of minute flaky minerals. This preferred orientation gives the slate a *cleavage*, that is to say, allows it to be split into thin parallel-sided sheets, a property made use of in the production of roofing-slates (Fig. 13.1). Any fine-grained parent rock whose composition allows it to produce flaky minerals such as chlorite or mica can develop a slaty texture on metamorphism; most *slates* are of pelitic composition, the new flaky minerals being derived from clay-minerals. Slates of this type are made up of chlorite and the white mica sericite with minute quartz grains derived from detrital quartz; organic matter in the parent rock gives specks of graphite and often also cubes of pyrites (FeS_2).

It is obvious that some types of parent rock can never give rise to slaty derivatives even when they are metamorphosed under suitable conditions: these are the rocks whose composition does not allow for the development of flaky minerals. Sandstones, for example, have so little alumina and magnesia that neither chlorite nor sericite can appear in large amounts. The metamorphic derivatives of these rocks formed in a slate-producing environment have only a very crude cleavage but often show signs of cataclasis. Limestones give rise to fine-grained *marbles* which again may have cataclastic textures. Basic igneous rocks are converted into *greenschists*, with a parallel texture less regular than that of slates, and composed largely of chlorite, epidote and albite, often accompanied by the amphibole actinolite and by calcite. Ultrabasic rocks such as peridotites are converted into *talc-schists*.

We can see from what has been said above that several different kinds of rock may be developed under the same conditions of temperature and stress. The differences are due to variations in the primary characters of the rocks subjected to metamorphism. The essential uniformity of the metamorphic environment is expressed by saying that all the rock-types show the same *grade of metamorphism*. The slates, greenschists and fine-grained psammites and marbles referred to above are all of the *slate grade*.

Rocks of schist grade. *Schistose textures* are exhibited by moderately coarse-grained rocks in

Fig. 12.3 *Mica-schist containing porphyroblasts of staurolite (an iron aluminium silicate)*

Fig. 12.4 *Amphibolite made largely of hornblende (dark—notice conspicuous cleavage) and plagioclase feldspar*

which the preferred orientation of flaky or elongated minerals produces a parallel texture or *schistosity*. *Schists* differ from slates in that they are coarse enough for the grains to be distinguished with the naked eye. They are normally rich in micas or in needle-shaped minerals such as amphiboles. They often contain conspicuous large grains or *porphyroblasts* of a different order of size from those making the groundmass (Fig. 12.3).

The coarse grain-size of the schists indicates that their metamorphism was accomplished at higher temperatures than that of slate-grade rocks; the strong preferred orientation of minerals suggests the operation of directed stress. The *schist grade* of regional metamorphism, produced under conditions of moderate temperature and strong stress, can be illustrated by the common derivatives of pelitic rocks, the *mica-schists* (Fig. 12.3). The rocks generally contain biotite, muscovite and quartz together with other silicate minerals rich in Al_2O_3 such as garnet, staurolite, kyanite, sillimanite, cordierite or andalusite. These aluminous silicate minerals tend to form conspicuous porphyroblasts; when one or two such species are abundant, their names are tacked on to the rock-name—a rock may thus be called garnet mica-schist, kyanite-schist, and so on.

Other rocks metamorphosed in conditions of the *schist grade* give rise to derivatives whose textures are, as a rule, either schistose or of the granular type dealt with below. Basic igneous rocks are converted into rocks made largely of plagioclase feldspar and hornblende and often containing minor amounts of garnet, pyroxene or epidote (Fig. 12.4). Such rocks are known as *amphibolites* when their texture is granular, schistose amphibolites or *hornblende-schists* when it is schistose. Psammitic rocks and limestones usually give granular textures as described below.

Granular textures are shown by rocks in which the dominant minerals are roughly equidimensional and may be developed over quite a wide range of metamorphic conditions. The two most common equidimensional minerals are quartz and calcite and, accordingly, granular metamorphic textures are most often seen in *quartzite* and *marble*, derived respectively from quartz-sandstone and limestone. Impurities in these rocks may give rise to inequidimensional metamorphic minerals scattered through the rock. These minerals may assume a parallel orientation which produces a rough parting or schistosity and increases in their abundance may provide a transition to true schists.

Quartzite, derived from sandstone, generally has a granular texture, the abundant quartz grains being welded into a mosaic. A little mica is almost always present and there may be in addition feldspars, garnet, hornblende or epidote, depending on the impurities in the original sandstone and the grade of metamorphism. Pure *marble* consists of a granular mosaic of calcite, fine-grained in the slate grade, coarser and more sugary-looking in the schist grade. Impurities give rise to lime-silicate minerals which, in the schist grade, may include pyroxene, epidote, olivine or hornblende together with plagioclase feldspar, lime-garnet or quartz.

Rocks of gneiss grade. *Gneissose textures* are characterised by large grain-size and by the arrangement of the component minerals to produce a rough banding. This streaky or banded arrangement of minerals gives a type of parallel structure known as *foliation* (Fig. 12.5); it may be accompanied by a rough schistosity where flaky or elongated minerals are common.

The majority of *gneisses* are produced by metamorphism under conditions of fairly high temperature and pressure and rocks of the *gneiss grade* are therefore formed in the hottest parts of mobile belts. The characteristic segregation of different minerals into streaks parallel to the foliation is connected with the increasing chemical mobility of material at high temperatures and, in some instances, with the presence during metamorphism of increasing quantities of pore-fluid. Most gneisses contain *feldspar*, which is produced partly by high-temperature reactions among the original components and partly as a result of metasomatism connected with the migration through the pore-fluids of sodium or potassium. Furthermore, many gneisses are penetrated by enormous numbers of granitic or pegmatitic veins, or are intimately mixed with granitic material (Fig. 4.11). The metasomatically produced feldspathic gneisses and the veined or granite-rich gneisses fall in the category of *migmatites* or mixed rocks (p. 53).

Pelitic gneisses are coarse, roughly foliated rocks made up largely of feldspar and quartz with muscovite or biotite; in addition, aluminous silicates such as garnet, kyanite or sillimanite are often present. The feldspar is usually gathered into little lenses or streaks: when these are eye-shaped, the rock is called *augen-gneiss* (German *Auge*, eye). *Psammitic gneisses* commonly form massive pinkish rocks made up essentially of quartz and feldspar. Impure calcareous rocks give *calc-silicate gneisses* in which calcite, pyroxene, hornblende, Ca-rich plagioclase and other lime-silicates are arranged in irregular streaks or patches.

We should mention finally that although most

Fig. 12.5 *Gneisses: banded gneiss* (left) *and augen-gneiss* (right)

gneisses are the products of high-grade regional metamorphism, gneissose textures can also be produced by the cataclastic modification of coarse-grained parent rocks, especially granitic rocks. The effect of stress on such rocks is to break down the less resistant minerals such as quartz into streaks of small grains, leaving the more resistant feldspars as large porphyroclasts. In this way; a *cataclastic augen-gneiss* is produced. It is distinguished from augen-gneisses of the type mentioned above by the indications of cataclastic breakdown and by the presence of low-grade minerals, such as chlorite, not appropriate to the gneiss grade of metamorphism.

7. Zones of Regional Metamorphism

In the last few paragraphs, we have grouped the products of regional metamorphism into three grades, each governed by a different metamorphic environment. These are, from low-grade to high-grade, the *slate grade*, the *schist grade* and the *gneiss grade*, typified by the slates, the schists and the gneisses. In a general way, the series from low to high grade expresses an increase in temperature, pressure and chemical activity during metamorphism. By mapping the distribution of rocks showing different grades of metamorphism, we can establish a series of *metamorphic zones* which reflect the conditions existing deep in the earth's crust during a period of metamorphism. Examples taken from Britain are shown in Figs. 12.6 and 16.9 which illustrate the distribution of metamorphic zones formed in the Scottish Highlands during the Caledonian orogeny, about 450 million years ago (see Chapter 16).

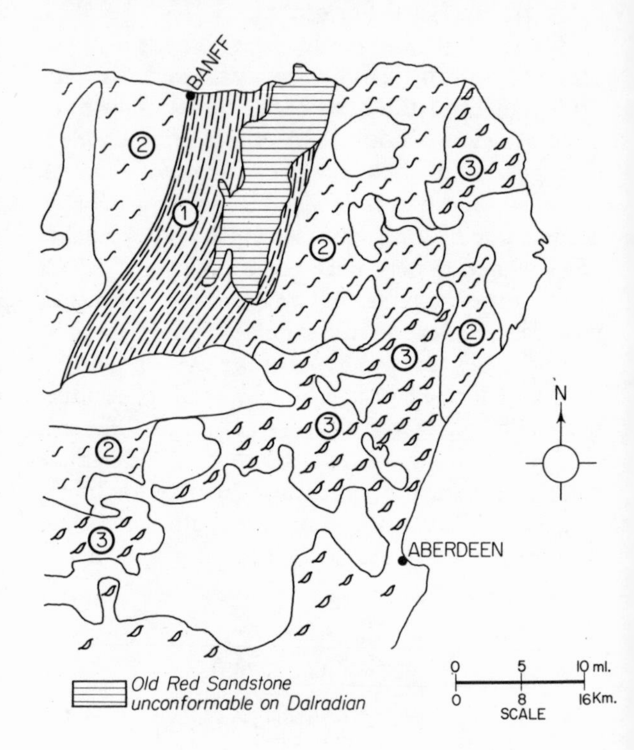

Fig. 12.6

Zones of regional metamorphism in north-east Scotland. The grade of metamorphism increases from slate grade (1) *through schist grade* (2) *to gneiss grade* (3). *Gabbro and granite intrusions left blank*

13

GEOLOGICAL STRUCTURES AND MAPS: 1

Folds and Cleavage

In this chapter a beginning is made not only on the study of structures produced by earth-movements but also on the interpretation of geological maps. The main types of structure are classified as folds, cleavages, thrusts, faults and joints and the first two types are described. Attention is given to the relationship between the three-dimensional shapes of the structures and the patterns of outcrops produced by the structures at the earth's surface.

1. Types of Crustal Movement

At several points in earlier chapters we have mentioned evidence which indicates that the crust has undergone deformation of various kinds. This evidence ranges from the direct proof provided by the displacement of rock-masses during earthquakes to the indirect testimony of the occurrence of rocks of marine facies on dry land or of plutonic rocks at the earth's surface. When all the indications of crustal movement are considered together, it is found that they provide evidence of disturbance of the crust in more than one geological setting. We may refer first to the earth-movements which take place in the *mobile belts* which have been mentioned in earlier chapters (pp. 6–14). In these long, relatively narrow zones of the crust, the originally flattish layers of sediment sooner or later become crumpled, intensely disturbed and often metamorphosed and invaded by granites. The *fold-belts* so produced are elevated bodily to make mountain ranges and the whole process of development and crustal movement in the mobile belts is termed *orogenesis* from the Greek *oros*, mountain. The second general type of crustal movement affects wide areas and is therefore styled *epeirogenesis* from *epeiros*, mainland or continent. Epeirogenic upheaval, warping and depression of huge portions of the crust usually take place without the violent small-scale disturbances which accompany orogenesis, and the effects of epeirogenesis are shown mainly by regression or transgression of the sea, by the development of raised beaches, the rejuvenation of rivers and similar phenomena.

2. Structural Geology and Geological Maps

In this and the succeeding chapter we are mainly concerned with the interpretation of the shapes and arrangements of rock-masses—their *structure*—as records of the effects of earth movements. As we have seen already (Chapter 1), sedimentary and igneous rocks have *primary structures* appropriate to their mode of origin. Subsequent deformation may modify these structures and superimpose others to produce a pattern of *secondary structures* controlled by the pattern of stresses in the crust. To understand the meaning of the primary and secondary structures, the geologist has to think of the geometrical shapes of rock-masses—beds of sediment, lava flows, igneous intrusions and so on—in three dimensions. The information on which he must rely comes from direct observation of rocks in the field, from records of mines and boreholes and, especially, from geological maps.

Geological maps show the areas over which rocks of different kinds and ages appear at the surface of the land, and so provide pictures of the geological structure as it is exposed to view by erosion. Many maps also show the *topography* or form of the ground by means of contours or other conventional symbols, so that the reader can envisage the forms of the rock-masses in relation to the ups and downs of hills and valleys. Many different kinds of information can be recorded on geological maps. Some maps are drawn to show the distribution of all types of rock including the purely *superficial* deposits such as alluvium, peat or blown sand. Others are drawn to show only the *bedrock* which extends continuously under the superficial blanket: in Britain, these two types are called 'drift' and 'solid' maps respectively. Other maps are constructed to serve special purposes—they may show the distribution of valuable ore-deposits, for example—and all may be useful in the study of structural geology.

The student looking at a geological map would be wise to think about the way such a map is made and the evidence on which it is based. The distribution of different rock-types can only be established on the spot and the geologist must go over the area himself to identify all the exposed rocks and examine the way in which they are arranged; he may be greatly helped, however, by looking at *aerial photographs* which not only show the nature of the country and the positions of rock *outcrops* but often reveal the pattern of the structure. On the *field-map* the geologist records all kinds of geological information which may be useful for understanding the structure. The essential information is later transferred to a *clean-copy* from which the final map is drawn for publication. In this chapter and the next we reproduce simplified geological maps to illustrate some of the geological structures described.[1]

[1] To supplement these maps the British student would do well to obtain a copy of the Geological Survey's *Geological Map of the British Isles* on the scale of 25 miles to the inch. This map (which we will call the *25-mile Map* for short) is a decorative piece of work and is sold for only a few shillings. Maps on larger scales, four miles or one mile to the inch, can be obtained which cover most of Britain and the student should try to see at least the sheets which cover his home area: a love of maps is in the blood of geologists.

3. The Importance of Datum Surfaces

In order to decide how any particular rock-mass has been disturbed, it is obviously necessary to have some idea of what its form and attitude were before disturbance took place. We can most easily acquire this information by reference to the primary structures in the rock. By far the most important structures for this purpose are the bedding-planes of sedimentary rocks; other *surfaces of reference* or *datum surfaces* are the tops and bottoms of lava-flows and the contacts of igneous intrusions, especially of dykes and sills.

In the vast majority of sedimentary rocks, the *bedding-planes* are formed in an almost horizontal position (the reasons for this are discussed on page 108); we are therefore justified in concluding that bedding-planes which are steeply inclined or strongly curved or broken have suffered secondary disturbance. So much depends on the application of this simple deduction that it is important for the student to be sure that he can recognise bedding-planes in any attitude and can distinguish them from other common plane structures such as cleavages and joints (Fig. 13.1). As a rule, bedding can be identified by the fact that successive beds separated by the bedding-planes are never quite identical, differing slightly in thickness, composition, grain-size, colour or fossil content, but this criterion is not entirely foolproof, especially when applied to metamorphosed sediments.

Since bedding-planes are originally more or less horizontal, it follows that each bed enclosed between two bedding-planes had originally an upper and a lower surface. Where no disturbance has taken place, the bed will obviously be right-way-up, with the upper surface on top; but earth-movements may turn beds upside down and it is therefore important to be able to establish the 'way-up' of sedimentary rocks.

Where the rocks contain fossils, the way-up may be found by applying William Smith's law of strata identified by fossils—the top of any bed or group of beds must lie on the side towards which younger fossils are found. Where fossils of stratigraphical value are lacking it is necessary to

Fig. 13.1
Recognising bedding: the bedding-planes are folded, and are crossed by a cleavage dipping towards the left

fall back on the structure of the sediments themselves. The use of *current-bedding* and *graded bedding* as means of distinguishing the original tops of sedimentary beds is explained on page 179 (Fig. 15.1).

4. The Main Types of Structure

When the rocks of the earth's crust are subjected to appropriate stresses, they will either *bend* or *break*, with the production of *folds* or *fractures*. Which kind of structure is formed depends on the condition of the rocks and the rate at which deformation takes place. Most rocks are *brittle* under surface-conditions and tend to break under stress, though they may yield very slowly by bending. At deep levels where temperatures and pressures are high, the majority of rocks become more *ductile* and can be deformed without breaking. We find, therefore, that structures of different kinds are formed in different conditions and we can recognise five such types of structure as follows:

(i) *Folds*, produced by curvature of the primary surfaces of reference.
(ii) *Cleavages*, very closely spaced secondary parting-planes produced during deformation.
(iii) *Thrusts*, fractures formed in close connection with folding.
(iv) *Faults*, fractures by which primary structures are broken and displaced.
(v) *Joints*, fractures which part primary structures but do not displace them.

5. Dip and Strike

In order to use all the information provided by our datum planes we need a simple method of recording the orientation of these surfaces with reference to the points of the compass and the vertical and horizontal planes. The method adopted is to record the *dip* or inclination of the plane and its *strike* or horizontal direction (Fig. 13.2).

The *dip* is defined as the angle and direction of the *maximum* inclination of the bedding or other

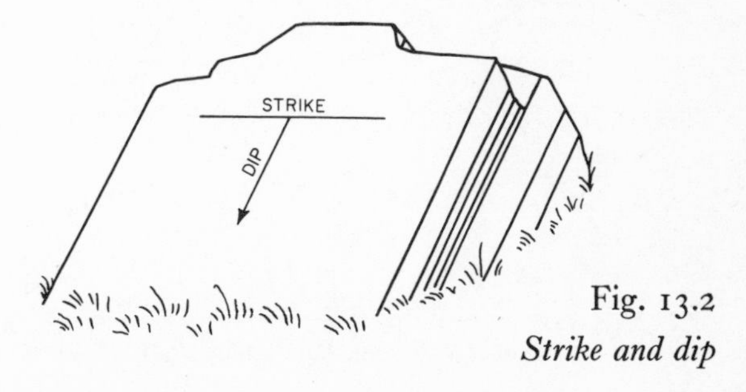

Fig. 13.2
Strike and dip

Fig. 13.3
The strike of beds illustrated by Devonian sandstones in a wave-cut platform, Cromarty, Scotland

geological plane, measured from the horizontal. The *strike* is the bearing of a horizontal line drawn on the plane; it is at right angles to the dip and can loosely be described as the general run of the beds across level ground (Fig. 13.3). The directions of strike and dip are measured in the field by compass and are usually recorded in terms of a 360° bearing from true north (allowance must be made for the magnetic declination at the place where the measurement is taken). The angle of dip is measured by a clinometer, one type of which is shown below. A simple clinometer can be made, with a little ingenuity, by mounting a celluloid protractor on a card and suspending a movable arm at the centre point. The strike and dip measurements made by the geologist are usually recorded on the geological map by symbols such as are shown in Fig. 13.4, the amount of dip being written in figures alongside each symbol.

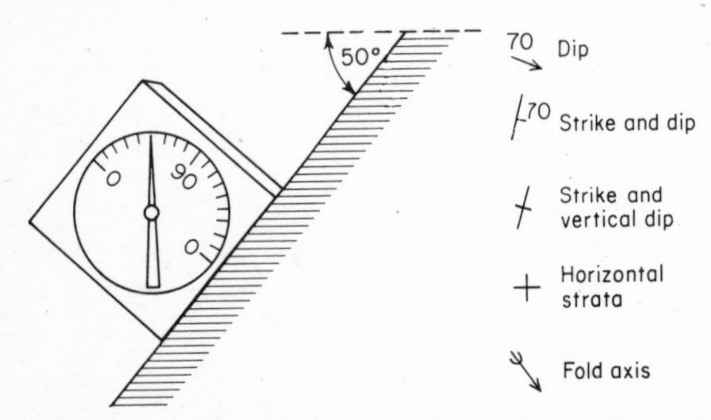

Fig. 13.4 *A clinometer* (left) *and some of the common symbols used on geological maps*

6. Shapes of Outcrops in Relation to Dip and Topography

The orientation of a bed or group of beds will affect the *shape of its outcrop* at the surface of the land. When beds are *horizontal*, they make very large outcrops where the land-surface is nearly flat. Where the land is undulating, the geological boundaries will be horizontal and will therefore be parallel to the topographical contours—high beds will form cappings on hills and low beds will appear in valley-bottoms. The hill-capping outcrops are examples of *outliers*, outcrops of younger rocks entirely surrounded by older. The reverse arrangement—outcrops of older beds surrounded by younger—gives *inliers*. The outcrop-patterns of horizontal beds are illustrated by the photograph in Fig. 13.5 and by Fig. 14.13, a simplified map of part of West Yorkshire where horizontal Carboniferous strata are seen. Examples of outliers to be seen on the 25-mile Map include the numerous outliers of Tertiary rocks resting on Chalk of the Chiltern Hills and in Hertfordshire.

Where beds are *vertical*, geological boundaries run in straight lines which show no deflections as they cross hills and valleys—the outcrop of a vertical bed thus forms a band crossing the map as if laid on with a ruler.

We can now consider *inclined beds* dipping at a constant angle (Fig. 13.6). Where such a bed crossing a valley dips upstream, its outcrop makes a V pointing upstream—the steeper the dip, the more obtuse the angle of the V. A bed which dips downstream usually forms a V pointing downstream: but where the angle of dip is lower than the gradient of the stream it forms a narrow V closing upstream at an angle more acute than that formed by the contours. The patterns of inclined beds crossing ridges are the reverse of those of the same beds in valleys—a ridge, from this point of view, is a valley turned upside down. Finally, the geological boundaries which run parallel to the topographical contours are necessarily *parallel to the strike direction*, since contours represent horizontal lines. All these relationships

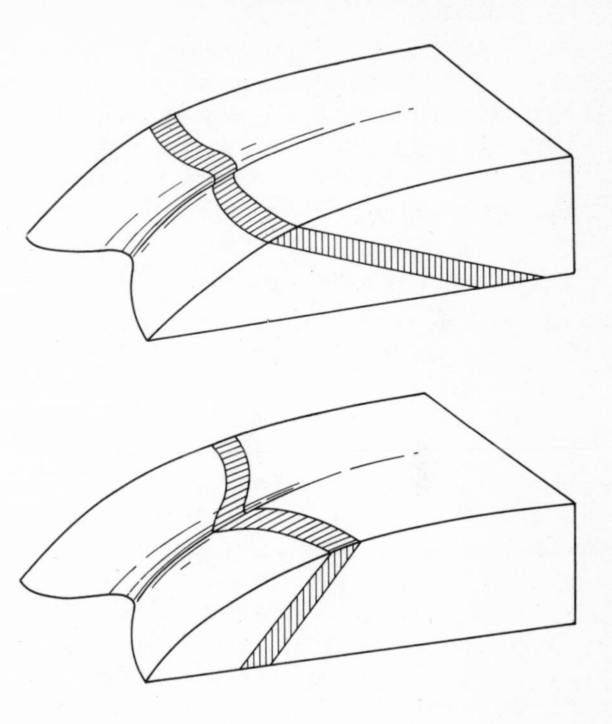

Fig. 13.6
Outcrop-patterns of inclined beds crossing valleys

Fig. 13.5 *Outcrop-patterns of horizontal beds illustrated by Carboniferous Limestone, West Yorkshire*

should be verified by the student—one can use sheets of cardboard to represent beds and any small irregular surface can provide model hills and valleys; it is also useful to imagine oneself standing in a gully and to predict the course of outcrops of beds of differing attitudes.

7. Thickness of a Bed and Width of Outcrop

The outcrop of a bed of given thickness varies in width according to the dip of the bed and the slope of the ground (Fig. 13.7). A vertical bed cropping out on a horizontal surface has an outcrop-width equal to the thickness of the bed. An inclined bed cropping out on a horizontal surface gives an outcrop wider than the thickness of the bed—the lower the angle of dip the wider the outcrop. The true thickness can be obtained by the geometrical construction shown in Fig. 13.7. On ground which is not horizontal, the outcrop-width is affected by both the dip and the topography; the construction needed to obtain the real thickness is shown in the figure.

8. Stratum-contours

Stratum-contours are lines drawn parallel to the strike of a bed at stated heights and provide a way of representing the form of the bed in three dimensions, just as topographical contours represent the form of the land-surface.

The strike of a bed, by definition, is the bearing of a horizontal line in the plane of the bed. If, therefore, a geological boundary crosses the same topographical contour on opposite sides of a valley or ridge, the horizontal line joining the two points of intersection must be a strike-line. Its height is given by the height of the contour crossed and it can therefore be used as a stratum-contour. Where the same geological boundary crosses several topographical contours, several stratum-contours can be constructed. *For a bed of constant dip* they will all be parallel straight lines and their spacing will depend on the angle of dip—the steeper the dip, the more closely crowded they will be. Stratum-contours of folded beds are curved and we shall not consider them here. We give instead an illustration of one of the uses to which the construction of stratum-contours can be put, an example of the so-called *three-point exercise*.

In Fig. 13.8 there is given a map of a hilly area in which a series of constantly dipping beds is known to occur. The upper surface of one particular bed crops out at the points A, B and C, situated on the 700-foot, 300-foot and 100-foot topographical contours. From C to A this geological boundary rises 600 feet, and by dividing the line AC into six equal parts we can fix points at 100-foot intervals along it. Point P on AC is at 300 feet, as is Point B: PB therefore represents the stratum-contour at 300 feet O.D. Since the dip is constant other stratum-contours will be parallel to PB and can be drawn at 100-foot intervals through the points fixed on AC. When stratum-contours have been inserted on the map, it is possible to complete the outline of the geological boundary represented by the top of our bed—this boundary must appear at each

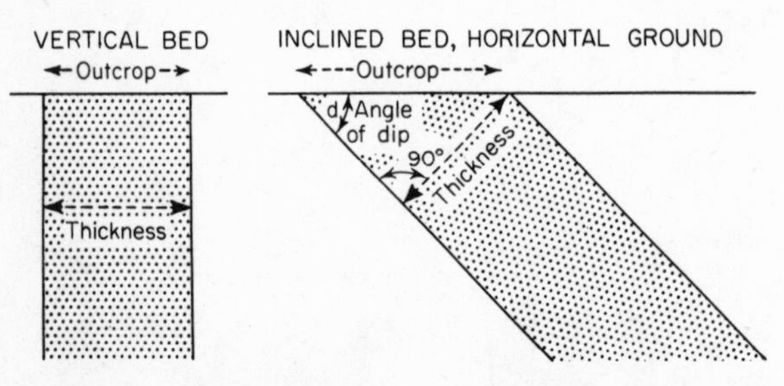

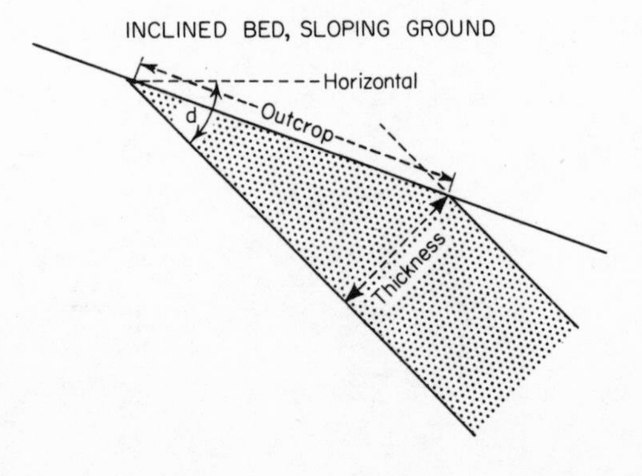

Fig. 13.7 *Width of outcrop in relation to dip*
Thickness, t = outcrop-width, w/cosec d − s
where s is the angle of slope

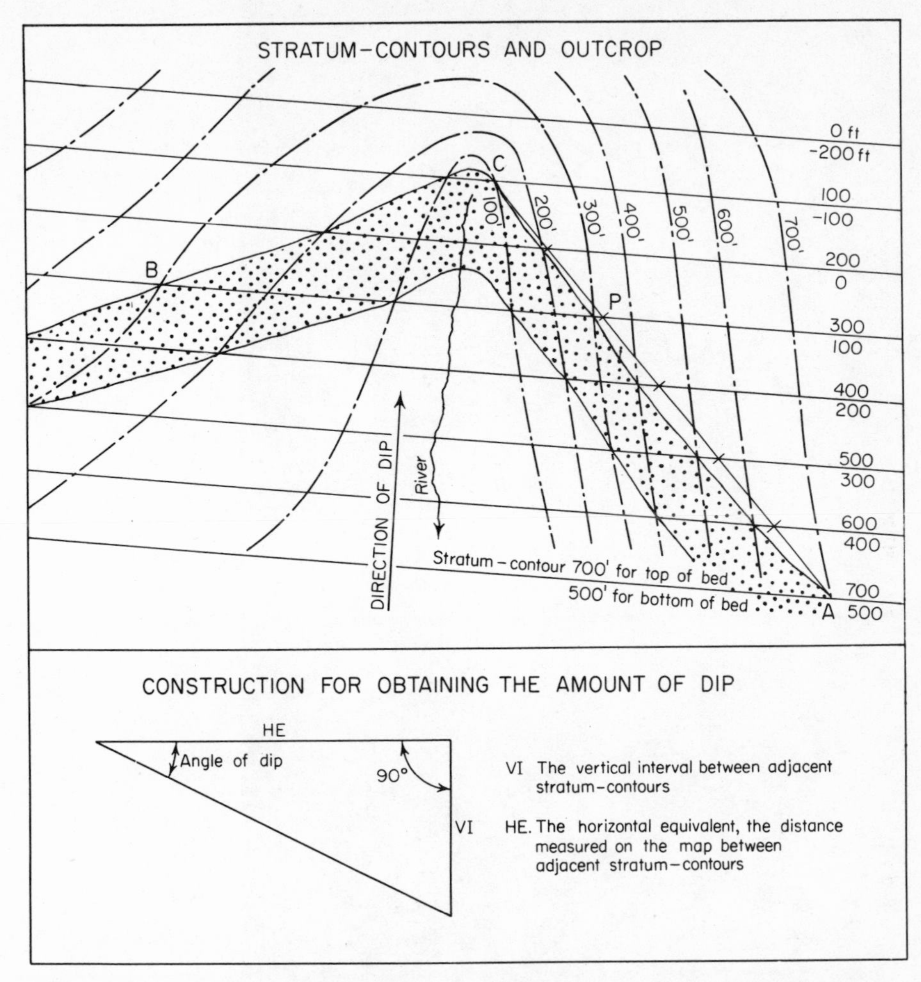

Fig. 13.8 *Stratum-contours used to plot the outcrop of a bed and to calculate the dip. The construction given below is a general one, not specifically related to the example*

point of intersection of equal topographical and stratum-contours and its course can be drawn in by joining all such points as shown in the figure. The direction and amount of dip can be found by a further construction. The dip direction is of course perpendicular to the stratum-contours. To find the amount of dip, draw a line of length equal to the distance on the map between two adjacent stratum-contours, measured in the dip direction; drop a perpendicular from one end equal in length to the vertical interval (in this example 100 feet) between the contours and complete the triangle. The dip (angle from the horizontal) can then be read off with a protractor.

Finally, if the thickness of the bed whose upper surface has been plotted is known (in the example, it is taken as 200 feet) the outcrop of the lower surface of the bed can be plotted by adjusting the values of the stratum-contours and joining their intersections with the appropriate topographical contours. Alternatively, if the outcrop of the base of the bed is known, the thickness can be found by stratum-contouring upper and lower surfaces and comparing the heights at a given point. The student should set himself other three-point exercises, experimenting with different kinds of topography and exposure, both at the surface and in boreholes.

A

B

Fig. 13.9
Folds: anticline (A) *and syncline* (B) *in Upper Palaeozoic rocks, south-west England*

9. Escarpments and Dip Slopes

So far, we have dealt with examples in which there was no direct relationship between the geological structure and the topography. If we think back to the ways in which erosion takes place, however, it will become obvious that such a relationship must often exist—hard beds may resist erosion and stand up to form hills while soft beds are worn down into hollows. Consequently, many small ridges and valleys carved by differential erosion are *strike-features* following the horizontal trend of tilted or vertical strata.

A succession of gently inclined beds of different resistances gives rise to a *scarpland* type of country of asymmetrical ridges; the steep side or *escarp-*

ment in each ridge represents the weathered-back edge of the bed, the longer and gentler side the *dip-slope* made by its upper surface, the two features together make a *cuesta*. Where the dip is steep, the ridge becomes a symmetrical *hogsback*. The 'edges' of Shropshire made of Silurian limestones are examples of escarpments and still more conspicuous examples are seen in the Weald of South-east England (Fig. 13.12). The north side of the Weald is made of a succession of clays, sandstones and chalk belonging to the Cretaceous system. The resistant formations are the Chalk, Lower Greensand and Hastings Beds, all of which dip northward and form scarps facing south with dip-slopes facing north; the less resistant Gault Clay and Weald Clay form hollows running east and west as strike-features. Some distance west of our line of section, the dip of the Chalk steepens and the escarpment and dip-slope pass into the Hog's Back.

10. Folds

Folding is shown up by the curvature of the primary surfaces of reference which may be obvious in the field (Fig. 13.9) or which may be revealed by the course of the outcrops of beds shown on the geological map (Figs. 13.12, 13.13) or may be shown up by variations in the amount and direction of dip. Three geometrical varieties of folds are distinguished, anticlines, synclines and monoclines. An *anticline* is, in its simplest form, a fold whose two sides or *limbs* dip outward away from one another. The *core* consists of rocks which are older than the rocks of the *envelope* wrapped round the core (Fig. 13.9A). A *syncline* is a fold in which the limbs dip towards one another or in which the core is younger than the envelope (Fig. 13.9B). A *monocline* is a step-like flexure joining two areas of more or less horizontal strata.

The two fold-limbs of a single bed meet in a line which is termed the *axis* of the fold. The limbs in successive layers meet along a surface passing through the axis in each layer; this is the *axial plane*. The fold-axis may be horizontal, like the ridge of a house-roof, or it may be inclined, in which case the fold is said to *plunge* or *pitch* in the direction towards which the axis descends.

The shape of a fold in *profile*, that is, in sections perpendicular to the fold-axis, depends on whether the axial plane is vertical or inclined, on the angle between the limbs and on the degree of curvature of the limbs. The terms used to describe different forms are explained in Fig. 13.10.

We may group folds also on a different basis—according to the way in which they were made (Fig. 13.11). *True folds* are produced by actual *bending* of the strata, as one might bend a thick pad of paper; experiment will show that as the pad bends the individual sheets (representing successive beds) slide on one another a little to allow distortion to take place. *Shear-folds* are produced without any bending by means of an infinite number of minute displacements along secondary *cleavage-planes* developed at an angle to the bedding. This kind of folding can be illustrated by drawing a straight line to represent a bedding-plane across the side of a pack of cards and allowing the cards to slip past one another until this line is distorted into a 'fold'. The partings between the cards represent the cleavage and it should be noticed that they are parallel to the axial plane of the fold produced. Finally, *flow-folds* are produced by plastic movements of soft and mobile rock material—they can be imitated in treacle without much difficulty. As a generalisation, it can be said that true folds are most

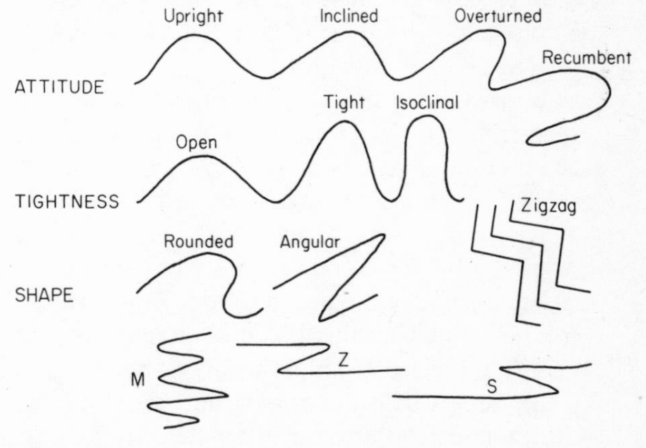

Fig. 13.10
Folds in profile: some of the possible variations

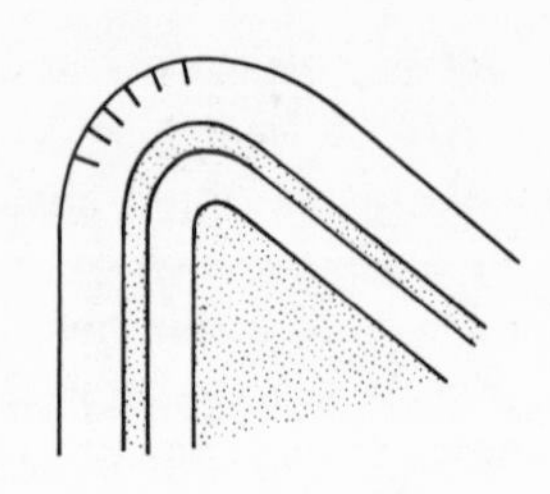

Fig. 13.11 *Two types of folding illustrated by specimens and diagrams. True folding* (above); *shear-folding* (below)

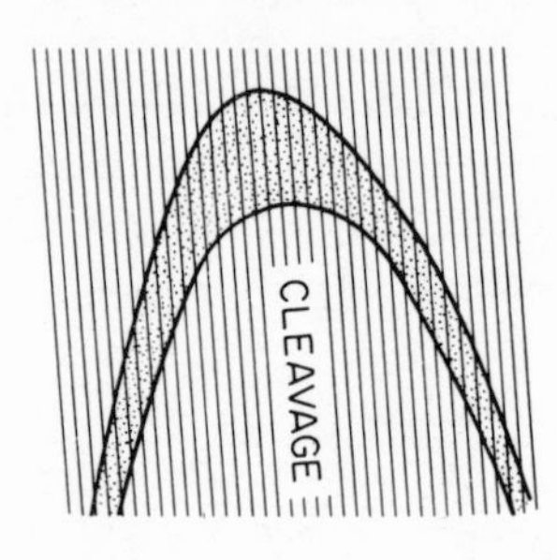

commonly developed in rocks which are hard and strong, especially in the upper parts of the crust; shear-folds in rocks which are being metamorphosed; and flow-folds in hot metamorphic or migmatitic rocks, or in watery clays, rock-salt or glacier ice.

The largest folds seen in the earth's crust are measurable in tens and hundreds of miles. Mountain-ranges such as the Alps are carved out of piles of enormous folds heaped up within orogenic belts. From these great structures, we can proceed downwards to folds which occupy a mountain-side, a single outcrop or a hand-specimen. Such smaller folds commonly mimic the style and orientation of the large folds formed by the same earth-movements and observations of the minor structures in the field may help in the interpretation of structures which are too large to be seen directly.

11. Outcrop-patterns Related to Folds

The Wealden anticline. The basis for the understanding of large folds is of course provided by the geological map. The first example we shall mention is the large simple fold beautifully shown on the 25-mile Map and revealed by the landscape of South-east England. This is the *Wealden anticline* whose axial plane runs nearly east and west through the Weald and whose limbs extend to the Channel coast in the south and the outskirts of London in the north (Fig. 13.12). A traverse from north to south across the *north limb* of the anticline shows, in order, the Chalk of the North Downs, Gault Clay, Lower Greensand and Weald Clay, all dipping northward: the escarpments of Chalk and Greensand form strike-features which reveal the structure even on a topographical map. In the centre of the anticline lies the Hastings Sand with fold-inliers of Jurassic strata, somewhat broken up by faults, appearing from beneath it. On the *south limb* the Hastings Sand is followed southward by Weald Clay, Lower Greensand, Gault Clay and the Chalk of the South Downs, all dipping southward. Again, the ridge of the Downs marks the strike of the Chalk. When the two Chalk outcrops of the North and South Downs are traced westward, they are seen to turn towards each other and unite in the region of Salisbury Plain at the axis of the fold. The junction of the fold-limbs westward indicates that the anticline has a *westerly plunge* which carries the rocks of the core down below the surface in this direction and allows the Chalk envelope to curve right round the fold-hinge.

If the student has available the 25-mile Map, he will notice that the Wealden anticline extends eastward into northern France. In the Boulonnais, the Chalk outcrops on the northern and southern limbs once more unite, revealing that the fold axis here plunges eastward: if we think of the whole structure at once, it clearly makes a flat *dome*. Further examination of the 25-mile Map shows that the Wealden anticline is one of a *set* or family of folds with similar orientations (Fig. 17.11). To the north of it lies the synclinal *London basin* with a core of Tertiary rocks and an envelope of Chalk. To the south lies the *Hampshire basin*, also a syncline with a Tertiary core, and still farther south the *Isle of Wight monocline*. Other large folds readily recognisable on the 25-mile Map are (1) the syncline of the Lanarkshire–Stirlingshire coalfield in Scotland, (2) the anticline of Carboniferous rocks in the southern Pennines, and (3) the syncline of the South Wales coalfield. The student should locate these structures, determine the orientation of the axial plane and draw sketch-sections where possible.

Interpretation of outcrop-patterns. We can build up more detailed pictures of fold-structures from maps by applying the methods of investigation discussed in Sections 6–8 of this chapter. The attitude of a folded bed can be established from the outcrop-patterns by the methods

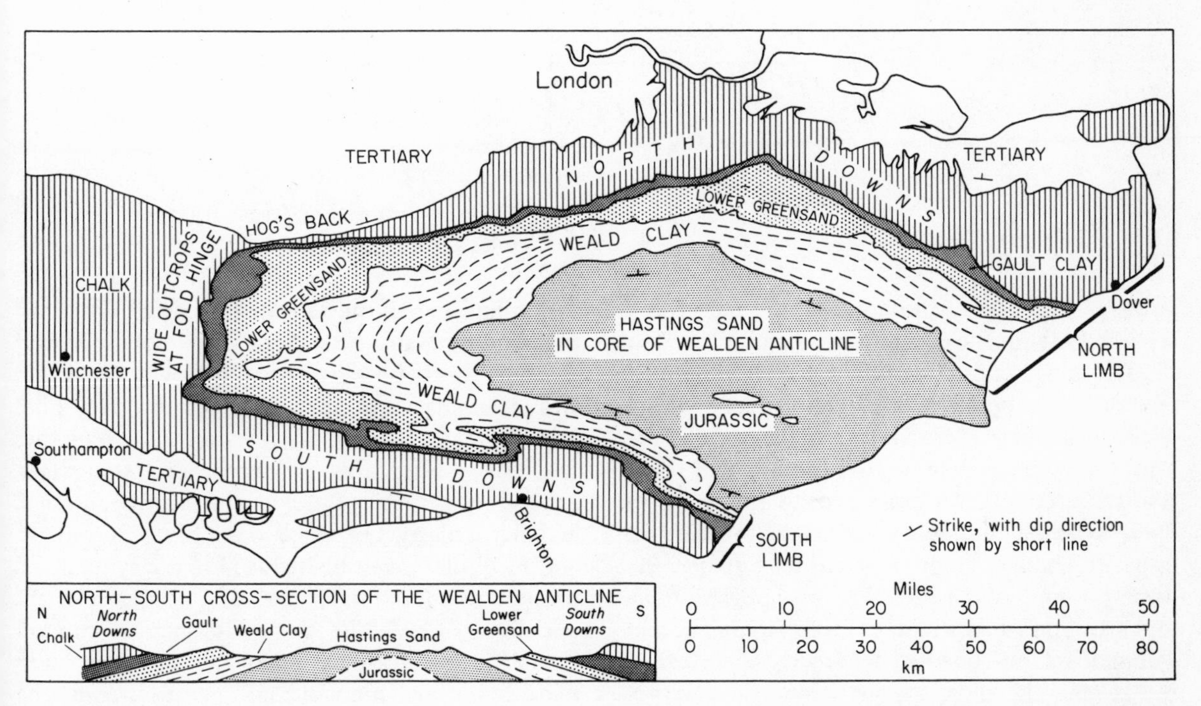

Fig. 13.12 *The Wealden anticline*

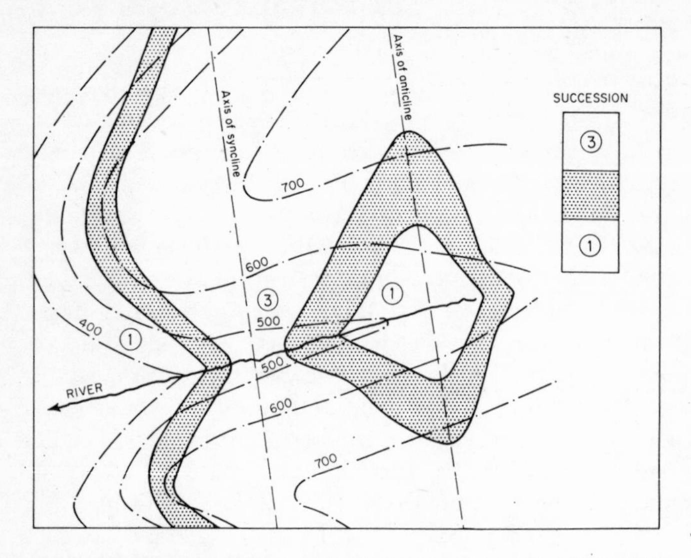

Fig. 13.13
Outcrop-patterns of folded strata: folds on horizontal axes

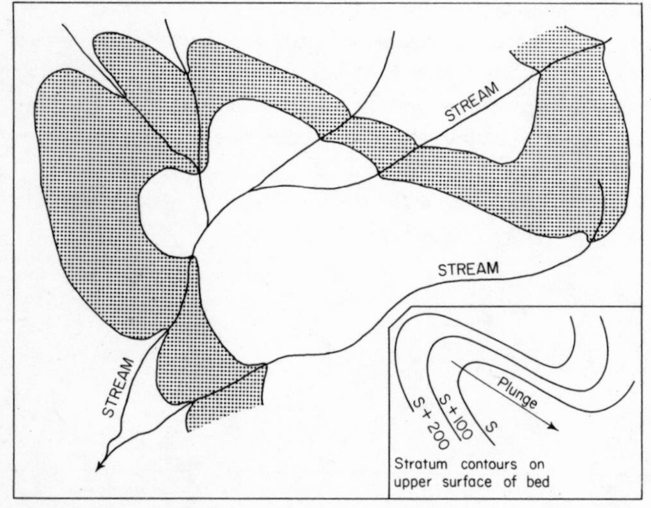

Fig. 13.14
Outcrop-patterns of folded strata: folds on inclined axes

already explained, provided that we think separately about the outcrops on each limb and at the fold-hinge. Thus, the outcrop of a folded bed will form V's pointing in directions appropriate to the local dip as it crosses a number of valleys—a little thought will show that the V's will point towards each other from opposite limbs of an open syncline and away from each other on opposite limbs of an open anticline. Further, the *outcrop-widths* of the same bed on opposite fold-limbs will be equal where the fold is upright and symmetrical but unequal where it is inclined—the steeper the dips, the narrower the outcrop; the outcrop-width is often greatest of all at the fold-hinge as may be seen in Fig. 13.12. Vertical fold-limbs are not deflected by the topography.

These lines of reasoning can be tried out on the two imaginary maps of Figs. 13.13 and 13.14. The first shows strata folded on *horizontal axes*—the folds have no plunge or pitch. By noting the general behaviour of the geological boundaries and particulariy the directions of V's in valleys and low hills, it can be deduced that a syncline is present in the western part of the area and an anticline towards the east. The general run of the fold-limbs can be established by eye and, with this information in mind, stratum-contours should be drawn for each fold-limb *separately* (the same geological boundary may cross a contour-line on both limbs of a fold and a line joining these two intersections would not be a stratum-contour). The stratum-contours will be found to be parallel straight lines, since the fold-axes are horizontal: their values and spacing can be used to determine the direction and angle of dip as already described. The student should finally test his understanding of the structure by drawing a sketch-section perpendicular to the strike; how to draw an accurate section is explained below.

We deal more briefly with *pitching or plunging folds* on *inclined axes*, examples of which are represented in Fig. 13.14. In this figure, the form of the topography is only indicated by the pattern of streams. The pattern of V's indicates that there is a syncline in the west and an anticline in the east. The direction of plunge is shown by the orientation of V's at the fold-hinges and it should be noted that the *syncline opens out* in the direction of plunge whereas the *anticline closes* in this direction. The changes in width of outcrop of the marker bed show that the common limb linking the syncline and anticline is steeper than the second limb of either fold. All the observations are summarised in the inset figure which shows approximate stratum-contours: this inset should be compared with the main map and an attempt made to imitate the form of the folds by folding a sheet of paper.

Cross-sections

The production of a vertical cross-section accurately drawn from data supplied by geological and topographical maps is one of the most useful ways of illustrating the geological structure: we will consider the procedure here because folds are the structures which respond best to this method. The line of the cross-section is usually drawn perpendicular to the trend of the fold-axis and one must therefore begin by establishing the general shape of the fold in the ways outlined above. Ideally, the horizontal and vertical scales should be the same, but where the section would become too cramped it may be better to exaggerate the vertical scale.

The first step is to draw in the *profile of the land-surface* along the chosen line of section. Mark on a strip of paper all the points at which the line of section crosses or touches a topographical contour, stream or recorded height: at the same time mark the points at which the line crosses geological boundaries or comes near to dip symbols. On another sheet, draw a horizontal line to represent sea-level. Transfer the measurements of heights to this base-line and for every point thus plotted draw to scale a vertical line upward to the height recorded by the measurement. The tops of all vertical lines can then be joined up by a smooth line which represents the land-surface (take care that the profile shows a valley-bottom at each point where the section crosses a stream).

The points where *geological boundaries* cross the line of section should then be marked off on the profile and a note made on it of all dip observations. If the section is drawn to the same vertical and horizontal scales, the inclination of strata at each point where the dip is known can be measured with a protractor and a line with this inclination drawn beneath the topographical profile. *Where the dip is constant*, geological boundaries can then be drawn parallel to the dip-lines. *Where the dip is not constant*, construct a perpendicular to each line of dip and project geological boundaries so that they cross each successive perpendicular at right angles (Fig. 13.15). If the section is drawn with an exaggerated vertical scale, the angles of dip must be increased and the section must be sketched in by eye.

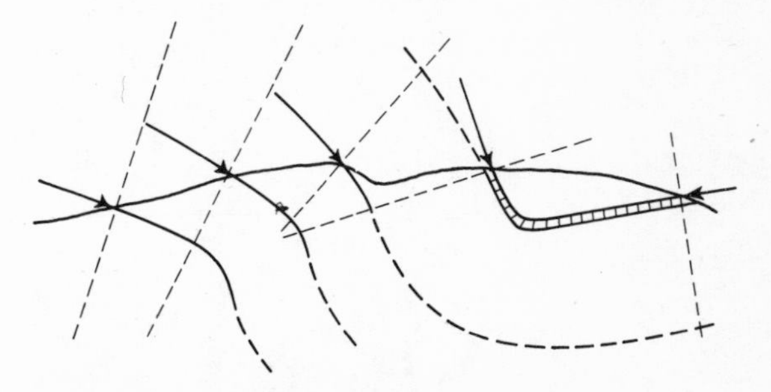

Fig. 13.15 *Plotting a cross-section where the dip is not constant*

Chapter thirteen

12. Cleavage

Cleavage has already been mentioned in connection with the formation of *shear-folds* and in our account of *metamorphic rocks of the slate grade* (p. 151). As we have seen, it is a purely secondary structure, a series of very closely spaced parting-planes which usually cut obliquely through the primary bedding. It is developed under conditions of moderate pressure and at temperatures well above surface-temperatures.

Slaty cleavage or flow-cleavage results from the parallel orientation of minute flaky minerals of metamorphic origin. It is the characteristic structure of the *slates*. From observations of slaty rocks it is found that fossils, pebbles and other small bodies have been flattened in the direction perpendicular to the cleavage and extended in the cleavage-planes—this flattening is a clear illustration of the effects of *compression* by earth-movements. The cleavage-planes themselves are parallel to the axial planes of shear-folds formed in association with them. The close connection which exists between the orientation of cleavage and that of the related folds makes it possible to use field-observations of cleavage in the interpretation of geological structure—its strike corresponds to that of the axial plane, its inclination reveals the asymmetry of the structure and its intersection with the primary bedding is usually parallel to the fold-axis. The student should verify these relationships by examining examples of shear-folds or by experimenting with a pack of cards for a model.

Fracture-cleavage is a mechanically produced type of cleavage in which the closely spaced partings are not associated with a preferred orientation of minerals. It is less regular than slaty cleavage and seldom yields usable roofing-material. Fracture-cleavage may be associated with shear-folds, when it is roughly (though seldom exactly) parallel to their axial planes, and is also formed where rocks have slipped past each other along thrusts and faults.

14

GEOLOGICAL STRUCTURES AND MAPS: 2

Thrusts, Faults and Joints

The study of geological structures begun in Chapter 13 is completed by an examination of thrusts, faults and joints.

1. Thrusts

Fold-thrusts and clean-cut thrusts. When tangential compression is driven beyond the limits that can be taken up by folding under particular environmental conditions, the rocks of the crust break and the effect of stress is then taken up by movement on gently inclined planes of rupture. Such dislocations associated with folding are termed *thrusts* and the actual planes of movement are *thrust-planes*.

Folds develop into thrusts either by breaking along the line of sharpest curvature—that is, near the fold-hinge—or by the thinning-out and final elimination of one limb as illustrated in Fig. 14.1. In mobile belts where enormous folds are piled up one above another (p. 164) the middle limb connecting a syncline and anticline is often replaced by a thrust so that, finally, a pile of detached sheet-like fold-limbs separated by thrusts is built up. In the Alps, for example, very large recumbent folds have been driven northward on thrusts, one above the other to make the pile out of which the Helvetic Alps are carved (Fig. 14.2). Thrust-sheets of this sort are called *nappes* (cf. napkin, a little sheet).

Another type of thrust is not directly related to the development of folds but appears as a clean break carrying an apparently undistorted wedge of rocks—we can suppose that this type develops when the rocks are too brittle to begin folding in time. The classic examples of *clean-cut thrusts* are seen in the *Moine Thrust-zone* of the North-west Highlands of Scotland (Fig. 14.3). This great group of dislocations was formed along the zone of weakness where the strongly disturbed rocks

Fig. 14.1
An overturned fold passing into a thrust

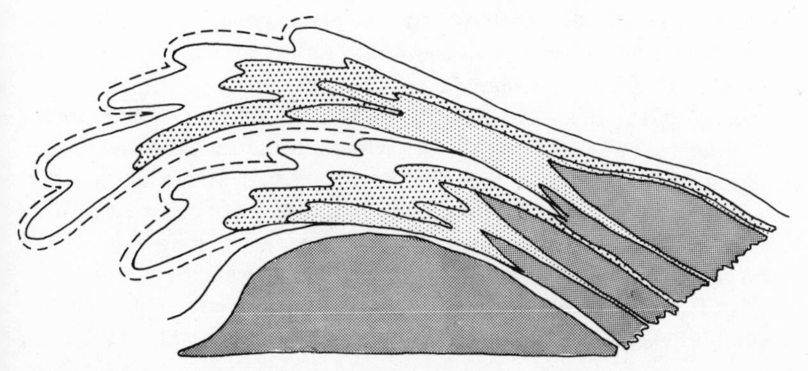

Fig. 14.2
Nappes in the Alps produced by combined folding and thrusting

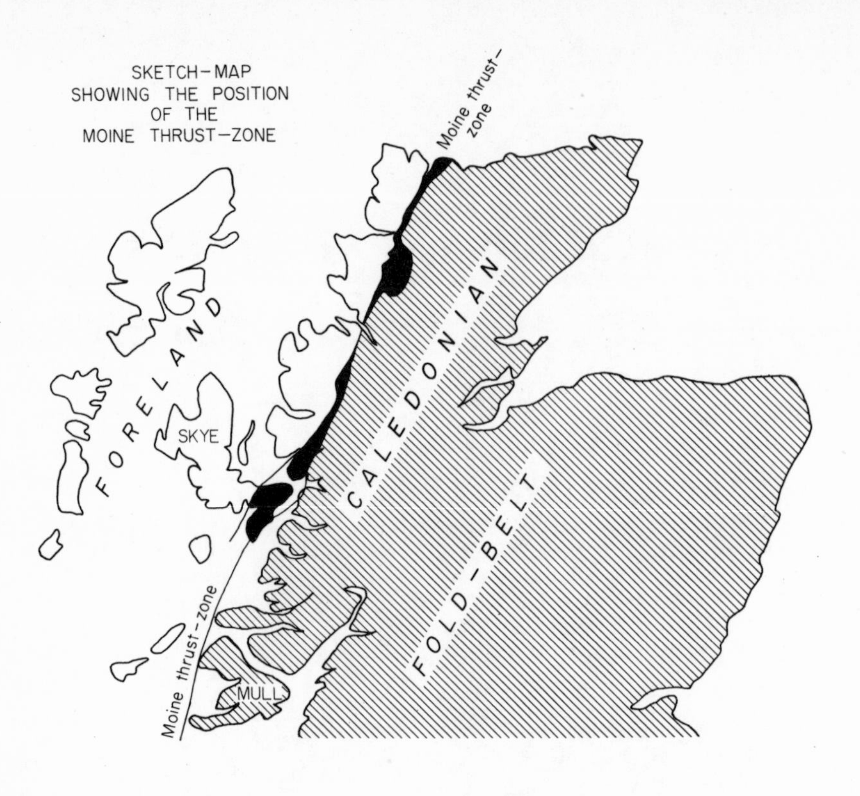

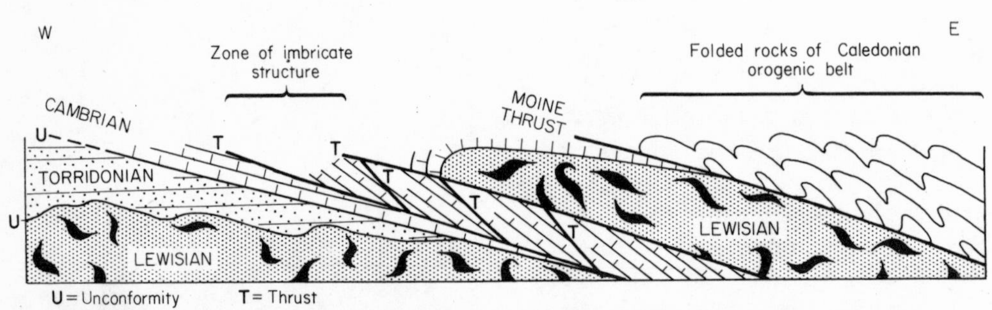

Fig. 14.3 *The Moine thrust-zone, north-west Scotland: diagrammatic map and section*

of a *mobile belt* (the Caledonian belt, see Chapter 16 for details) joined the rigid and unmoved rocks of the *foreland* or stable area to the west. In the thrust-zone, a number of large thrusts and many smaller ones were developed, striking parallel to the margin of the mobile belt and dipping eastward towards it. On these thrusts, slices of the rocks of the foreland and western part of the mobile belt have been carried westward and piled up one above the other: it can be shown that some slices must have travelled at least ten miles and their actual displacements were probably very much greater. Along and near the thrust-planes themselves, the transported rocks have been broken and ground down to very fine-grained *mylonites* (p. 150).

The North-west Highlands. The simple section of the Moine thrust-zone which illustrates the features described above (Fig. 14.3) shows at the same time a series of other structures which provide a record of a long history of geological events in North-west Scotland. It is worth turning aside from our study of fractures to examine these other features as an exercise in the *interpretation of geological structure in terms of geological history*. In the western part of the section (left) can be seen the rocks of the stable foreland. The oldest of these, lying at the base of the succession, are those labelled Lewisian, metamorphic rocks of gneiss grade which record, by their complicated structure, evidence of two important periods of Pre-Cambrian orogenic activity. Resting on the Lewisian gneisses with an irregular base are arkosic sandstones of the Torridonian formation. They are almost horizontal and are not metamorphosed; clearly they are

6. Development of Moine thrust-zone

5. Folding and metamorphism in Caledonian orogenic belt

4. Deposition of CAMBRIAN and lower ORDOVICIAN unconformably on Torridonian and Lewisian: plane of marine erosion at base

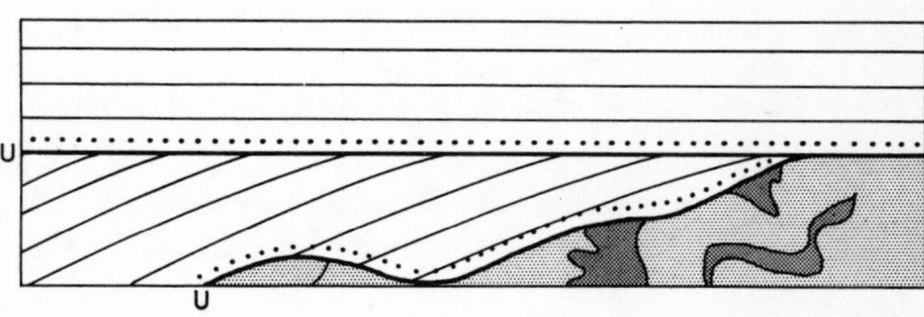

3. Tilting of Lewisian and Torridonian

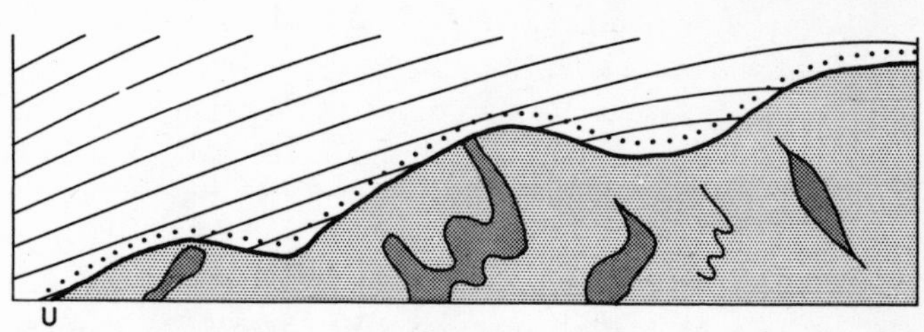

2. Deposition of TORRIDONIAN unconformably on Lewisian: old land-surface at base

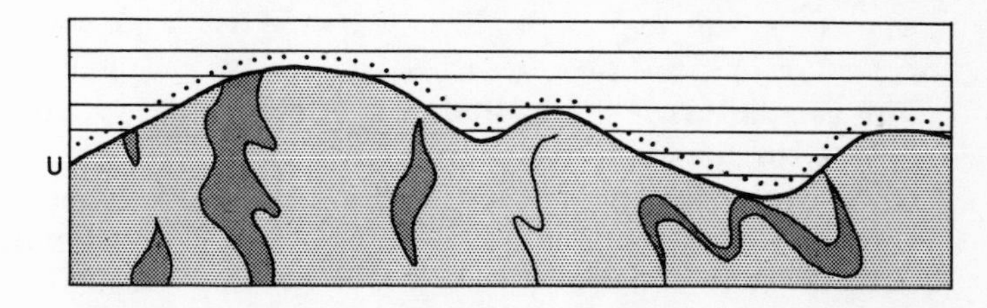

1. LEWISIAN GNEISS produced by several periods of folding and metamorphism

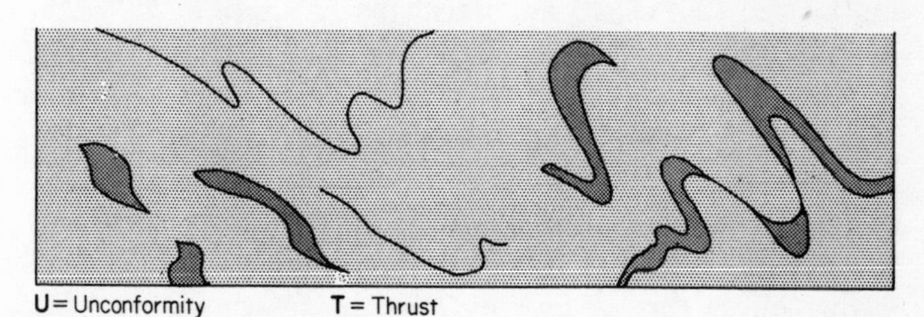

Fig. 14.4

Stages in the geological history of north-west Scotland

separated from the Lewisian by a great unconformity and it can be deduced that the Lewisian rocks, formed far down in the crust, must have been elevated and deeply eroded (partly by subaerial agents, with the production of an irregular land-surface) before the Torridonian sediments were deposited. Resting, with a second unconformity, on both Torridonian and Lewisian is another series of sedimentary rocks, pure quartzites followed by dolomitic limestones containing marine fossils of Cambrian and early Ordovician age. This series rests on a planar surface of marine erosion cut across the older rocks by the advancing Cambrian sea. But before this plane was cut, the Torridonian and Lewisian rocks must have been tilted, since the plane crosses the bedding-planes of the Torridonian at an angle.

At some time after the deposition of the Cambrian succession on the foreland, the deformation which produced the thrust-zone began. The bedded Cambrian sediments broke along a swarm of small fractures and piled up in slices like tiles on a roof to give what is termed an *imbricate structure* (Latin *imbrix*, a tile). The underlying Lewisian was displaced by larger thrusts and driven westward on top of the younger Cambrian. Finally, folded and metamorphosed rocks produced farther east in the interior of the mobile belt were carried over all on the still higher *Moine Thrust*. This history, which should be checked step by step against the section (Fig. 14.3), is schematically summarised in Fig. 14.4 and is discussed again in Chapter 16.

Outcrop-patterns controlled by thrusts. Returning to the subject of thrusts, we must deal briefly with the ways in which they affect the arrangement of outcrops at the earth's surface. Thrust-planes are normally inclined at gentle angles and, as with any other gently dipping surfaces, it is possible to determine their strike and dip by the way they run over hills and valleys —a thrust, for example, may be expected to make a V upstream or downstream in just the same way as a bedding-plane of similar attitude. The rocks lying beneath a thrust may be revealed in 'thrust-inliers' or *windows* by erosion and those lying above it may be preserved in isolated 'outliers' or *klippen* (singular *klippe*).

Reference to Fig. 14.3 shows that the main effect of the thrusts marked in the figure is the *repetition of strata* at more than one level—the Lewisian is seen, for example, not only in its natural position at the base of the succession but also resting on a thrust at a higher level. When a geological map shows that a particular formation appears at more than one level in a succession with constant dips, the presence of thrusts may be suspected.

2. Faults

Definitions. A fault is a fracture or dislocation in the crust along which there has been *displacement* of the rocks on one side relative to those on the other. This general definition embraces the category of *thrusts* dealt with in the last section as well as the remaining types of fault considered below; we have singled out thrusts for separate treatment largely because of their close connection with folding.

The surface on which movement takes place during faulting is the *fault-plane* which may be vertical, steeply inclined or (as with the thrusts) gently inclined. The direction and amount of inclination is recorded by the *dip*, the angle between the fault-plane and the horizontal (Fig. 14.5): the horizontal trend of the plane is its *strike*. The intersection of a fault with the ground-surface is known as the *fault-line* or *fault-trace*. The upper side of an inclined fault, and the rocks which lie above it, are referred to as the *hanging wall*; those below it are the *foot-wall*.

Still further terms are needed to define the relationship between the fault and the strata which it displaces: *dip-faults* strike parallel to the local direction of dip of the beds, *strike-faults* are parallel to the strike and *oblique faults* cut across both strike and dip directions.

Movement on a fault may be in any direction, horizontal, vertical or some combination of both. The *displacement* or *slip* is the sum of all the effects of movement and is recorded by the relative positions on either side of the fault of two originally contiguous points: the *vertical component* of the slip, taken by itself, is known as the *throw* of the fault (Fig. 14.5).

Normal, reverse and transcurrent faults. Faults can be classified according to the kind of movements that have taken place on them into normal faults, reverse faults and transcurrent or strike–slip faults. The *normal faults* (originally so called because they are the normal type found in British coalfields) are those in which the hanging-wall rocks have moved *down the dip* of the fault-plane. Consideration of Fig. 14.5 will show that this movement results in a *lengthening* of the crust across the fault.

Small normal faults are extremely common in almost all geological situations. Large normal faults occurring in groups produce a considerable effect of lengthening and are especially common in the more stable areas, outside the mobile belts. Groups of faults arranged so that alternate dislocations dip in opposite directions produce the effect of *block-faulting* illustrated in Fig. 14.6: the crust is separated into high blocks or *horsts* between outward-dipping faults and low blocks, *troughs* or *graben* between inward-dipping faults.

Rift-valleys are topographical depressions produced by the subsidence of long narrow segments of the crust between normal faults. These valleys may continue without interruption for hundreds and even thousands of miles. They are among the largest structures of the crust and it is thought by many geologists that their formation is connected with the deep-seated processes which are responsible for breaking up continental masses and causing continental drift to take place. At the present day, a huge system of branching rift-valleys extends through Africa from the Zambesi to the Red Sea and the Jordan Valley (Fig. 14.7), continuing for more than one-sixth of the earth's circumference; and a second system extends from north to south the length of the Atlantic Ocean. We realise the extraordinary nature of the structures when we remember that the floor of the Dead Sea is some 2,600 feet below the level of the Mediterranean Sea (Fig. 14.7).

Reverse faults, the second group, are those on

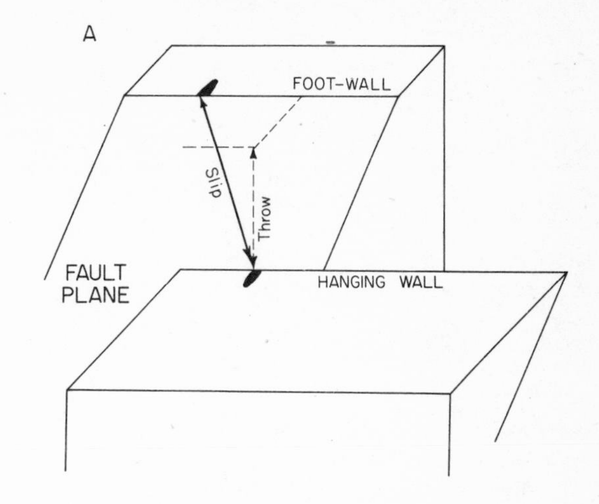

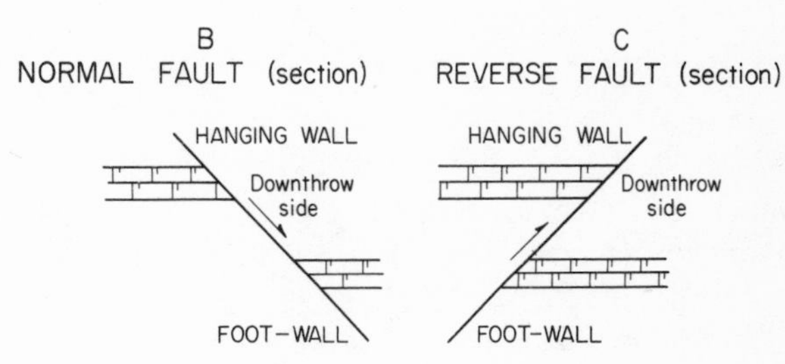

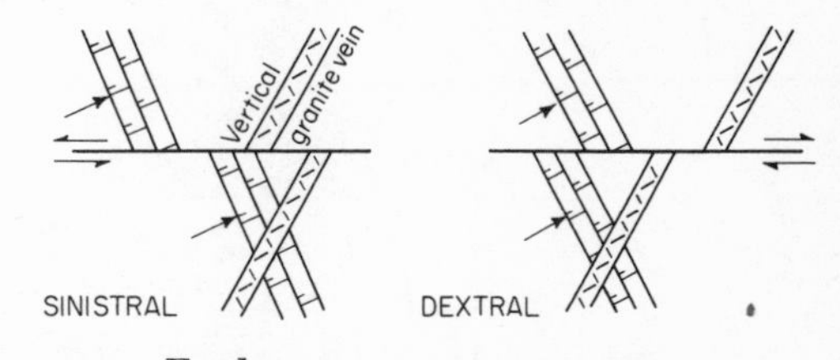

Fig. 14.5 *Faults*

A. The elements of a fault

B, C and D. The three main types of fault

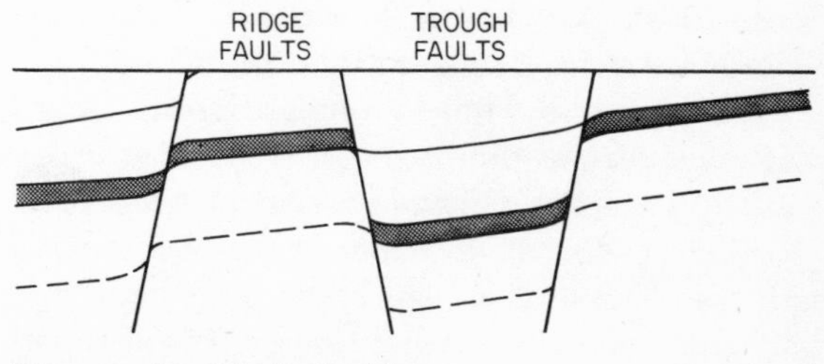

Fig. 14.6 *Block-faulting*

Chapter fourteen

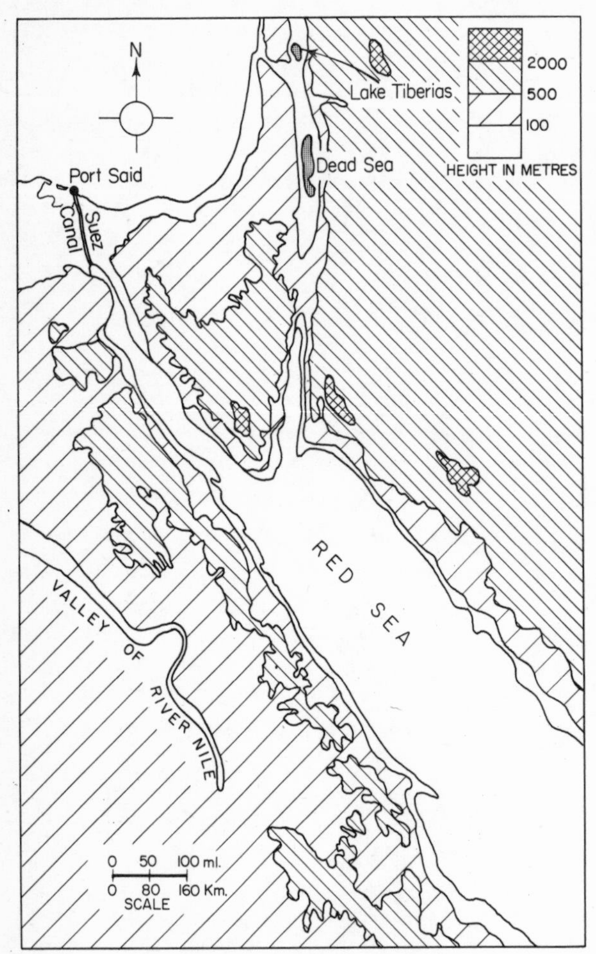

Fig. 14.7 *A rift-valley: topographical map of part of the Red Sea and the Jordan Valley showing the long narrow depression made by the rift-valley*

which the rocks of the hanging wall move *up the dip* of the fault-plane. They result in *shortening* across the fault and in duplication of strata: reverse faults with low dips are *thrusts* which have already been dealt with (Fig. 14.3).

The faults of the third group are *wrench-faults*, *tear-faults*, *strike–slip-faults* or, best, *transcurrent faults* on which *horizontal* movement takes place (Fig. 14.5). The fault-planes are almost vertical and the effect of faulting as seen on a map is to shift rocks laterally, often for many tens of miles. Transcurrent faults are important structures in the Scottish Highlands, where they have a north-easterly strike and are usually *sinistral* (i.e. the rocks are seen to have been displaced to the left as one looks across the fault: displacement to the right is *dextral*). The largest of the Highland wrench-faults is the *Great Glen Fault* which runs along the hollow occupied by Loch Ness and the Caledonian Canal. The displacement of a granite intrusion suggests that movement on the fault amounts to more than 60 miles (Fig. 14.9).

The movement of rock-masses during faulting often leads to the breaking-up of rocks along the fault, with the production of *fault-breccia*, a jumble of angular fragments often cemented together by clayey *gouge*. The fault-plane itself may be smoothed and polished by friction and may be scratched or grooved by the material moving against it. The scratches or *slickensides* are parallel to the direction of slip, though they usually record only the late stages of movement. Other small-scale effects of faulting include the bending of strata against the fault by drag of the moving rock-mass. Sometimes, an ore-bearing vein may be bent and broken and the fragments distributed along the fault: observation of the 'drag of ore' may establish the direction of displacement and make it possible to locate the faulted continuation of the vein.

Faults on maps. The real proof of faulting is supplied by the geological map. As the map is being made, the geologist finds that the outcrop of a particular set of beds or of an igneous intrusion unexpectedly comes to an end, that a well-defined fold or other structure disappears, or that there is an abrupt change in strike or dip. As mapping progresses, it may become apparent that certain beds are either displaced or repeated or, alternatively, are missed out ('cut out') at the place where they would be expected. These discontinuities indicate that faulting has taken place. In addition, the fault-line may be marked by breccia or gouge, and may be excavated by erosion into a long hollow, as in the example of the Great Glen Fault already mentioned (Fig. 14.9). Finally, where faulting is of recent date, the land-surface itself may be displaced. Vertical fault-movements may produce a *fault-scarp*, a steep slope joining the land on the upthrown and downthrown sides. A *fault-line scarp* is a slope at a fault which results simply from differential erosion where the rocks on either side of the fault are of different resistances.

The *fault-trace* or outcrop of the fault itself follows a course which is controlled by the attitude of the fault and the topography. The attitude of the fault can be established by using the lines of reasoning discussed in Chapter 13 (p. 159). We may recall that the trace of a vertical fault will not be deflected by variations in the topography, that of a steep fault will be only slightly affected and that of a gently dipping fault such as a thrust will be strongly affected.

We may consider next the effects of faulting on bedded rocks arranged in various ways; in the diagrams illustrating this topic (Fig. 14.8) the complications due to topographical irregularities are got rid of by showing the outcrop-patterns as they would appear on a plane surface.

Normal or reverse faulting of *horizontal beds* brings higher and lower layers against each other, the throw being equal to the change in level of any bed. The effects of faulting on *inclined beds* vary according to the relationship between the direction and throw of the fault on the one hand and the direction of dip of the beds on the other. A *dip-fault*, whether normal or reverse, displaces the outcrop of an inclined bed laterally so that younger beds on the downthrow side are brought against older beds on the upthrow side. A *strike-fault* cuts out beds where the downthrow is in the same direction as the dip of the strata; but it repeats beds where the downthrow is in the direction opposite to the dip.

Transcurrent faults produce lateral displacements of all structures regardless of the attitude of these structures: their effects are illustrated by the displacement of a granite body by the Great Glen Fault (Fig. 14.9).

The effects of faulting on *folded strata* can, with a little thought, be predicted from what has already been said, since we can establish the effect on each fold-limb separately (Fig. 14.10). A fault at right angles to an open fold is, more or less, a dip-fault with respect to each limb and the outcrops of beds will be displaced accordingly. As the figure shows, the width of a syncline is increased on the downthrow side and that of an

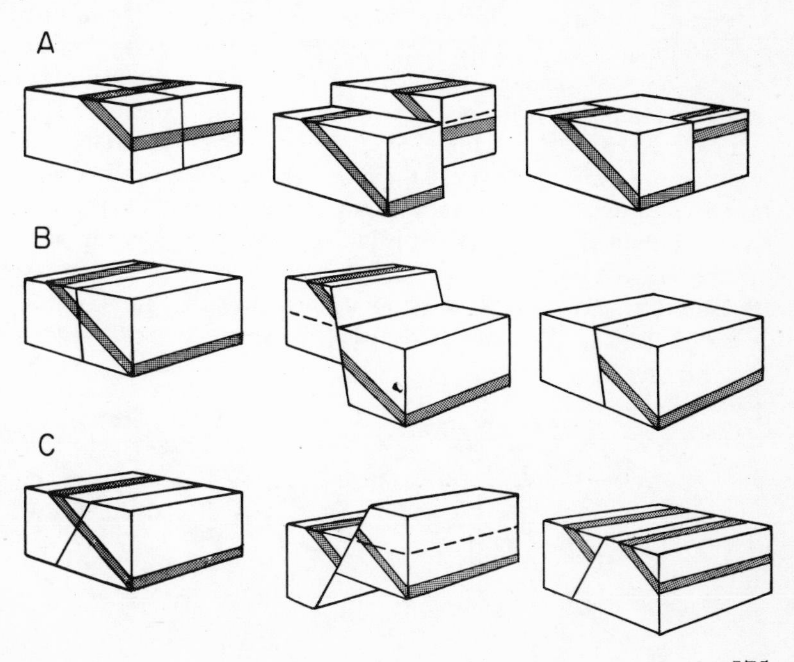

Fig. 14.8

Outcrop-patterns of faulted beds: the left-hand diagrams show the relationships of beds and faults, the central diagrams show the effects of faulting, and the right-hand diagrams the outcrop-pattern revealed after erosion

A. Dip-fault
B. Strike-fault with downthrow in dip direction
C. Strike-fault with downthrow against the dip

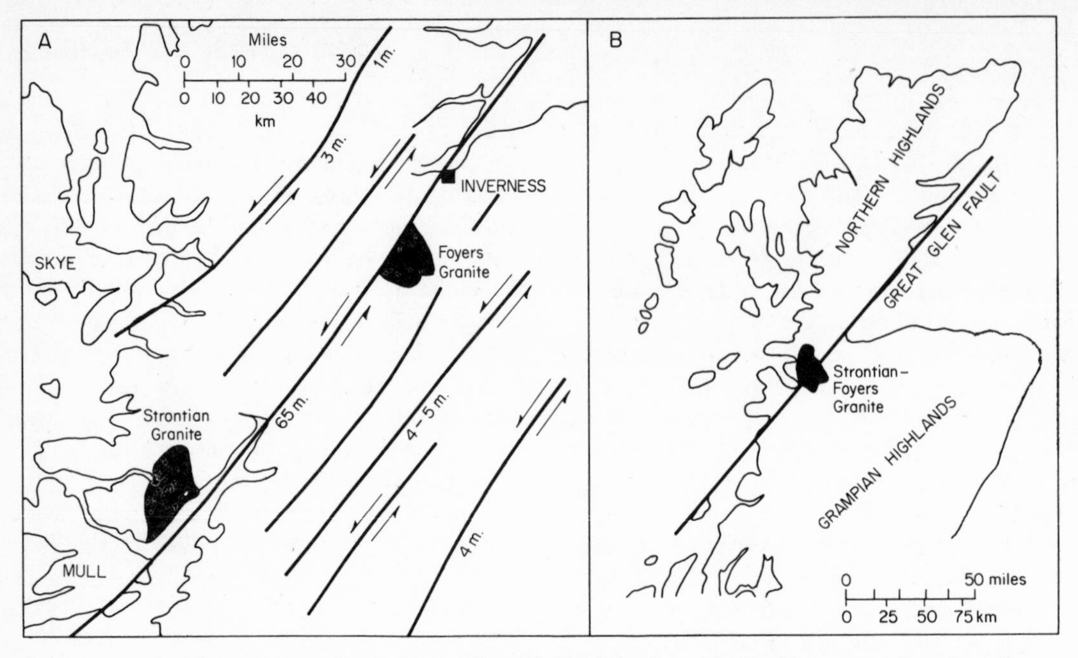

Fig. 14.9 A. *Transcurrent faults in the Scottish Highlands, with displacements given in miles.* B. *The regions on either side of the Great Glen Fault restored to the positions which they occupied before faulting took place*

anticline is correspondingly decreased. A transcurrent fault shifts both fold-limbs in the same direction without altering the width of the fold.

3. Joints

A *joint* is a parting which tends to separate a once-continuous mass of rock into two parts but which does not displace either part. In rocks exposed at the surface, joints are usually numerous and often control the manner of erosion—it is important to learn to recognise them or they may be mistaken for bedding-planes. Many joints are widened, stained or emphasised by the effects of weathering, the process being carried furthest in limestones where solution along joints produces *grikes* (Fig. 5.4). Because fluids can penetrate along joints, their presence or absence may have important economic effects. Well-jointed strata can act as reservoirs for water or oil and can allow the passage of these fluids—the main *aquifer* of the London basin is the Chalk which transmits water via open joints (p. 64). On the other hand, well-jointed rocks in the floor of a reservoir may allow water to leak away, and strongly jointed rocks may be too unstable to provide secure foundations for dams.

Joints result from many kinds of stresses and their arrangement varies according to their origin. A group of parallel joints is called a *joint-set* and several intersecting sets make a *joint-system* (Fig. 14.11).

Shrinkage-joints are produced during the dry-

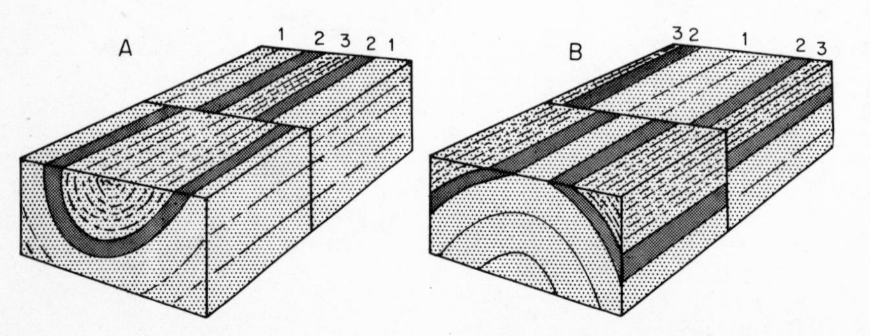

Fig. 14.10
Displacement of syncline (A) *and anticline* (B) *by faults. The nearer portion of the block in each diagram is on the downthrow side*

Fig. 14.11
Joints: a rectangular joint-system in the granite of Land's End, Cornwall

ing-out of sediments and, much more spectacularly, during the cooling and contraction of igneous bodies. The best example is provided by the *columnar jointing* of sheet-like bodies of fine-grained igneous rocks as seen in the Giant's Causeway of Northern Ireland and in Staffa in the Western Isles of Scotland. As these homogeneous rocks cooled, shrinkage-cracks tended to develop around regularly spaced centres towards which the tensional stresses converged. Interference between neighbouring centres led to the production of a honeycomb pattern of joints outlining columns set perpendicular to the cooling surface and having on average six sides (Fig. 14.12).

Jointing in large igneous intrusions is again related to the shapes of the intrusions. Where the bodies are roughly circular in plan, radial and concentric joints are frequently developed. These, together with horizontal joints which are also often developed, outline rectangular blocks and may cause the igneous rock to weather with a wall-like pattern. Such an arrangement is shown by the *tors* of the Dartmoor and Cornish granites. The joint-patterns of igneous bodies control the shape and size of the solid blocks which may be

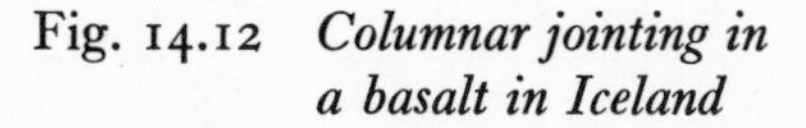

Fig. 14.12 *Columnar jointing in a basalt in Iceland*

used for building and, before quarrying granites or similar rocks, it is necessary to know that they are not so much broken-up by joints as to be unsuitable for use as building stone.

The *jointing of coal* is well-known in practice, since it controls the way mines must be laid out and the size of the easily obtained coal-fragments, but its origins are not really understood. Most coal-seams show a well-developed, closely spaced vertical joint-set called the *cleat* and a second less conspicuous set at right angles to it called the *end*. These two sets combine with the bedding to produce the characteristic cuboidal shape of ordinary house-coal.

Joints related to earth-movements are developed in most areas where folding or faulting has taken place, but surprisingly little is known about their arrangement. Very commonly, a joint-set is formed at right angles to the axes of folds and other joints may be roughly parallel to the fold-axes.

4. Geological Structure as a Record of History

Every aspect of the geological structure revealed in any area is a record of some event in the history of that area. The arrangement of the sedimentary rocks in layers, with younger following older in a long succession, reflects the history of deposition. The presence of unconformities indicates interruptions in the period of deposition. The shapes of igneous bodies in relation to the sedimentary beds tell us of their mode of formation. Finally, the distortion of the primary structures by folds or displacement of these structures by faults reflects the working of crustal stresses. The kinds of structures present suggest whether the forces at work were orogenic forces acting in mobile belts—systems of tight folds, nappe structures and overthrusts are characteristic—or whether they were the milder forces concerned with epeirogenic movements—broad open folds and faults may predominate. Finally, the sculpturing of the land-surface reveals what erosional forces have been at work. We shall see all these topics illustrated in the next three chapters, but since practice alone can make the interpretation of structural evidence easy we end this chapter with a final exercise.

In Fig. 14.13 is reproduced a geological map of a part of west Yorkshire. The student should attempt to draw a sketch-section from north to south and should write an account of the geological history revealed by map and section. He should check the evidence relating to the following conclusions:

1. Pre-Carboniferous folds plunging to the south-east in the older rocks.
2. Pre-Carboniferous faulting.
3. Unconformity between the Carboniferous Limestone and the older rocks.
4. Post-Carboniferous faulting.

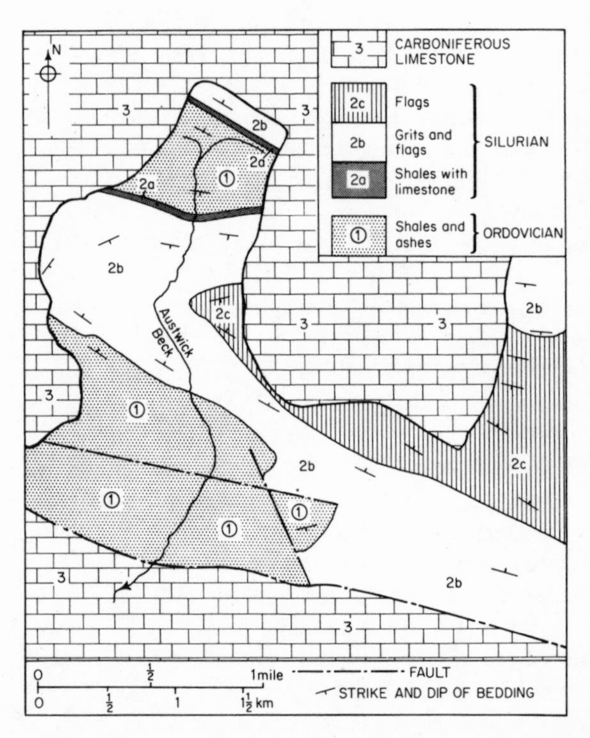

Fig. 14.13 *Geological map of part of west Yorkshire (after King and Wilcockson)*

15

HISTORICAL GEOLOGY: 1

General considerations

This chapter serves as an introduction to the study of historical geology. It summarises the evidence from which the history of deposition, deformation, intrusion and metamorphism can be established and also deals with the geological time-scale and its divisions.

1. Introduction

In the earlier chapters of this book we have been concerned with the general nature of geological processes and with the rocks and structures produced by them. With this information as a background, we can go on in the next few chapters to examine the geological history of portions of the earth's crust. Since any study connected with history, whether human or geological, is concerned with time, the first essential for the geologist is to be able to arrange the geological events recorded by the rocks and structures of any area in their proper order. He must, in other words, be able to determine the *relative ages* of these rocks and structures. Many of the criteria by which age-relationships can be established have been mentioned at appropriate places in earlier chapters. They are brought together and summarised in the next few paragraphs.

2. Timing in Sedimentary Rocks

The stratigraphical succession. The fundamental law governing the interpretation of sedimentary sequences is William Smith's *Law of Superposition* (p. 2); provided that there has been no secondary disturbance, any bed in a pile of sedimentary rocks must be younger than the bed below and older than the bed above. The same law governs the interpretation of lavas and pyroclastic rocks which, like sediments, are formed on the earth's surface.

The Law of Superposition cannot be applied directly to strata which have been tilted or folded by earth-movements, because such secondary disturbances may have turned the strata upside down or tilted them into a vertical position. The forms of the primary structures, however, often enable us to identify the *original top* of a bed or lava-flow, whatever its attitude may be (Fig. 15.1). The structural differences between the tops and bottoms of beds provide a built-in record of the '*way-up*' of a sequence of strata. Where the strata are right-way-up, the original tops are still on top and the strata are said to *young* or to *face* upward. Where the strata are wrong-way-up or *inverted*, the strata face downwards.

The application of the Law of Superposition (or, for folded sequences, the use of 'way-up' criteria) makes it possible to establish the order in which a number of beds or groups of beds were deposited. As we saw in Chapter 10, a sequence of groups from the oldest to the youngest gives the *stratigraphical succession*.

Correlation. The stratigraphical succession at any locality provides a record of the history of deposition at that locality. To compare this record with that of any other region it is necessary to be able to match or *correlate* strata of the same age. This may be attempted in various ways. In the simplest circumstances, a group of strata in one locality may be connected with a group in the other by a *continuous outcrop*; the geologist can walk on the same group of rocks from one locality

Fig. 15.1
Way-up: vertical beds of current-bedded sandstone face towards the right

to the other and their identity is self-evident, though it may not imply a precise equivalence in time. A second method of correlation depends on matching the characters of groups in the two localities—a succession of, say, sandstone followed by limestone in one may be equated with a similar succession in the other. The dangers of this method of *lithological correlation* will be fairly obvious. The facies of a sediment depends not on its age but on its environment of deposition and at any particular moment of time sandstone may be deposited in one area, limestone in another and clay in a third. Moreover, since the number of possible types of sediment is limited, rocks such as sandstones and limestones recur again and again throughout the stratigraphical succession.

A more satisfactory method of correlation depends on the *presence of fossils*. The assemblage of fossils contained in any sedimentary rock varies according to the stage of evolution reached at the time when the rock was deposited. Organic evolution involves the continuous and progressive change of all groups of animals and plants. It has never stopped or been reversed, and for this reason the fossil populations of successive groups of strata show differences which are consistent all over the world. They can therefore be used in the correlation of sedimentary successions, a fact which was expressed by William Smith in his *Law of Strata identified by Fossils* (p. 2). The fossil record covers only the last 600 million years of earth-history—this period is known as the *Phanerozoic* ('evident life') part of this history—and it is therefore only for rocks deposited in Phanerozoic times (less than a fifth of geological time) that correlation by means of fossils is possible.

Chronological divisions based on fossils. The changes in fossil populations marking stages in the progress of organic evolution are used as a basis for dividing this long span of 600 million years. Four great time-divisions or *eras* are recognised on this basis: working backwards in time these are the Cainozoic ('modern life'), the Mesozoic ('intermediate life'), and the Upper and Lower Palaeozoic ('ancient life'). The Cainozoic era is often called the Tertiary. Each era is further divided into a number of geological *periods* distinguished by smaller differences in fossil assemblages. The deposits of each period constitute a *geological system* and, as we have already seen, the sequence of systems arranged in order from the oldest to youngest makes the *stratigraphical column* (Fig. 10.10).

Each geological period was so long that thousands of feet of sediment could be laid down and, before detailed correlations can be made, it is obviously necessary to make further subdivisions. The smallest divisions of the succession made on the basis of fossils are the *zones* and are identified by means of *zone-fossils* (p. 120). To be useful, a zone-fossil must be fairly common within its own zone and rare or absent in older and younger zones—that is, it must be a short-lived species. A zone-fossil should also have as wide a geographical distribution as possible and should occur in a variety of different sedimentary facies. Some of the more important groups from which zone-fossils are drawn are the graptolites, corals, trilobites, ammonites and foraminifera. Examples are illustrated in Chapter 11.

Strata which are more than 600 million years in

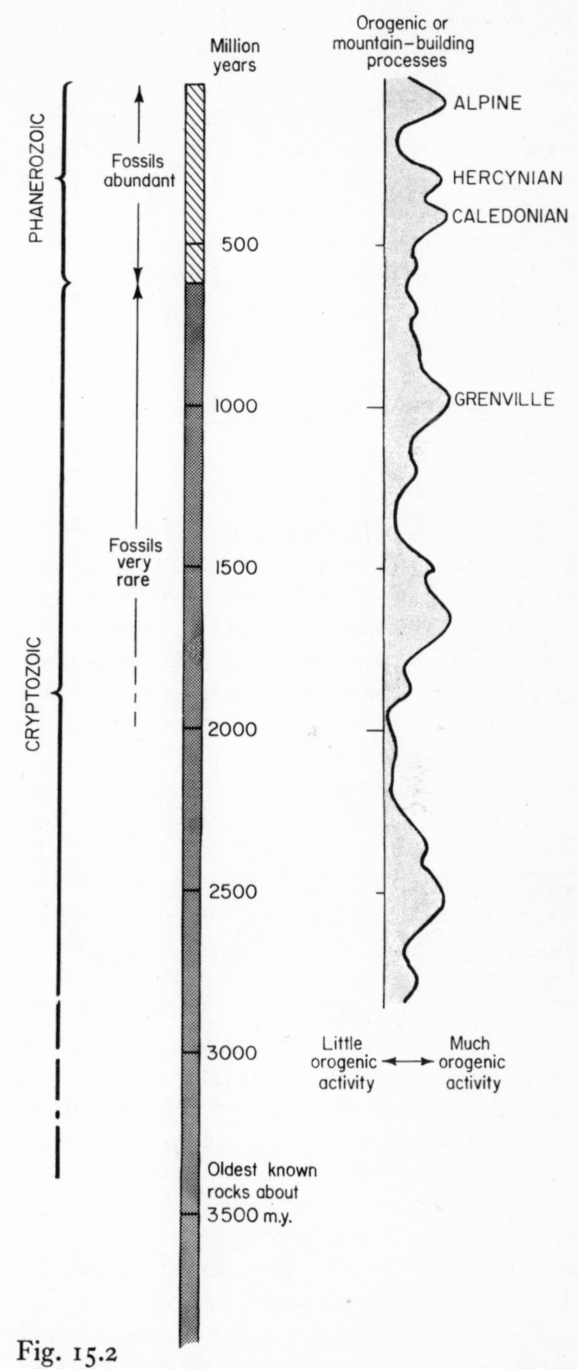

Fig. 15.2
The geological time-scale, with periods of orogenic activity important in the Northern Hemisphere shown diagrammatically

age, that is, older than the Cambrian Period, belong to the *Cryptozoic* ('hidden life') portion of earth-history. They seldom have any fossils at all and such fossils as they do contain have little value for correlation. It is therefore impossible to establish world-wide correlations by means of fossils for these Pre-Cambrian rocks. Yet the span of Pre-Cambrian time covers more than four-fifths of the whole of geological time (Fig. 15.2). For this Cryptozoic part of earth-history, the stratigraphical record is less useful than the record of deep-seated happenings which is discussed in the next few paragraphs.

3. Timing of Earth-movements

Earth-movements are recorded by the fact that the rocks affected by them are disturbed—that is, the rocks are displaced along faults, folded or distorted. Obviously, rocks which were not in existence at the time when movement took place could not have been affected, and the date of a phase of movement must therefore lie between the age of the youngest disturbed rock and that of the oldest undisturbed rock.

In addition, earth-movements indirectly affect the processes of erosion and deposition at the earth's surface and may therefore be recorded in the stratigraphical succession. Movements which lead to uplift may bring deposition to a halt and cause erosion to begin. When, at some later date, deposition is resumed new sediments are laid down *unconformably* on the eroded remains of the older series (p. 118). An *angular unconformity* (Fig. 10.9) separates a folded or tilted succession from an undisturbed younger succession and thus provides a record of earth-movement. An example is shown in Figs. 14.3 and 14.4, where an unconformity separates the violently disturbed gneisses of the Lewisian from the unmetamorphosed Torridonian and Cambrian.

The structures produced during one phase of earth-movement can often be dated relative to those of another phase. In principle, the criteria are very straightforward; an older structure will be deformed or displaced by a younger structure. Thus, a fold may be displaced by a younger fault, or it may be distorted by a younger fold. By such evidence, it is possible to build up a *history of deformation* in the crust.

4. Timing of Igneous Intrusion and Metamorphism

We now come to consider other geological happenings which take place in the depths of the crust. *Intrusive igneous bodies* are, as we have seen (p. 44, Fig. 4.2), necessarily younger than the country-rocks which they invade, and the youngest rocks of the intruded series give an upper limit for the age of the intrusion. A lower limit may be provided by the occurrence of younger strata resting unconformably on the eroded surface of the intrusion, or by conglomerates containing boulders of the igneous rock (Fig. 15.3). Similarly, an *episode of metamorphism* must be younger than the age of the youngest metamorphosed rock and older than the oldest rock resting unconformably on the metamorphic series or containing boulders of the metamorphic rocks.

Episodes of intrusion or metamorphism may also be dated in other ways. *Radiometric age-determinations* of igneous or metamorphic minerals (p. 18) give the ages of these minerals in terms of millions of years and so enable them to be placed in their correct position in geological time. Structural evidence may establish the age of intrusion or metamorphism relative to a phase of earth-movement; for example, an intrusive dyke which cuts across a fold or fault is younger than the movement producing the structure, whereas a dyke which is itself folded or faulted must be older than the movement. The age-relationships of earth-movement and metamorphism may be a little more difficult to unravel because, as we have seen, the two processes often go on simultaneously. Metamorphism contemporaneous with movement is often demonstrated by the fact that the metamorphic minerals are arranged to produce cleavages or other structures formed in response to deformation (p. 151). Deformation after metamorphism is shown by fracturing and bending of the metamorphic minerals.

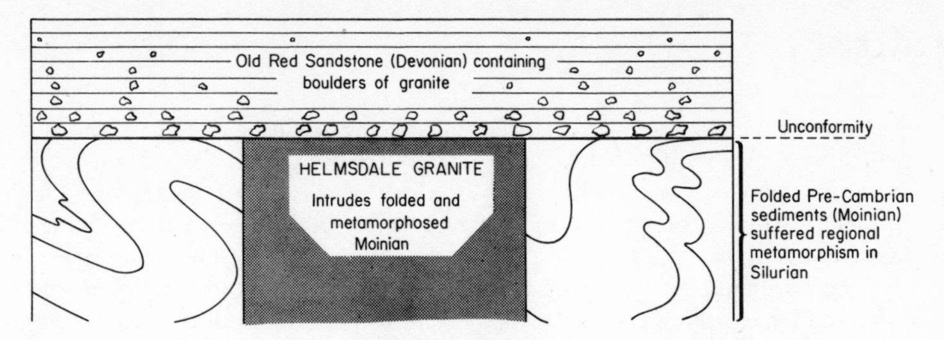

Fig. 15.3 *Geological dating of an igneous intrusion illustrated by the Helmsdale Granite of Scotland. The granite is younger than the folding and metamorphism affecting Pre-Cambrian sediments (last metamorphism is Silurian) but older than the overlying Old Red Sandstone.*

5. The Pattern of Geological History

By making use of the criteria summarised above it becomes possible to fit together the history of deposition at the earth's surface and the record of events that have taken place in the depths of the crust. Although much still remains to be discovered, enough is known about these relationships to show that geological events do not follow one another haphazardly. Certain patterns of events have been repeated a number of times in the course of earth-history and it appears that these consistently related events must have a common cause. The most important of the recurring patterns of events is that connected with the formation, growth and ultimate upheaval of *mobile belts* or *orogenic belts* (Fig. 15.4).

The mobile or orogenic belts are, as we have seen, long narrow tracts within which the crust is subject to abnormal activity, both tectonic and igneous. A newly developed mobile belt usually has a tendency to subside—as O. T. Jones has put it, it gets a 'sinking feeling'—and therefore acts as a trap in which considerable thicknesses of sediment may accumulate. Such a subsiding belt is known as a *geosyncline*. Some *oceanic trenches* of the present day (p. 14) may represent mobile belts in this stage of evolution. As development proceeds, volcanic activity may begin from centres in or near the belt and volcanoes may build up new islands or mountains as in the *volcanic island arcs* of the present day.

Subsidence in the geosyncline is frequently interrupted by earth-movements of other kinds. Compression of the geosynclinal filling takes place when the more stable *forelands* on either side approach each other; such compression gives rise to folding and thrusting and may force parts of the filling up above sea-level, thus leading to erosion and the development of unconformities. The rapid erosion started off by upheaval in turn supplies quantities of clastic sediment to the remaining basins of deposition. The later stages of geosynclinal development are therefore often marked by the accumulation of thick series of coarse detrital sediments; in the prevailing conditions of crustal instability, the sediment is often redistributed by turbidity currents with the production of turbidites (p. 88).

The infilling of the geosyncline, perhaps aided by other processes taking place in depth, forces the basement underlying the geosynclinal succession downwards to great depths. The whole thickness of the crust is therefore disturbed by the development of the mobile belt, and its lower parts begin to grow unusually hot. The prolonged unrest culminates in one or more violent episodes of *mountain-building* or *orogenesis* during which the narrow mobile belt, already distorted by long subsidence and beginning to be softened by heat, is squeezed between the stronger and more stable forelands. The geosynclinal filling and its basement are *folded and sliced up by thrusts* and may at the same time begin to undergo *metamorphic changes* and become partly molten. Sooner or later, the whole deformed belt is heaved up to produce a range of mountains and the sea is expelled from the mobile belt. The molten material generated deep in the crust is squeezed up through the folded rocks to produce *intrusions of granitic composition.*

The newly formed mountain-belt or *orogenic belt* is rapidly attacked by erosion with the production of vast quantities of detrital debris. Much of this is carried by rivers to inland basins within the mountains or to great alluvial plains and deltas at their feet. When the mountain-building episode is over, the orogenic belt gradually loses its mobility and settles down into a more stable area. The same kind of activity, however, continues in a new mobile belt and so the building of an orogenic belt is repeated in another area.

The life-time of an orogenic belt is usually several hundred million years in length, and although there are many variations in detail, the general pattern of events outlined above and summarised in Fig. 15.4 can be recognised in the history of most belts, whatever their age. The recurrence of this pattern provides a natural way of dividing the immense length of geological time. The culminating phases of metamorphism and granite intrusion which take place during the development of each belt can be dated in terms of millions of years by radiometric methods. Thus, a number of fixed points are established on the time-scale (Fig. 15.2) which can be used as points of reference—for example, a succession of sedimentary rocks may be shown to have been deposited in the interval between two periods of orogenic activity, and can thus be placed in its correct position in the span of geological time. This scheme of division of the time-scale is still very rough, for the techniques of radiometric dating have only been used on a large scale for about fifteen years. It does, however, provide a framework into which events ranging through the whole of geological time can be fitted.

The most important orogenic episodes recognised in the history of Western Europe and North America are shown, with their approximate dates in Fig. 15.2. It may be said that four such episodes of mountain-building have taken place during the period covered by the fossil record. These are the *Caledonian orogeny* which ended at the end of the Lower Palaeozoic, the *Hercynian orogeny* ending at the end of the Upper Palaeozoic, the *Laramide orogeny* (important in North America but almost unnoticed in Europe) in the Cretaceous period and the *Alpine orogeny* in the mid-Tertiary. The effects of the Caledonian and Hercynian orogenies, as well as those of at least two Pre-Cambrian orogenies, are clearly recorded in the geological history of the British Isles and we shall use these great phases of tectonic activity as landmarks in our study of this history.

6. The Oldest Known Rocks

The oldest rocks which have been recognised at the time this book was written occur in two regions. In Southern Rhodesia, metamorphic minerals in rocks derived from impure sandy and clayey sediments have been dated at about 3,400 million years; the sediments themselves must of course be older than this. In the Kola region of the north-west U.S.S.R., minerals in granitic and migmatitic rocks have been dated at 3,500 million years. The interesting thing about these very old rocks is that, so far as is known, there is nothing very unusual about their composition or structure. Sedimentary rocks, similar to those investigated in Southern Rhodesia and metamorphosed in similar ways, can be found in many younger sequences; and the granites and migmatites of Kola can be matched in younger orogenic belts. It is reasonable to conclude that geological processes were already working in much the same way as they have done throughout geological history by the time when these rocks were formed. James Hutton, the geologist who formulated the principle of uniformitarianism, came to the conclusion that the geological record revealed 'no trace of a beginning, no prospect of an end'. This conclusion, published in 1788, is still valid.

7. The Record of Organic Evolution

Although very few fossils have been found in Pre-Cambrian strata (that is, in rocks older than 600 million years), it seems certain that living

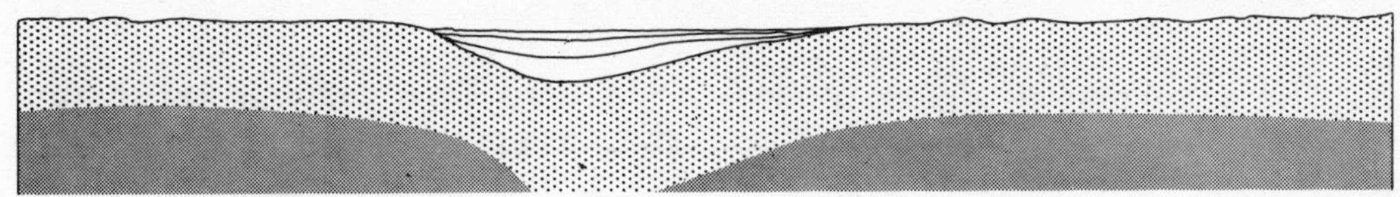

1. The geosyncline is initiated and begins to fill with sediment

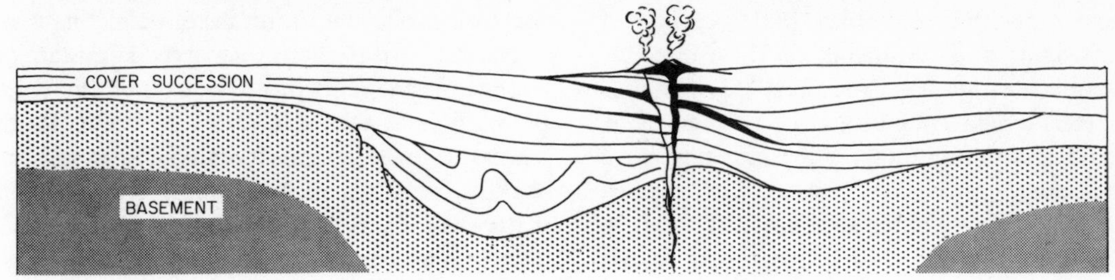

2. The geosyncline accumulates a thick pile of sedimentary and volcanic rocks

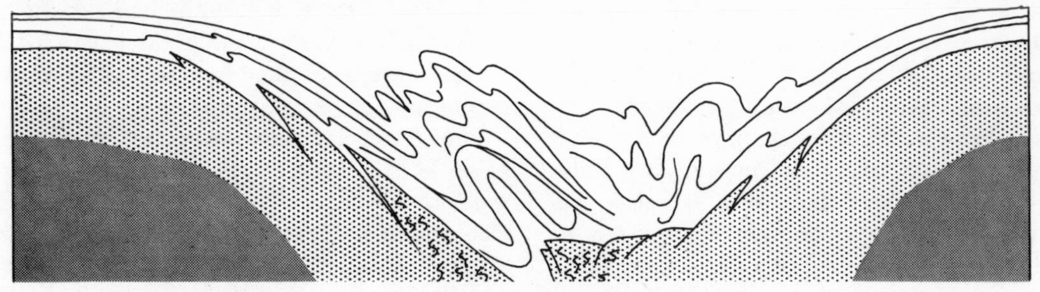

3. The geosynclinal filling and its basement are folded and fractured in the mobile belt, metamorphism begins

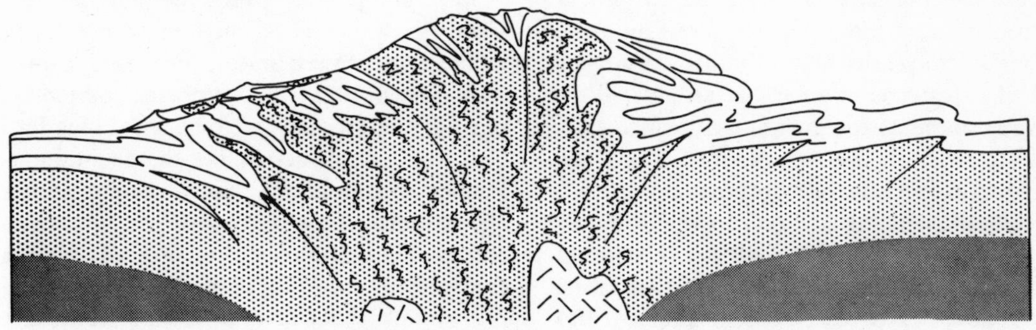

4. Further folding, migmatisation and metamorphism are followed by upheaval and the production of a mountain belt

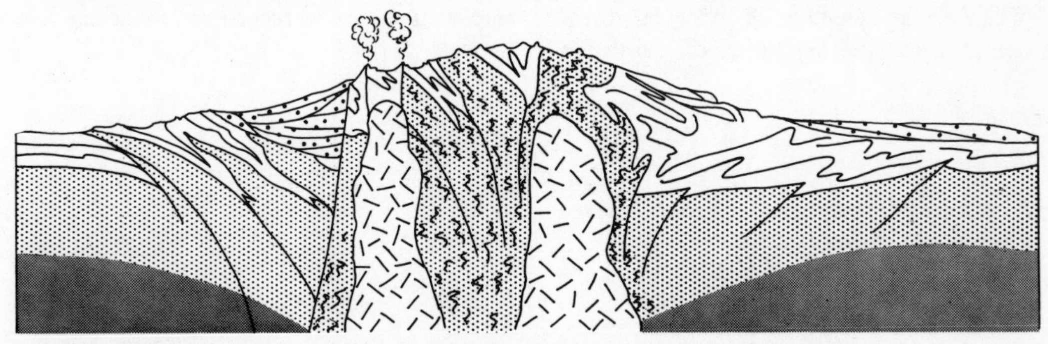

5. Granitic bodies intrude the fold-belt, erosion attacks the mountains and late-tectonic sediments accumulate

Fig. 15.4 *Stages in the evolution of a mobile belt*

organisms of some kind were already in existence in Pre-Cambrian times. In Finland, pelitic sediments known to be more than 1,800 million years in age contain finely divided carbon whose isotopic composition resembles that of carbon of organic origin; and sediments of the same age contain obscure sac-like structures, a few inches in diameter, which are outlined by carbonaceous films, and are thought to be fossils; they have received the appropriàte name of *Corycium enigmaticum*. In Pre-Cambrian sediments between 1,000 and 600 million years in age, fossils are occasionally present, though they are excessively difficult to find. The most widespread types are nodular masses of calcium carbonate sometimes showing a concentric layering. These *stromatolites* are calcareous algae (p. 146). At a very few localities, Pre-Cambrian fossils have been found in considerable variety. One such locality is in South Australia, where shales not far below the base of the Cambrian contain organisms resembling jellyfish, colonial coelenterates and worms, as well as a number of curious forms unlike any Phanerozoic animals. With the exception of the stromatolites, most known Pre-Cambrian fossils appear to represent soft-bodied organisms devoid of calcareous or siliceous armour. This fact largely accounts for their scarcity since soft-bodied creatures, as we have seen, can only be fossilised in exceptional circumstances.

The oldest Phanerozoic strata in which fossils occur abundantly are those of the Cambrian system. The appearance of the Cambrian fauna seems to have taken place rather suddenly and it is probable that some important change in the habits of life or the structure of living creatures took place at about the beginning of Cambrian times. It is fairly generally agreed that this change was connected with the evolution of protective shells and skeletons which provided material more suitable for fossilisation than the soft bodies of Pre-Cambrian animals. The reasons why such hard parts were evolved simultaneously in several independent phyla are not known. One possibility is that their development followed the widespread colonisation of the floors of the shallow seas where protection from turbulent waters and from competing organisms was particularly necessary. Another suggestion which has been considered is that lime did not become available in sea-water until late in the Pre-Cambrian; in view of the great length of Pre-Cambrian time revealed by radiometric measurements this suggestion seems rather improbable.

The Cambrian fauna—the first of which we have detailed knowledge—included members of most of the main phyla of the animal kingdom. Coelenterates, brachiopods, molluscs, worms, arthropods and echinoderms were all represented and it is evident that these groups had already been differentiated from one another in Pre-Cambrian times. The phylum Chordata, which includes the Vertebrates, probably came into existence during the Cambrian period. Since most of the divisions of the animal kingdom had thus been established before the beginning of the Phanerozoic, it follows that we cannot learn much about their origin from the fossil record. This record does, however, provide a remarkably complete picture of the way in which many of the groups changed and became diversified in the course of their evolution. Some of the landmarks in the fossil record are mentioned in later pages.

16

HISTORICAL GEOLOGY: 2

The Pre-Caledonian and Caledonian Foundations of the British Isles

This chapter covers the early part of the geological history of Britain in which the foundations of the geological structure were laid. The Pre-Caledonian basement rocks are dealt with as a preliminary to our main topic, the building of the Caledonian orogenic belt. We contrast the histories of sedimentation, igneous activity, earth-movement and metamorphism in the mobile belt with those in the stable forelands, and examine the structure of the mobile belt.

1. The Geological Position of the British Isles

The British Isles lie near the north-western margin of the continent of Europe (Fig. 16.1). To the east, a continental region extends for thousands of miles, only covered here and there by shallow epicontinental seas. The main geological structures of Britain continue without interruption beneath the North Sea and English Channel to join up with structures in the rest of Europe—from a geological point of view, Britain is simply an outlying part of Europe and the seas which isolate her are of quite recent origin.

To the west, the arrangement is very different. At the continental slope beyond the continental shelf which fringes the western coasts, the sea-floor drops to oceanic depths to make the basin of the North Atlantic. There is no evidence that the geological structures of the continent continue out beyond the margin of the continental shelf—indeed, the Caledonian and Hercynian mobile belts which cross the British Isles appear to be

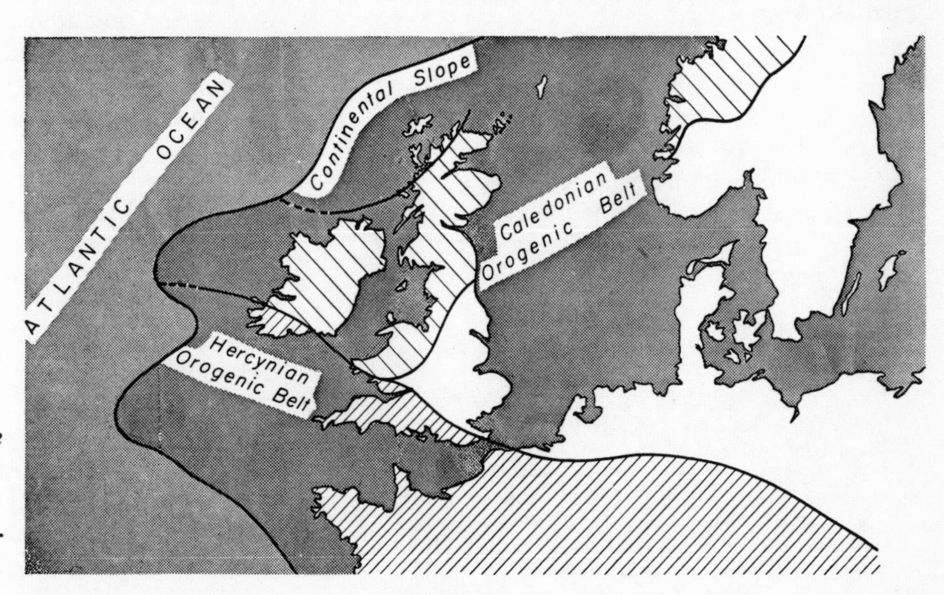

Fig. 16.1
The continental slope off western Europe cuts across the Caledonian and Hercynian mobile belts

cut off short at the continental margin. This apparent interruption of structures is one of the arguments for continental drift (p. 15). The land masses of North America and Greenland were probably at one time joined to Western Europe but later split off and drifted westwards, carrying with them the continuations of the broken mobile belts.

The oldest rocks which have been recognised in Britain are at least 2,600 million years old. Our survey of the geological history of the region must therefore cover in outline the whole period from the time when those immensely ancient rocks were formed up till the present day when the modern beach and flood-plain deposits are being laid down. Since it is obviously impossible to deal in detail with every episode in this long history, we shall concentrate on the geological happenings which seem especially important in themselves, or which illustrate some problem of general geological interest. The student in Britain should be able to fill in the gaps, especially where his home district is concerned, by means of the *Regional Guides* of the Geological Survey and the *Guides* issued by the Geologists' Association, details of which are given below.

2. Outline of the Geological History of the British Area

If, according to the methods outlined in the previous chapter, we use the main episodes of *orogenic activity* as landmarks in geological history, we can divide the record as follows, beginning with the oldest event. The map given in Fig. 16.2 is constructed on this method and should be carefully studied during the reading of this and the following chapter.[1]

[1] Further details of the geological structure of many regions of the British Isles can be obtained from two series of publications:

(i) *British Regional Geology:* a set of handbooks published by the Geological Survey of Great Britain which can be obtained at prices of around 10–15s. per volume from H.M. Stationery Office or through booksellers.

(ii) *Excursion Guides of the Geologists' Association* which cover classic localities in Britain and can be obtained at a price of a few shillings from the Publishers, Benham and Co. Ltd, Sheepen Road, Colchester.

1. OLDER PRE-CAMBRIAN EVENTS: the *older Pre-Cambrian* rocks, between 2,600 million and 1,000 million years in age, were subjected to orogenic disturbances at several periods before the inception of the Caledonian orogeny and are separated by major unconformities from the younger rocks.
2. THE CALEDONIAN EVENTS: the *younger Pre-Cambrian* and *Lower Palaeozoic* rocks were subjected to the *Caledonian Orogeny* which ended about 400 million years ago. The Caledonian mobile belt ran across Britain from north-east to south-west; the areas which lay beyond its limits and made the Caledonian foreland can be seen in north-west Scotland and Central England. An unconformity generally separates rocks affected by Caledonian movements from the overlying Upper Palaeozoic and younger strata.
3. THE HERCYNIAN EVENTS: the Upper Palaeozoic strata were deformed by the *Hercynian Orogeny* which ended about 300 million years ago. The southernmost parts of England and Ireland lay within the Hercynian mobile belt, and the rocks of these regions were strongly disturbed by folds and thrusts. The remaining parts of Britain lay outside the mobile belt in the more stable foreland and were only gently folded. An unconformity separates rocks affected by Hercynian movements from the overlying blanket of little-disturbed Mesozoic and Tertiary rocks.
4. ALPINE EVENTS: the blanket of *Mesozoic* and *Tertiary* rocks accumulated in the foreland of the Alpine mobile belt. It is affected in the south of England by the outermost ripples of the *Alpine Orogeny* but is only gently folded.

Since the topography is dependent to some extent on the character of the bedrock, the geological map (Fig. 16.2) reflects the natural geographical divisions of the British Isles. The toughened Pre-Cambrian and Lower Palaeozoic

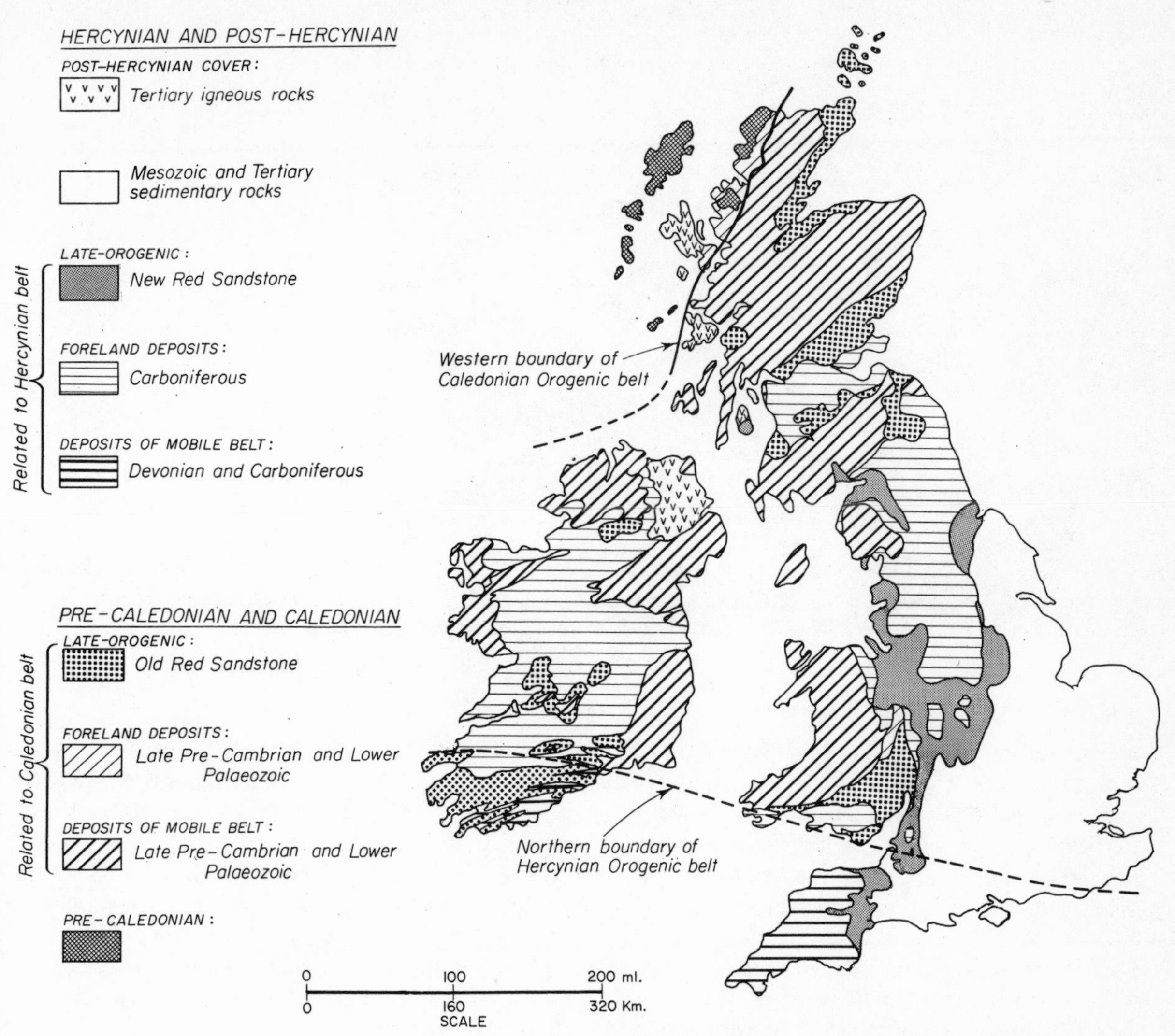

Fig. 16.2 *Simplified geological map of the British Isles*

rocks make the rugged mountainous country of the Scottish Highlands, the Southern Uplands, the Lake District, North and Central Wales and the eastern and north-western parts of Ireland. The somewhat less hard Upper Palaeozoic tends to give bleak but not rugged country, as in the Pennines and south-west Ireland, while the softer Mesozoic and Tertiary weather to give rolling hills and broad valleys in which most of the best farming land of Britain is situated.

The shaping of the British Isles as a distinct group of land-masses was not completed until late in geological history. It was not until the end of the Caledonian orogeny that even the foundations of the structure were laid, and not until the end of the Mesozoic era that most of the British region emerged above the sea-level. In the remainder of this chapter we shall deal with the *laying of the foundations*, leaving the later history of Britain for the next chapter.

The geological history of the British area is unusually full and varied. On this account we pay most attention to the events recorded in the British rocks—but the same methods of interpreting these records apply to the rocks of the whole world.

Fig. 16.3
The Lewisian Gneiss: a banded gneiss outcrop, north-west Scotland

3. The Pre-Caledonian Basement

The oldest groups of rocks which can be recognised are those which were already in existence before the Caledonian mobile belt began to develop and which made the *basement* on which Caledonian sediments accumulated. Since deposition in the Caledonian belt began in late Pre-Cambrian times, this means that the *Pre-Caledonian groups* include all *Pre-Cambrian rocks older than about 1,000 million years*. It is as well to remember that their history covers a period longer than all the subsequent geological periods put together. The principal regions in which Pre-Caledonian rocks can be seen are those, on either side of the Caledonian mobile belt, which did not become involved in the Caledonian orogeny. They are the regions of the *North-west Highlands of Scotland* which will be dealt with in some detail, and *Central England* which will be mentioned more briefly.

The North-west Highlands and Outer Hebrides

The western seaboard of Scotland from Skye northwards, with the islands of the Outer Hebrides, belong to the *north-western foreland* which remained stable throughout the Caledonian orogeny (Fig. 14.3). In this foreland, the Pre-Caledonian rocks are grouped together under the general name of *Lewisian Gneiss*. They are strongly metamorphosed, coarsely crystalline, streaky gneisses (Fig. 16.3) resulting from the transformation, under high temperatures deep in the crust, of very ancient sedimentary successions and igneous intrusions.

We do not know much about the primary characters of the Lewisian rocks, because secondary metamorphic and structural modifications have almost destroyed them. A good deal is known, however, about the history of their metamorphic and structural changes. Putting together the evidence we can summarise this history as follows:

(i) *Before 2,600 million years*. Sedimentation followed by intrusion of basic and ultrabasic igneous rocks—giving the parent rocks.

(ii) *2,600–2,200 million years. The Scourian episode*; several phases of metamorphism and migmatisation converting the parent rocks into streaky gneisses composed in varying proportions of feldspars, quartz, pyroxenes, hornblende and biotite. These gneisses make the *Scourian complex* (Fig. 16.4) and have been dated by radiometric analysis of the metamorphic minerals.

(iii) *About 2,200 million years.* Intrusion of a swarm of *dolerite dykes* into the Scourian Gneisses and, possibly, deposition of a new sedimentary sequence.

(iv) *1,600–1,150 million years. The Laxfordian episode* of several phases of metamorphism, migmatisation and folding affecting the rocks of the Scourian complex, the post-Scourian dolerites and post-Scourian sedimentary rocks.

A few comments on this long history may be made. First, it should be noted that the oldest event of which there is evidence is one of sedimentation—this event must have taken place before 2,600 million years, the date of the first phase of Scourian metamorphism. The pre-Scourian sediments are known to have included pelitic and psammitic rocks and limestones not obviously different from sediments of later date. For sedimentation to have taken place, there must previously have been erosion and transport of weathered material and we can therefore be certain that the surface-processes concerned in these activities had already begun. Our record of geological history in Britain reveals, in Hutton's phrase, no vestige of a beginning.

A second point of interest is that the deep-seated processes of metamorphism and deformation recorded in the Scourian and Laxfordian episodes appear to have been similar in their effects to the processes accompanying orogenic activity in later geological times. The dyke-swarm formed in the interval between the two episodes was intruded when the crust was in a much more stable and rigid condition, capable of forming long parallel fractures up which the basic magma rose. Thus, there is evidence that periods of crustal mobility alternated with periods of stability, just as they have done in later geological times.

Central England

In the *south-eastern foreland* of the Caledonian orogenic belt, the Pre-Caledonian basement is exposed at a few localities where the cover of younger rocks has been removed by erosion. Chief among these *Pre-Cambrian inliers* are those of Shropshire, Charnwood Forest and the Malvern Hills (Fig. 16.5).

In *Shropshire*, a thick volcanic formation composed largely of acid lavas and tuffs is seen in a number of isolated hills such as the Wrekin and Caer Caradoc. This formation, known as the *Uriconian*, is overlain by slates, flags and greywackes (the *Stretton Series*) which in turn are covered unconformably by red sandstones and

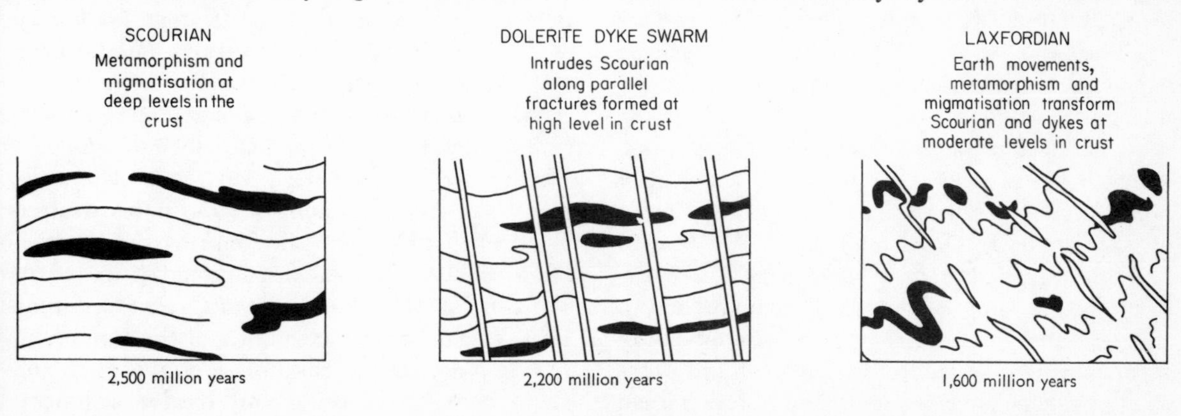

Fig. 16.4 *Three stages in the history of the Lewisian*

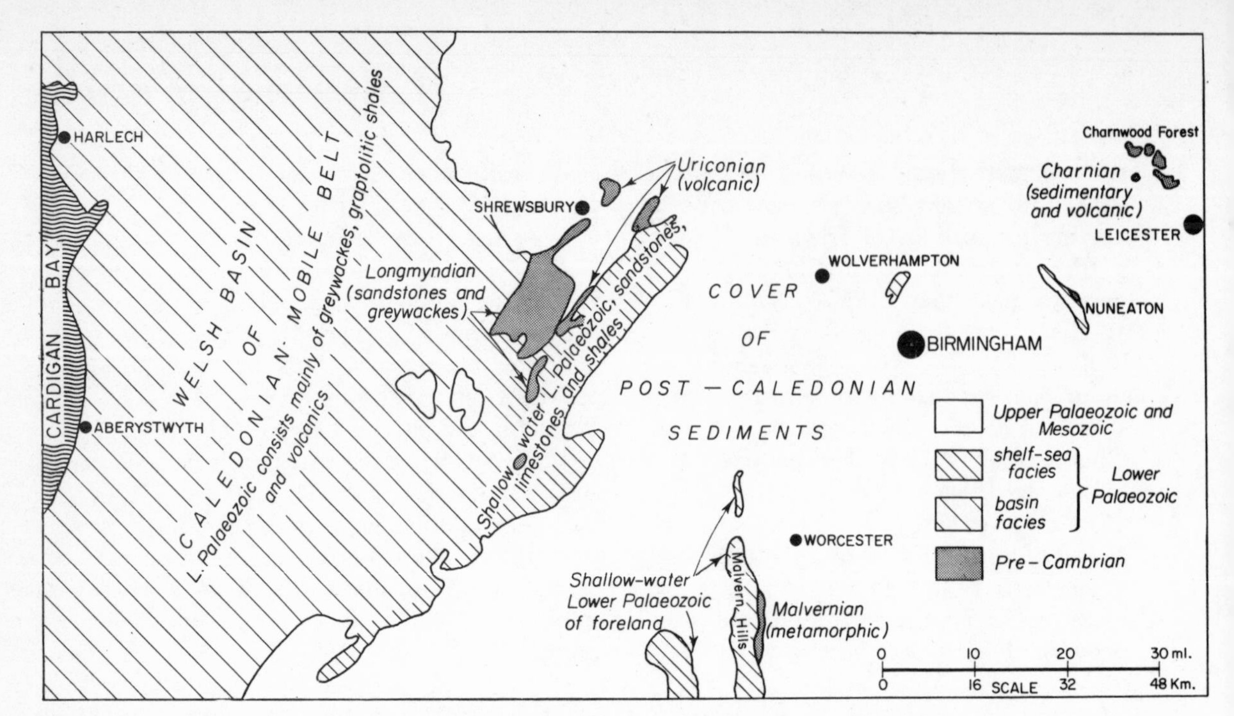

Fig. 16.5 *The south-eastern margin of the Caledonian mobile belt: parts of the south-east foreland and Welsh basin showing the facies of Lower Palaeozoic sediments. The inliers of the Pre-Cambrian basement are marked and some of the Pre-Cambrian formations in these inliers are named.*

conglomerates (the *Wentnor Series*). Both these sedimentary divisions crop out in the Longmynd and are collectively called *Longmyndian*. All the members of this succession are folded and are covered unconformably by Lower Palaeozoic.

In *Charnwood Forest*, volcanic rocks similar to the Uriconian are interbedded with sandstones and shales to make a *Charnian* formation, unconformably covered by Triassic. An interesting feature of the Charnian is the occurrence of a Pre-Cambrian fossil—*Charnia*—which appears as frond-like bodies sometimes attached to bladders interpreted as floats.

In the *Malvern Hills* the Pre-Cambrian rocks, which make a long narrow ridge, are highly metamorphosed gneisses unlike any other rocks of the south-east foreland. These *Malvernian* gneisses may belong to a fairly early Pre-Cambrian period, but evidence is accumulating to suggest that the volcanic and sedimentary formations of Shropshire, Charnwood and other smaller inliers are not older than 1,000 million years.

4. Building the Caledonian Orogenic Belt

The development of the Caledonian mobile belt began in late Pre-Cambrian times, about 1,000 million years ago, and lasted until the Devonian Period, about 400 million years ago, when the whole belt was heaved up to make a mountain-range often called the *Caledonides*. Throughout this long time sedimentation, igneous activity and earth-movement in the British area were all controlled by the way in which the mobile belt was developing. For us, therefore, *the history of the late Pre-Cambrian and early Palaeozoic is the history of an orogenic belt and its forelands*.

When mobility began, in late Pre-Cambrian times, the first part of the belt which developed a 'sinking feeling' and began to accumulate sediment was the zone alongside the north-western foreland in the region now making the Scottish Highlands and north-west Ireland (Fig. 16.6). We can call this the *Highland Zone*. Some time later, at or after the beginning of the Cambrian, basins lying to the south-east of the Highland Zone began to subside and receive sediment; there were several such basins, partly separated from one another by ridges, which received thinner sequences of sediment. Because the various parts of the mobile belt developed somewhat independently, we will deal with their histories separately. They can be listed under three headings as shown below and in Fig. 16.6. Their stratigraphical successions are compared in Fig. 16.7.

(i) The *Highland Zone*.
(ii) The *Southern Uplands*, with their continuation in Ireland.
(iii) The *Welsh Basin*, with the Lake District, south-east Ireland and the Isle of Man.

5. The Highland Zone and the North-west Foreland

A. Stratigraphical succession. When the Caledonian geosyncline began to develop in the Highland Zone, the land-surface was made of crystalline *Lewisian Gneisses* which had been deeply eroded in the period following the Laxfordian episode. The first sediments of the Caledonian sequence were therefore *deposited unconformably on the Lewisian basement.* The stratigraphical succession in the *Highland Zone* itself belongs to three major groups, as given in the table on p. 194—the rocks of these groups acquired secondary metamorphic characters which are dealt with later:

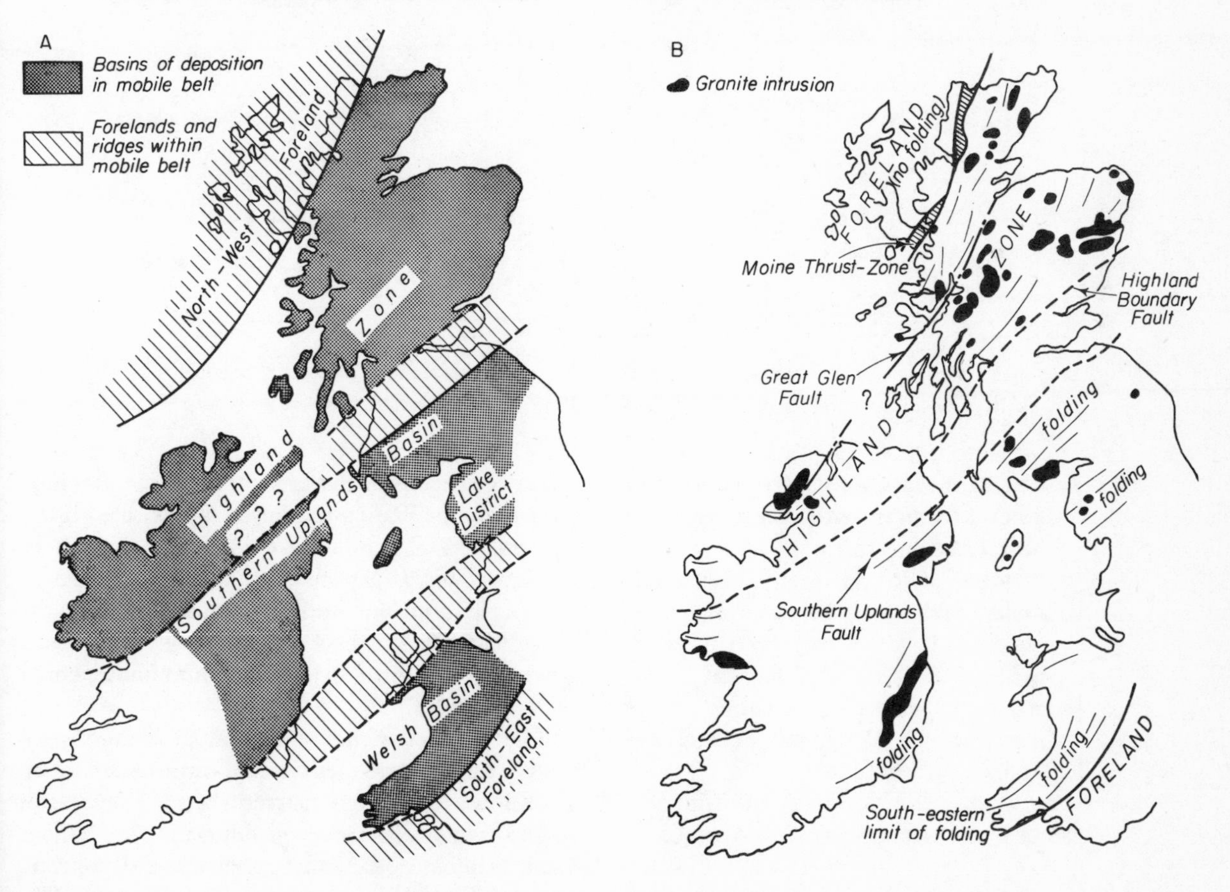

Fig. 16.6 A. *The principal regions of deposition in late Pre-Cambrian and Lower Palaeozoic times*
B. *Structures of the Caledonian orogenic belt (lines show trends of folds and related structure) and the distribution of Caledonian granites*

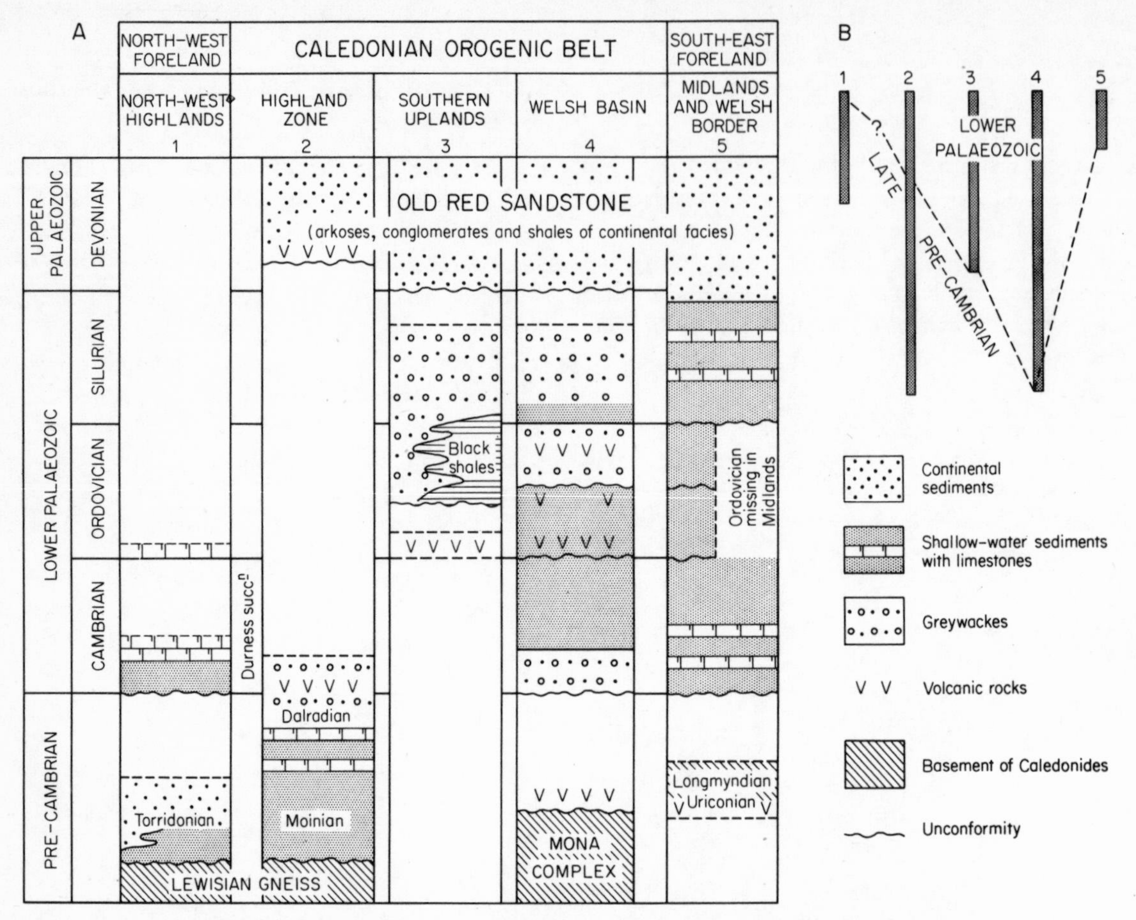

Fig. 16.7 A. *Stratigraphical successions in the Caledonian mobile belt and its forelands*
B. *The successions drawn roughly to scale to illustrate variations in thickness*

PRE-CAMBRIAN, possibly passing up into CAMBRIAN:

3. UPPER DALRADIAN: *greywackes* and other *turbidites*, with some layers of *basic lavas* and tuffs.
2. LOWER DALRADIAN: pure *sandstones*, *pelites* and *limestones* of orthoquartzite facies, with a layer of *tillite* believed to be a glacial deposit.
1. MOINE SERIES: *feldspathic sandstones* and *pelites*, deltaic or shallow-water in origin.

(unconformity)

BASEMENT OF LEWISIAN GNEISS

The characters of this succession tell us a good deal about the history of deposition. The Moine Series is a thick pile of poorly sorted clastic sediment derived from the erosion of a mountainous land-region. The psammitic layers are often current-bedded and were evidently deposited in shallow water. It is possible that the Series represents a gigantic delta built out by material brought down from the north-western part of the foreland where, as is shown below, a mountainous land-mass existed.

The *Lower Dalradian* consist of rather well-sorted sandstones, pelites and limestones. The psammites are often current-bedded and the assemblage may have been deposited in shallow water where waves and currents could concentrate sand and clay material in separate beds. The *tillite group* contains boulders and chips of granite and other rocks scattered through a sandy base.

Many geologists think that it was deposited by an ice-sheet which spread as shelf-ice (p. 96) over the sea in which the Dalradian sediments were accumulating. Similar deposits of about the same late Pre-Cambrian age have been found in Norway, Spitzbergen and Greenland; their distribution suggests that an ice-sheet comparable with those formed in the Great Ice Age covered much of the North Atlantic region.

The *Upper Dalradian*, the last group of sediments to be deposited before the folding of the Highland basin, consists of thick, badly sorted greywackes and pelitic rocks. Graded bedding and slump-structures are common and the sediments appear to have been laid down in fairly deep water by turbidity currents. Interbedded with them are basic pillow-lavas erupted under water. The appearance of turbidites and lavas in the Upper Dalradian suggests that the Highland Zone was becoming increasingly unstable and, in fact, the next thing which happened was a period of folding and metamorphism.

The stratigraphical succession so far dealt with was deposited *within the mobile belt*. It accumulated through the later part of Pre-Cambrian time and, probably, part-way through the Cambrian. Its total thickness was probably more than *eight miles*, and it is obvious that the basement in the mobile zone must have subsided so that the surface of the accumulating sedimentary pile remained just at or below sea-level.

In the *foreland* to the north-west of the mobile zone, the Lewisian basement remained much more stable; the succession laid down on it had a thickness of not more than *three miles* and is interrupted by an unconformity, indicating that the subsidence did not continue steadily. It consists of two members, the Pre-Cambrian *Torridonian* and the Cambro-Ordovician *Durness succession* whose general characters have already been discussed (pp. 171–2, Figs. 14.3 and 14.4).

The Torridonian formation is made almost exclusively of feldspathic sandstones and shales derived from rapidly eroded and incompletely weathered material and was probably deposited along the border of a mountainous region. In the northern part of the foreland, the Torridonian rests on a very irregular surface of Lewisian Gneiss which is shown by its topography to be a *pre-Torridonian land-surface*; the old stream valleys, scree-slopes and mountain crests can still be recognised under the blanket of Torridonian sediment. In the southern and south-eastern parts of the Torridonian outcrop, the succession is thicker than it is in the north and includes a higher proportion of shaly and silty rocks. This finer-grained assemblage may have been deposited farther from the mountainous source-region in a deltaic or shallow-water environment. It probably passed into the Moine Series of the Highland Zone.

The Lower Palaeozoic *Durness Succession* rests unconformably on the underlying Torridonian and Lewisian; the Torridonian strata had been gently arched and subjected to erosion before deposition began again. The basal unconformity of the Durness quartzite is a plane surface cut by marine erosion (Fig. 14.3) and the succession which rests on it is a typical series deposited during a period of marine *transgression*. A very thin conglomerate is followed by current-bedded quartz-sandstone and by beach sandstones showing the burrows of marine worms. These are followed by shaly beds laid down in a deepening sea and then by the thick Durness Limestone, deposited in waters free from detritus. The scanty fossils are mainly trilobites and gastropods.

B. Structure and metamorphism. When the main Caledonian orogenic movements began in early Palaeozoic times, the thick succession of the geosyncline was squeezed against the more stable foreland into systems of enormous folds (Fig. 16.8). At about the same time, the whole succession became hot and began to undergo metamorphism. The *main folds of the Highland Zone* are arranged with their axial planes running NNE–SSW or NE–SW, roughly parallel to the margins of the zone (Fig. 16.6). In the western parts of the zone, huge wedges of Lewisian gneiss derived from the basement were driven up into the Moinian succession, their thin ends pointing westwards towards the foreland. In the south-

Chapter sixteen

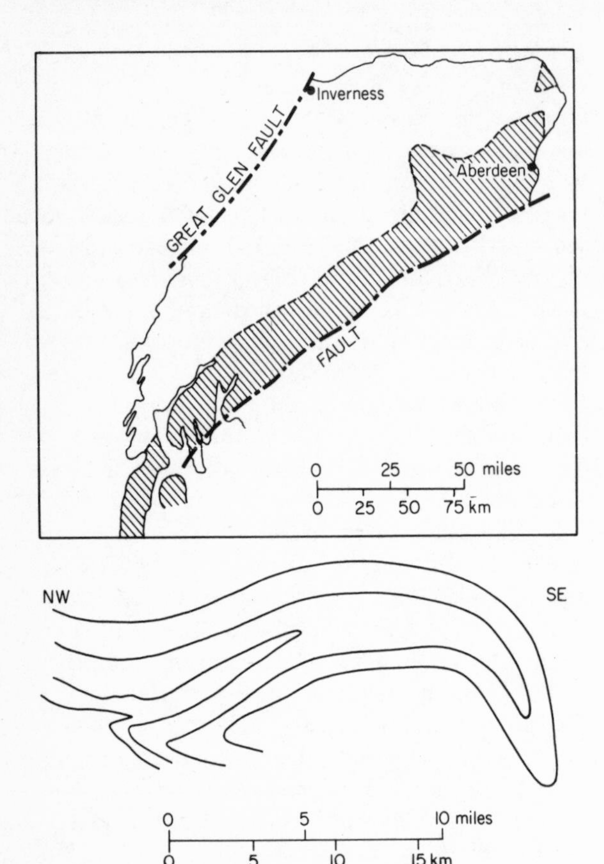

Fig. 16.8 *A recumbent fold in the Caledonian orogenic belt: the Tay nappe*

Above The shaded area represents the outcrop of the core and lower limb of the fold

Below Cross-section

eastern part of the zone, the Dalradian rocks were piled up in flat recumbent folds; the largest of these, called the *Tay nappe*, can be traced parallel to the fold-axis for more than 200 miles.

The *effects of metamorphism* in the Highland belt are shown by the conversion of Moinian and Dalradian sedimentary rocks into crystalline rocks. In the regions which reached the highest temperatures, the rocks are *gneissose* and are often impregnated with granitic material to form *migmatites*. The regions which remained cooler are characterised by *schistose* rocks or by *slates* recrystallised at still lower temperatures. The distribution of these three grades and of certain metamorphic minerals defines a series of *zones* of decreasing intensity of metamorphism (Fig. 16.9). Intense metamorphism is recorded in most of the central and lower parts of the pile of folded rocks, especially where migmatites are plentiful; in these parts, the rocks became very plastic and developed fantastic contortions during folding. Low-grade slates are seen near the margins of the fold-belt and near the top of the fold-pile.

While folding and metamorphism were going on within the mobile belt, the *foreland*, with its stable basement and relatively thin cover-succession, remained almost undisturbed and suffered no metamorphism. The contrast in behaviour between the rigid foreland and the hot softened rock-mass of the fold-belt set up stresses along the line of junction and, towards the end of the Lower Palaeozoic era, the rocks of the fold-belt were driven westward on strong thrusts over the rocks of the foreland (Fig. 14.3). This *Moine thrust-zone* described on pages 169–70 extends parallel to the fold-belt from the north coast to Skye and the Inner Hebrides.

C. Caledonian granites. The metamorphic rocks of the Highland Zone are intruded by many large bodies of granite (Fig. 16.6) which cut through or disturb the fold-structures described above. Some are known to invade unmetamorphosed Devonian rocks lying unconformably on the Dalradian, and it is evident that they were intruded towards the end of the period of orogenic activity. The prolonged heating and compression of the fold-belt provided conditions in which granitic magma could be formed in the lower levels of the crust and the partially molten material generated at depth was then squeezed upwards to make intrusive granites.

D. Late-orogenic sediments: the Old Red Sandstone. The period of folding and metamorphism in the Highland Zone culminated, at about the end of the Silurian period, in the *upheaval of the fold-belt to produce a mountain-range*. Rapid erosion followed and an enormous mass of coarse, badly sorted and incompletely

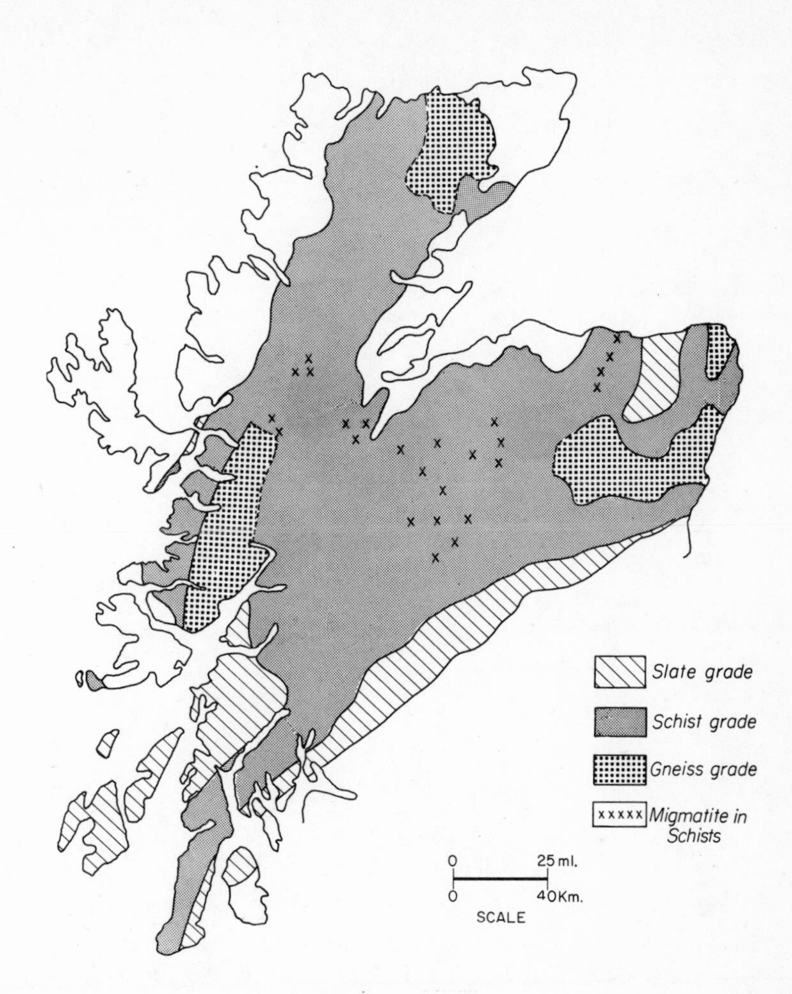

Fig. 16.9 *Zones of regional metamorphism in the Highlands of Scotland*

weathered detritus was produced. This detritus began to accumulate in low-lying areas within and on the fringe of the mountains to produce a formation of red or brown arkoses, conglomerates and sandstones, with occasional shales and calcareous shales. The whole formation, Devonian in age, is known as the *Old Red Sandstone*. It is of continental facies, representing the deposits of screes, alluvial fans, flood-plains and wide shallow lakes.

The main outcrops of Old Red Sandstone in the Highland region represent the infillings of two *intermontane basins* supplied with detritus by rivers from the mountains on their flanks (Fig. 17.1). The most northerly basin occupied the site of the *Moray Firth, Caithness, Orkney and Shetland*. Much of the sediment consists of fine sandstones, flagstones and muddy limestones laid down in great sheets of shallow water and containing many primitive fossil fish. The more southerly intermontane basin occupied the *Midland Valley of Scotland*, to the south-east of the Highlands. Here, thick beds of conglomerate were deposited in alluvial fans along the margin of the basin, while finer-grained sandstones and shales were laid down in its central part. Thick basic and intermediate lavas are interbedded with the sediments.

6. The Southern Uplands

To the south-east of the Highland Zone, folded Lower Palaeozoic rocks are seen in a broad north-east–south-west belt, extending through the Southern Uplands of Scotland and across

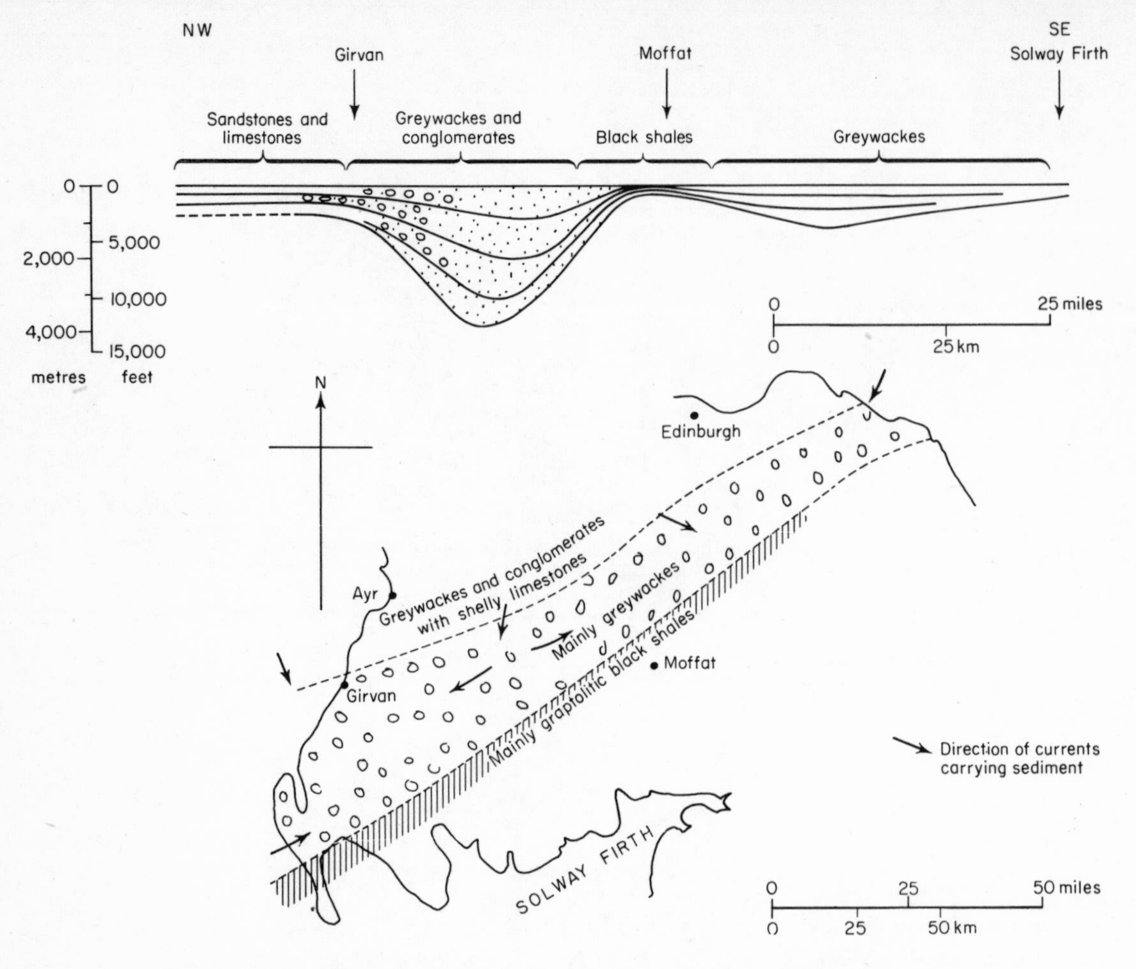

Fig. 16.10 *The Southern Uplands basin in the Middle and Upper Ordovician (based on Walton and Williams)*

Above Diagrammatic section showing variations of thickness and sedimentary facies
Below Diagrammatic map showing the facies variation and the pattern of palaeocurrents

Ireland. This belt consists of *Ordovician and Silurian* strata which appear to have been deposited in and along the margin of a deep marine basin. No Cambrian rocks are exposed, and it is probable that the basin did not begin to receive thick masses of sediment until the Ordovician. It thus began to subside at about the time when subsidence ceased in the Highland zone.

A. Sedimentary succession. The sedimentary succession in the Southern Uplands of Scotland shows remarkable *facies variations* from which it is possible to reconstruct the history of the basin. The oldest rocks visible, the *Lower Ordovician*, are basic pillow lavas associated with a little sediment. They are covered, often unconformably, by *younger Ordovician* sediments and in these we can recognise a number of different facies-belts, arranged parallel to the length of the basin, as shown in Fig. 16.10. We examine this Ordovician facies-variation more closely since it provides a good exercise in the geological method.

(i) In the *most northerly zone* as seen near Girvan, the succession consists largely of *limestones*, *sandstones* and *mudstones* laid down in shallow water and well sorted by waves and currents. The fossils are abundant shelly organisms, such as brachiopods, gastropods and trilobites. (ii) In the *next zone to the south-east*, the sediments are up to 14,000 feet thick and are mainly coarse-grained, *conglomerates* and *graded greywackes*. They represent turbidites deposited rapidly and with little sorting, and contain scarce shelly fossils. Between the turbidite-beds there are occasional layers of black shale containing graptolites. (iii) *Towards the south-east* the turbidites become finer-grained and very much thinner. Eventually they pass laterally into a very thin series of *black shales* rich in carbon. The bedding-planes of these shales are covered with *graptolites*, representing a pelagic fauna of animals drifting in the surface-layers of the sea.

This arrangement of the different belts suggests a possible palaeogeographical setting. To the north-west lay a shallow sea swept by waves and currents and populated by bottom-dwelling organisms. From time to time, the sea-floor was upheaved and erosion began which supplied much of the debris laid down in the next belt to the south-east. Here, the sea deepened rather abruptly—the floor may have been let down along active faults—and turbidity currents and slumping repeatedly brought in floods of coarse detritus. This piled up to make an enormous wedge, tapering south-eastwards away from the source of the sediment. Farther south-east still, the finest muds accumulated in deep still water beyond the reach of the turbidity currents, attaining a thickness of no more than 200 feet. The carbonaceous nature of the shales indicates that they were laid down under foul-bottom conditions in waters which were stagnant and deprived of oxygen. Few shelly organisms lived on this murky sea-bottom and the graptolite skeletons raining down from the surface layers therefore remained undamaged by scavengers.

The conditions established in the later part of the Ordovician period continued through the *early part of the Silurian*. Before the middle of the period, thick greywackes began to extend south-eastwards over the zone of black shales, so that the contrast between the greywacke and black shale zones disappeared. During these late stages, the basin seems to have been almost filled up by the influx of detritus.

B. Effects of the Caledonian Orogeny. Earth-movements disturbed the pile of strata in the Southern Uplands basin more than once—for example, after the eruption of the early Ordovician pillow-lavas—but the main phase of orogenic activity took place near the end of the Silurian period. At this time, the rocks were hardened, fractured and folded to produce great corrugations of NE–SW trend; but they did not suffer regional metamorphism, nor were they displaced by large thrusts as in the Highland Zone. After the main phase of folding, the rocks were intruded by granites (the Galloway Granites) very like those of the Highland Zone (Fig. 16.6). The presence of these granite intrusions suggests that the lower part of the crust must have been heated up and partially melted, although there was no metamorphism at the levels which are now exposed.

C. Old Red Sandstone. Erosion of the new mountain region of the Southern Uplands provided material which accumulated as *late-orogenic* Old Red Sandstone of Devonian age in the manner already discussed. Erosion debris was contributed from the Southern Uplands to the intermontane basin of the Midland Valley of Scotland and other intermontane basins may have developed on the south-east side of the Southern Uplands.

7. The Welsh Basin, the Lake District and the South-eastern Foreland

In the south-eastern part of the Caledonian belt, thick successions were laid down in *geosynclinal basins* occupying much of *Wales*, *south-east Ireland*, the *Lake District* and the *Isle of Man*. Deposition began at or even before the beginning of the Cambrian and continued, with interruptions, until about the end of the Silurian. At some time during this period, all the areas just mentioned may have been covered by a continuous sea, but at others islands and promontories divided them into a number of smaller basins. The *south-eastern foreland* remained stable throughout the period; at some times it was covered by a shallow shelf-sea and at others it stood above sea-level and contributed erosion debris to the geosyncline.

There are conspicuous differences between the succession laid down on the stable foreland and that in the adjacent Welsh basin (Fig. 16.5). On the foreland, the succession is thin, incomplete, of orthoquartzite facies and almost free from volcanic rocks. In the Welsh basin, it is very thick, largely of greywacke facies and associated with thick volcanic sequences. We illustrate these contrasts in the next few pages and in the diagram of Fig. 16.7.

Before the development of the basins, both the foreland and a large part of the future geosyncline appear to have constituted a low-lying land area. At the beginning of the Cambrian period the sea advanced over this land and deposited the first members of the Lower Palaeozoic succession unconformably on the underlying rocks. Evidence from many regions including north-west Scotland (p. 172) shows that enormous areas of the continents were flooded by shallow seas at the beginning of the Cambrian period and we may regard this *Cambrian transgression* as being due to a widespread *eustatic movement*, a rise in sea-level or a general depression of the lands.

A. Sedimentary successions. In the *Welsh basin* the *Cambrian* succession begins in the Harlech district with thick greywackes showing graded bedding and other features indicative of turbidite deposition. The greywackes are followed, in the higher parts of the Cambrian, by fine-grained flaggy sandstones and pelitic rocks which suggest a falling-off in the supply of detritus. Fossils are rather scarce and trilobites, with graptolites in the youngest beds, are the main types.

In the *foreland* region of Shropshire and the Midlands, a typical transgressive series follows the basal Cambrian unconformity (Fig. 16.7); it begins with a thin conglomerate followed by clean, well-sorted sandstones and passes up to sandstones interbedded with occasional limestones and shales, and finally to flagstones and shales. Fossils include shelly bottom-dwellers such as brachiopods and molluscs as well as trilobites and, in the highest beds, graptolites.

The ending of the Cambrian period was marked by earth-movements and uplift, as a result of which much of the basin and foreland rose above sea-level. The sea rapidly spread back over the Welsh basin where Ordovician sediments frequently rest unconformably on the underlying rocks; but most of the foreland region remained above sea-level and consequently Ordovician sediments are missing there.

The distinctive feature of the *Ordovician* succession in both Wales and the Lake District is the repeated occurrence of basic and acid *lavas and pyroclastic rocks* interbedded with marine sediments. These volcanic rocks form some of the most mountainous country, including the Snowdon and Cader Idris ranges.

The Ordovician sediments show conspicuous *facies-variations* according to the depth of water in which they were deposited and the distance from the land (Figs. 16.5, 16.11). (i) In the central parts of Wales the sediments are predominantly shales and greywackes. They are said to be of *graptolitic facies* because their only common fossils are graptolites which drifted in the surface-waters of the open sea. (ii) Near the eastern margin of the Welsh basin, and in the vicinity of islands within the basin, the sediments are largely sandstones, shales and limestones of *shelly facies* containing a fauna of bottom-living brachiopods, trilobites, molluscs, corals and calcareous algae (Fig. 16.11).

Towards the end of the Ordovician period, violent earth-movements once more interrupted deposition. The *Silurian succession* is therefore in many places unconformable on the underlying rocks. This succession is wholly sedimentary—volcanic activity had petered out towards the end of the Ordovician period—and its lower parts show facies-variations not unlike those already described in the Ordovician. (i) A graptolitic facies of thick greywackes and shales is seen in the central parts of the Welsh basin and similar shales occur in the Lake District. The greywackes, like those of the Cambrian and Ordovician, are turbidites and many of them show *slump structures*

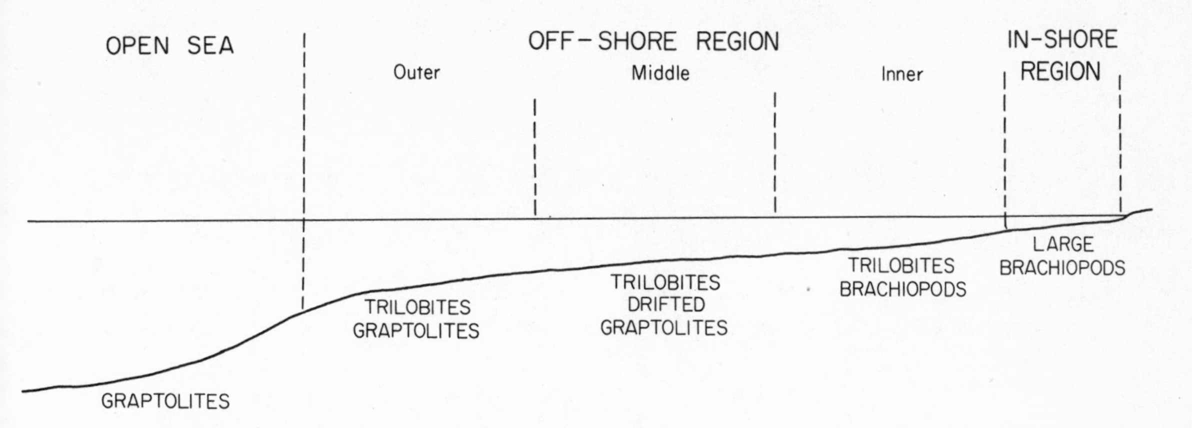

Fig. 16.11 *The Welsh basin in the Ordovician: diagrammatic section showing the relation of faunal facies to distance from the shore-line (after Elles)*

produced as a result of repeated slipping of sedimentary layers on the unstable sea-floor. (ii) Along the eastern margin of the Welsh basin and in Shropshire and the Midlands, the rocks are of *shelly facies*, consisting of clean sandstones, shales and highly fossiliferous limestones. The faunas include reef-building corals and calcareous algae, as well as echinoderms, brachiopods, trilobites and early cephalopods. These shallow-water Silurian sediments extend far over the stable foreland which, as we have seen, had remained above sea-level for most of the Ordovician period. The lowest beds are locally of *beach-facies* and can be seen, for example, around the margins of the Longmynd banked against ancient cliffs and sea-stacks.

Towards the end of the Silurian period, there began the phase of mountain-building earth-movements which finally expelled the sea from the mobile belt. The highest Silurian beds are not seen in either the Welsh basin or Lake District but, because the orogenic disturbance had very little effect on the foreland region, deposition went on in Shropshire to provide a continuous succession linking Silurian and Devonian. A progressive upward change from *marine to terrestrial facies* marks the regression of the sea. The youngest marine sediments are grey shales containing brachiopods and a few other organisms which were capable of living in brackish water. Near the top is the *Ludlow Bone Bed*, crammed with the scales and teeth of fish, together with scraps of arthropods and brachiopod shells. This strange deposit, only a foot thick, seems to mark an episode when organisms were killed off in great numbers, perhaps by a change in the salinity of the water. Above the Bone Bed, the deposits change to brown, yellow or red shales, marls and sandstones in which the majority of fossils are fresh-water arthropods and fish. It is probable that these deposits were laid down in *lagoons* cut off from the sea. They pass up into the terrestrial Old Red Sandstone.

B. The effects of the Caledonian Orogeny. By the time deposition ended in late Silurian times, an enormous thickness of Lower Palaeozoic strata, amounting in places to 40,000 feet, had been laid down in the Welsh basin and near-by parts of the mobile belt. Not more than 5,000 feet had been deposited over Shropshire and the Midland area, so that, as in the Highland region, there was a marked contrast in the level of the underlying basement. When folding began, its effects were concentrated in the mobile region and its intensity died away eastwards towards the foreland (Fig. 16.12). Within the Welsh area, the pile of Lower Palaeozoic strata was compressed to produce a number of broad anticlines and synclines whose axial planes generally run almost parallel to the eastern margin of the basin. The

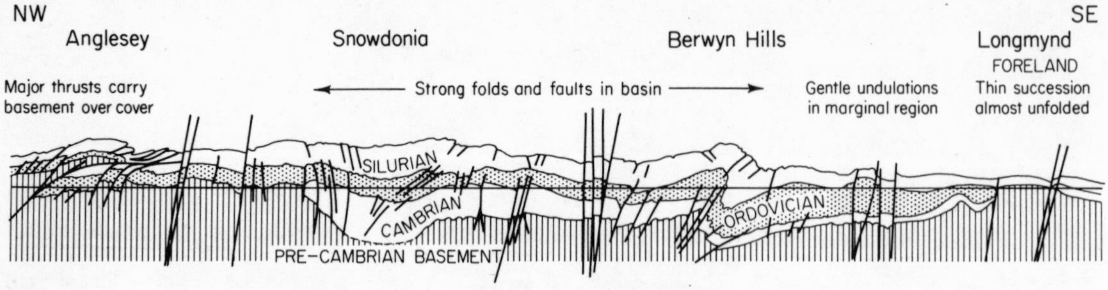

Fig. 16.12 *Caledonian folding and faulting in North Wales: diagrammatic cross-section (based on Shackleton)*

underlying basement was fractured and, in the region of Anglesey where the cover was thin, was driven up into the sedimentary pile as a thrust-sheet. These basement rocks of Anglesey and neighbouring parts of North Wales constitute the *Mona Complex*, an assemblage of gneisses, grits, quartzites, phyllites and volcanics, with intrusions of granite and gabbro. Little or no metamorphism accompanied deformation—in North Wales, the metamorphism was just sufficient to produce a strong slaty cleavage and to give the famous Welsh roofing-slates. *Granitic intrusions* similar to those of the Highland Zone and Southern Uplands are found in the Lake District and in Eastern Ireland but are lacking in Wales (Fig. 16.6).

C. Old Red Sandstone. Post-orogenic sediments of *Devonian* age were formed in the vicinity of the new mountains of Wales in much the same way as they were farther north. The bulk of these Old Red Sandstone deposits are seen in a crescentic zone extending south-westwards from Shropshire along the margin of the original basin of deposition (see Fig. 17.1). In this zone, the erosional debris from the new mountains was spread out in alluvial fans and in flood-plain, lake and delta deposits stretching southwards to the sea, which had now retreated to the latitude of the Bristol Channel. The Old Red Sandstone consists predominantly of red or brown sandstones, marls and shales containing fresh-water fish and arthropods. An unconformity within the succession indicates that slight earth-movements still continued.

8. Lower Palaeozoic Life

The Lower Palaeozoic fossil record provides a history of increasing variety and complexity in the groups which had been differentiated during the Pre-Cambrian. One very important group—the Chordates—may have appeared within the Cambrian period; the first graptolites are late Cambrian and the first vertebrates Ordovician. Among plants, none of the group of complex land-plants was in existence at the beginning of the era and the first pteridophytes or spore-bearing plants are found in Silurian rocks. This late development of land-plants had one very important consequence—throughout the Lower Palaeozoic era the lands were more or less bare of vegetation and, since there was therefore no food supply for them, they were also lacking in animal life. The Lower Palaeozoic faunas and floras are thus almost exclusively *marine or fresh-water*.

The *coelenterates*, already established in the Pre-Cambrian, included at least three groups of coral-like organisms—the *Tabulata*, *Rugosa* and *Stromatoporoidea*. These groups (Fig. 11.6) are almost confined to the Palaeozoic and first appeared as reef-builders in the Ordovician and Silurian. *Brachiopods* of the Lower Palaeozoic include a number of simple *inarticulate* forms, such as the primitive *Lingula*, which were more important at this early period than at any later time (Fig. 11.8). In addition, *articulate* forms allied to *Orthis*, *Leptaena*, *Productus* and *Spirifer* are common. Among the *Mollusca*, gastropods occur in Cambrian, Ordovician and Silurian rocks, *lamellibranchs* and *cephalopods* are present, though not of great importance, in the Ordovician and Silurian: the cephalopods were represented only by nautiloids such as *Orthoceras* (Fig. 11.13). *Echinoderms* belonged mostly to the division *Pelmatozoa* of fixed organisms.

The most important and characteristic Lower Palaeozoic fossils belong to two groups—the Trilobita and Graptolithina. The *trilobites* provide the *zone-fossils* used to divide the Cambrian succession as shown in Fig. 11.10—the primitive form *Olenellus* and its allies are characteristic of the Lower Cambrian, *Paradoxides* of Middle Cambrian and *Olenus* of Upper Cambrian. The *graptolites*, appearing at the end of the Cambrian, provide zone-fossils for the Ordovician and Silurian, successive zones being distinguished according to the number and arrangement of stipes in the colonies as already described (Fig. 11.18). Although both trilobites and graptolites lingered into the Upper Palaeozoic, their importance was very greatly reduced after the end of the Silurian.

17

HISTORICAL GEOLOGY: 3

Britain on the margin and fringes of Hercynian and Alpine orogenies

This chapter traces the evolution of Britain from the ending of the Caledonian orogeny to the Great Ice Age. During the building of the Hercynian orogenic belt, Britain lay on the margin of the mobile region: geosynclinal sedimentation and orogeny took place in the south, more restricted deposition and disturbance in the north. The Coal Measure forests were formed towards the end of this orogeny. In Mesozoic and early Tertiary times Britain remained more stable, receiving relatively thin sequences mostly of shallow-water sediments. Violent igneous activity broke out in the north and west during the early Tertiary, and in mid-Tertiary times the south was affected by the outermost ripples of the Alpine orogeny. In late Tertiary times, the most important process was the sculpturing of the land-surface by erosion and in the Pleistocene this surface was still further modified by the advance of great ice-sheets.

1. The Upper Palaeozoic Era: on the Margin of an Orogenic Belt

Before the building of the Caledonian mountains had been completed in the north-west, the development of a new mobile belt had already begun in the south. This was the *Hercynian orogenic belt* which occupied a broad east–west zone in central and southern Europe (Fig. 16.1). The first geosynclinal basins of deposition had been established there early in the Lower Palaeozoic and sedimentation continued, with interruptions produced by earth-movements, through most of the Upper Palaeozoic era. Late in the Carboniferous period, more widespread and intensive earth-movements led finally to the upheaval of a folded mountain-belt on the site of the mobile zone; in the last stages of the Upper Palaeozoic and the earliest parts of the succeeding Mesozoic era a great land area was formed occupying almost the whole of central and northern Europe. Erosional debris from this land collected to produce continental sediments comparable with those formed at the close of the Caledonian mountain-building period.

The evolution of the Hercynian orogenic belt was, in outline, very like that of the earlier Caledonian belt which has already been described, but British geologists obtain a rather different picture of the Hercynian mountains because Britain lay near the borders of the developing belt and not in its central parts. *South-west England and southern Ireland* were within the geosyncline and received thick sequences of marine sediments interbedded with volcanic rocks. *The remainder of Britain* belonged to the more stable foreland region (Fig. 16.1) and remained through much of the Upper Palaeozoic either just below or just above sea-level, receiving deposits of both marine and terrestrial facies.

The *structural and metamorphic effects* of the Hercynian orogeny were most intense in Spain, France and southern Germany in the central and southern parts of the mobile belt. In these areas schists and gneisses were produced by regional metamorphism. In southern Britain, metamorphism led only to the production of slaty rocks. *Granitic intrusions* were formed towards the end of the orogeny not only in central and southern Europe but also in south-west England. Connected with these granites were *ore-bearing veins* which have been mined as a source of tin and other metals for many centuries (p. 234).

Since the British area lay on the margin of the Hercynian mobile belt, the sedimentary rocks laid down in Britain show very conspicuous *variations in sedimentary facies*: each member of the Upper Palaeozoic succession changes its character as it is traced northward from the geosyncline into and across the foreland. From the variations in successive formations, we can reconstruct a history of the changes both in the environments of sedimentation and in the distribution of land and sea. *We shall take these changes as our main topic in discussing the Upper Palaeozoic.*

The Devonian Period

At the beginning of the Devonian period the great Caledonian mountain ranges stood high above sea-level in the northern and western parts of the British Isles. The sediments of Devonian age laid down in these regions were of *terrestrial facies*, and constituted the *Old Red Sandstone*. Devonian sediments of *marine facies* were confined to the developing *Hercynian belt* in Cornwall, Devon and the southernmost parts of Ireland. The origin of the Old Red Sandstone is, as we have seen, bound up with the history of the Caledonian orogenic belt and this formation was therefore discussed in the previous chapter (pp. 196–7). The marine Devonian, on the other hand, is a part of the Hercynian belt and as such is dealt with here.

The most southerly *marine Devonian* outcrops are seen along the south coast of the Cornish peninsula (Fig. 17.1). Here, the formation is entirely marine and largely pelitic, and is associated with basic *pillow lavas*. In the lower and middle portions of the succession, the majority of the fossils represent bottom-dwelling organisms, including brachiopods, gastropods, crinoids and corals. The occurrence of fragments of plants and of fresh-water fish-like vertebrates washed in by rivers in the lower beds suggests that land lay not far to the north. Towards the top of the Devonian succession, the free-swimming *goniatites* (Cephalopoda) become common, suggesting that the shore-line had moved northward and that the area now lay in the open sea.

Along the coast of North Devon further outcrops of Devonian rocks are seen (Fig. 17.1). These evidently lay close to the shore-line of the Devonian sea, for they consist of alternations of shallow marine, littoral and terrestrial deposits. The terrestrial sediments are of Old Red Sandstone facies and contain scraps of land-plants and teeth and scales of fresh-water fish. The marine sediments are shales and sandstones with occasional thin limestones. Their fossils include bottom-living marine organisms as well as plant- and fish-fragments derived from the neighbouring land. Some of the sandy and shaly beds, like the muds of many present-day tidal flats, are marked by innumerable worm-burrows. In contrast to the more southerly outcrop, there is little or no volcanic material.

To the north of the Cornish peninsula, the Devonian sediments are, as we have seen, almost entirely of continental facies, consisting mainly of sandstones, marls and shales laid down on the flood-plains of rivers fringing the Caledonian mountains. The alternation of sediments of this type with shallow-sea deposits in North Devon records small fluctuations in the position of the Devonian shore-line.

The Carboniferous Period

At about the beginning of the Carboniferous period an important change in palaeogeography took place. The sea now advanced northward over much of the low-lying foreland, depositing shallow-water marine sediments over most of Ireland and central and northern England. Even at this time, some portions of the old Caledonian mountains—by now deeply eroded—remained above sea-level and either received no sediment or continued to accumulate sediments of terrestrial facies.

The distribution of *Lower Carboniferous* rocks has already been discussed as an example of

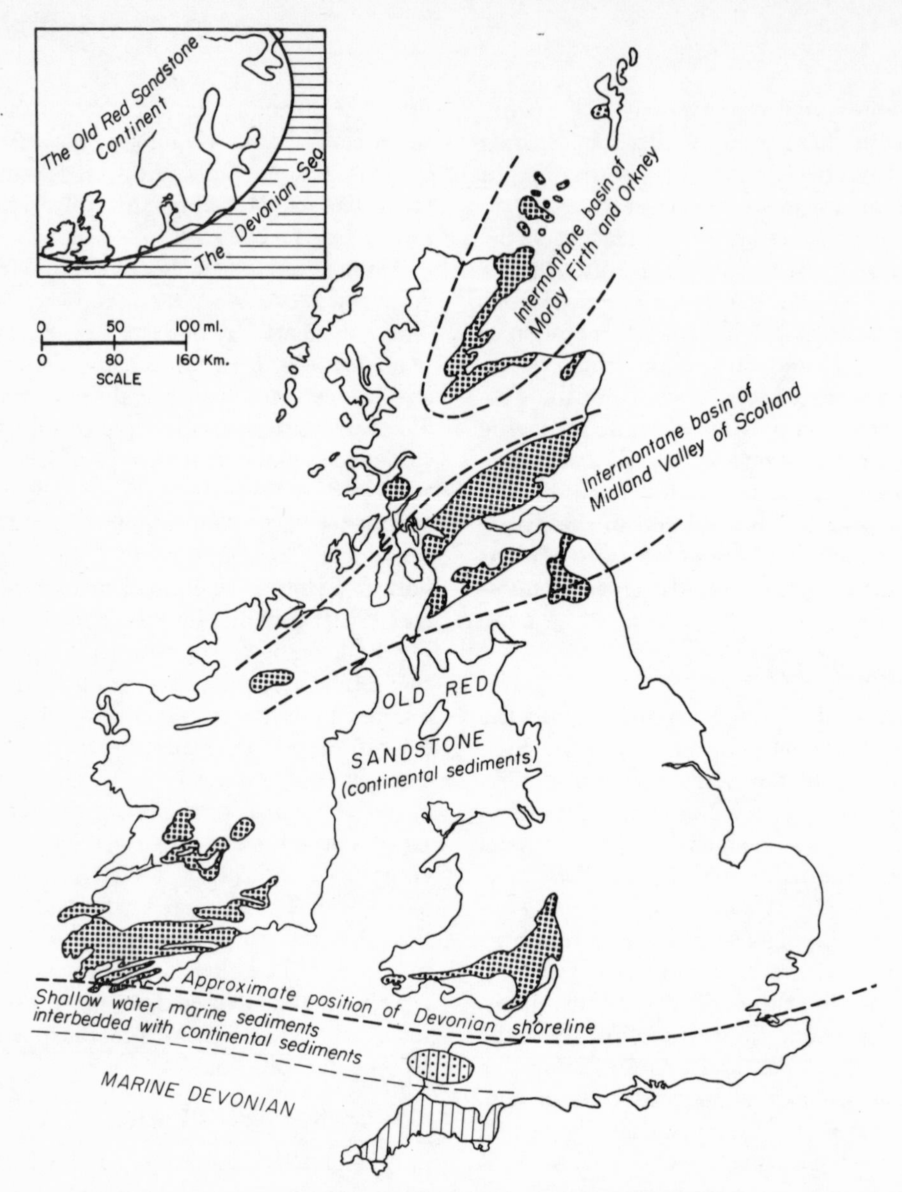

Fig. 17.1 *The principal outcrops of Devonian rocks showing the Old Red Sandstone of intermontane basins and piedmont areas in central and northern Britain, and the marine Devonian of the Hercynian mobile belt in the south*
(Inset) *The Old Red Sandstone continent of northern Europe*

facies variation (p.116, Fig. 10.7). We can now interpret these variations. The deep-water shales and sandstones of the most southerly region are deposits of the Hercynian *geosyncline*. Like the underlying Devonian, they contain occasional volcanic intercalations. The predominantly calcareous marine sediments of the Midlands and northern England are deposits of the shallow seas invading the *foreland*. 'St. George's Land', the island which stood above sea-level in Wales and parts of the Midlands, is a relic of the much larger land-area of Old Red Sandstone times.

Another relic occupied northern Scotland, and the Lower Carboniferous sediments fringing this area are sandy or muddy terrestrial deposits with which are associated basic lavas erupted from many small volcanoes.

The calcareous Lower Carboniferous formation of the foreland is known as the *Carboniferous Limestone* and is especially conspicuous in the Pennines and the Mendip Hills. Its fossils represent organisms which flourished in clean shallow seas, notably corals—including colonial reef-building tabulates and rugosa (Fig. 11.6)—crinoids, brachiopods, gastropods and lamellibranchs. Within the area covered by the Carboniferous Limestone, variations in the fauna, the thickness and the lithology—from shelly limestones to reef-limestones or muddy limestones—show that conditions were not entirely uniform and, moreover, that even the foreland was not altogether stable. Local movements on faults allowed some regions to subside more rapidly than others, with consequent differences in the conditions of sedimentation as illustrated for part of the Pennine region in Fig. 17.2.

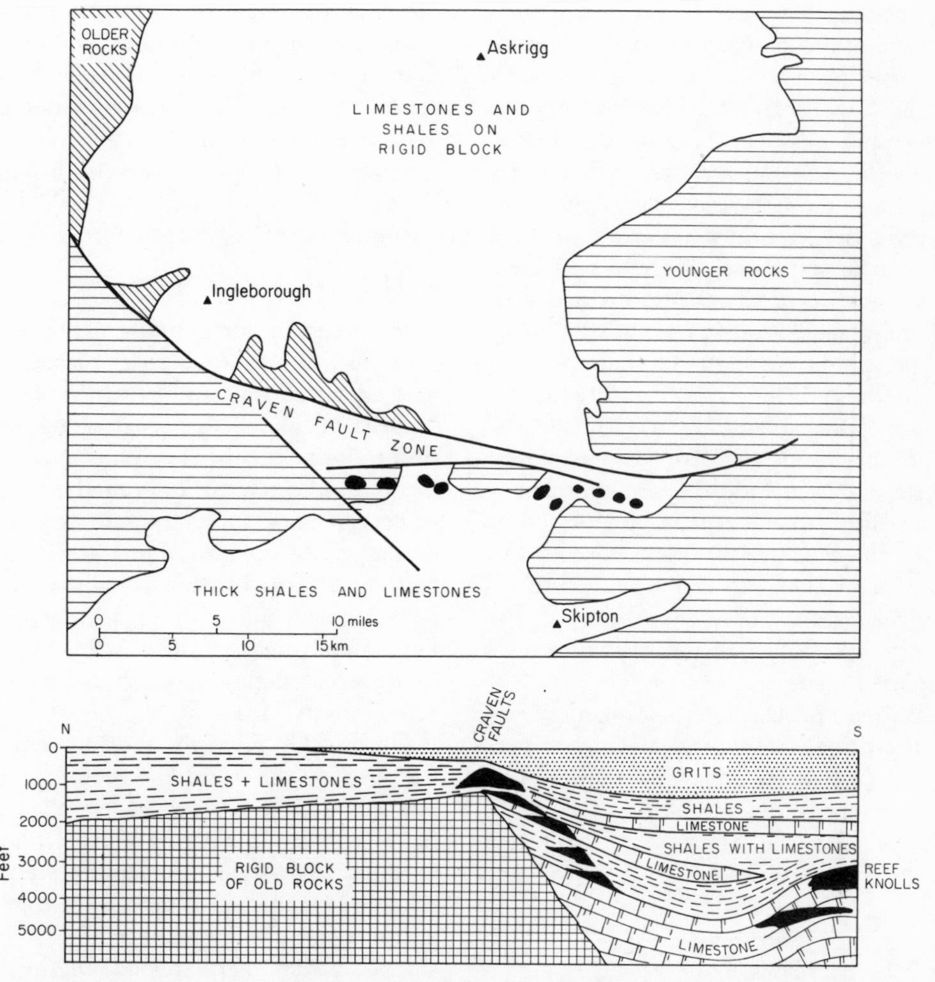

Fig. 17.2 *Facies variations in the Lower Carboniferous of the Craven Uplands, Lancashire and Yorkshire (based on Wray): diagrammatic map and section, length of section about fifty miles*

Chapter seventeen

At about the beginning of the *Upper Carboniferous*, new supplies of sandy and muddy sediment were delivered to the sea covering the foreland, probably as a result of uplift and erosion of the surviving land-areas. Limestone deposition came to an end and a great deltaic series of sandstones and mudstones—the *Millstone Grit*—was spread out over much of northern England and the Midlands. By the growth of this series the sedimentary pile was gradually built up above sea-level. Only in the *geosynclinal region* of Devon and Cornwall did uninterrupted marine deposition continue—here pelites, sandstones and greywackes containing occasional goniatites were laid down.

To the north of the geosyncline, in the *foreland region*, the principal Upper Carboniferous sediments are the *Coal Measures* laid down in flood-plains, deltas and swamps. This formation, the main source of coal in Europe, accumulated in a broad zone along the northern side of the Hercynian orogenic belt, not only in the British Isles but also in northern France, Belgium and Germany. It is made up of shales interbedded with occasional fine-grained sandstones and with coals. The fossil flora indicates that the plains were at times thickly forested (Fig. 11.22), and that fern-like and scrambling plants grew beneath the trees to produce a dense plant cover. Other characteristic fossils include the shells of fresh-water mussel-like lamellibranchs (important as zone-fossils), the skeletons of amphibia and fish and scraps of arthropods of various kinds, including primitive insects. The characters of the abundant vegetation and fauna suggest a tropical or sub-tropical climate.

The Coal Measure strata show in many places an orderly repetition of a sequence of three or four different kinds of sediment. This sedimentary *rhythm* is as follows:

TOP *Coal seam*
seat-earth
sandstone
shale
coal seam
seat-earth

This rhythm, repeated many times, records a sequence of events which began with the accumulation of sediment brought down by rivers, giving the shale and sandstone layers. In the next stage, subsidence of the surface almost ceased, and prolonged growth of forests allowed organic debris to pile up above the seat-earth or soil of the forest floor. This stage was brought to an end by more rapid subsidence which led to the burial of the forests beneath new layers of detrital sediment. Warping of the basin while subsidence was taking place led to rejuvenation of the rivers, allowing them to carry down sand as well as mud. Some of the shale layers which follow the coal seams contain marine fossils, showing that subsidence was rapid enough to drown the forests beneath the sea. These *marine bands*, usually less than twenty feet in thickness, can be individually recognised over great areas of Britain and the adjacent continent and are invaluable for correlation of the Coal Measures.

2. The Hercynian Orogeny

Mountain-building movements went on at intervals through the Upper Palaeozoic era in the parts of the Hercynian belt which lay to the south of Britain. The British part of the belt did not suffer violent disturbance until almost the end of the period. Much earlier in the Upper Palaeozoic, however, enormous transcurrent faults were developed, by northward pressure from the mobile belt, in the hard, brittle rocks of the old Caledonian mountains, which now made the foreland. The most important of these was the *Great Glen Fault* producing a lateral displacement of over 60 miles (see pp. 174, 175; Fig. 14.9).

During the main phase of orogenic activity at the end of the Carboniferous, the *geosynclinal sequence* of Devon and Cornwall was strongly folded and in some places metamorphosed to slate grade. Many thrusts and faults were formed near the northern margin of the orogenic belt especially in the region of the Bristol coalfield, in Pembrokeshire and in southern Ireland. On the *foreland*, the Upper Palaeozoic rocks were warped to produce a set of broad anticlines and synclines broken by

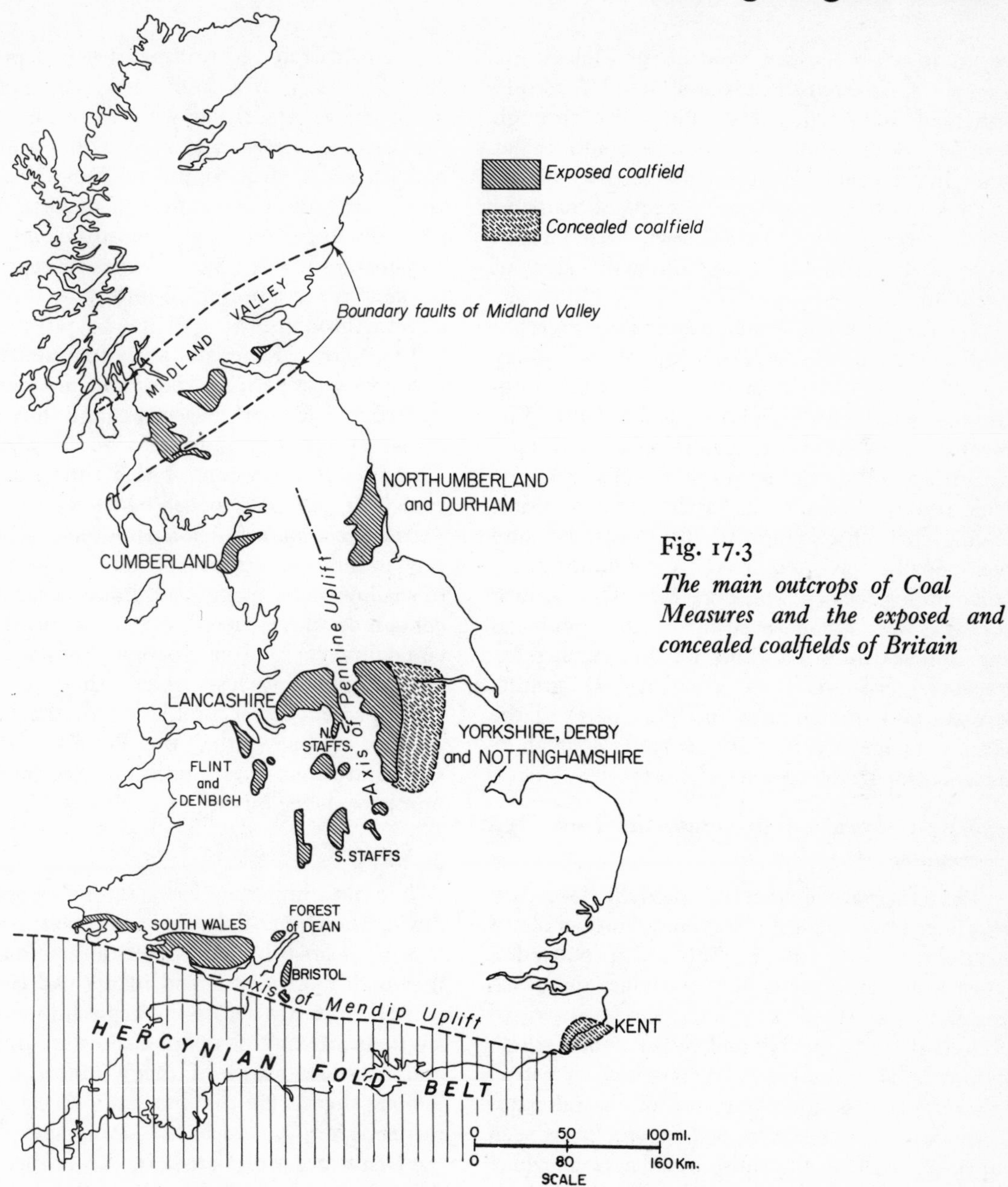

Fig. 17.3
The main outcrops of Coal Measures and the exposed and concealed coalfields of Britain

many faults. These structures are of great practical importance because, in the *period of erosion* which followed warping, the coal-bearing strata at the top of the Upper Palaeozoic were worn away on the crests of the anticlines and preserved only in the synclinal basins.

The *coalfields* of Britain are therefore mostly situated in these old synclines formed in the foreland during the Hercynian orogeny (Fig. 17.3). The main structure of the foreland was a wide low anticline running north and south along the line of the Pennines, flanked on either side by basins in which lie the coalfields of Lancashire, Yorkshire, Nottinghamshire and Derby. A second anticlinal ridge runs east and west through southern Wales and Somerset; the South Wales and Bristol coalfields lie on its northern flank. Farther north, the coal-bearing strata of Scotland are preserved in a basin let down between two old-established faults in the Midland Valley of Scotland.

Hercynian granites were intruded into the rocks of the mobile belt in the Cornish peninsula—they make the moorland areas of Dartmoor, Bodmin Moor, Land's End and the Scilly Isles. The magma which formed these granites carried small quantities of tin, copper, zinc and other elements which were concentrated in the residual fluids during the consolidation of the intrusions, and were finally deposited in veins penetrating the surrounding rock (p. 234, Fig. 18.2). The Cornish *tin mines* were worked even before the coming of the Romans to Britain and are now almost exhausted. Vapours from the residual granite magma also reacted with solidified parts of the granite to make soft, pure *china-clay* which is another important product of Cornwall.

3. Late-Orogenic Sediments: the New Red Sandstone

The upheaval which accompanied the Hercynian orogeny drove the sea from almost the whole of central and northern Europe and produced a great continental area of mountains and open plains. Erosion of the high lands provided abundant sandy, pebbly and clayey debris which accumulated to produce a formation of terrestrial arkoses, breccias, conglomerates, marls and clays. This formation, the *New Red Sandstone*, is seen especially well on the south coast near Torquay and in the Midlands where it makes fertile red-soiled land in Cheshire, Hereford and Worcestershire. The New Red Sandstone is in most respects similar to the Old Red Sandstone deposited at the end of the Caledonian orogeny. It is distinguished by the occurrence of rocks of *desert facies* including *dune-sandstones* of eolian origin and *evaporates* formed by the drying-out of inland lakes. We have already referred to these desert deposits (p. 105, Figs. 9.5, 9.6) and need not repeat the details here. Associated with them are breccias representing screes, and alluvial fan and flood-plain deposits, all of continental facies. Fossils are very scarce, the environment of deposition being inhospitably hot and dry. Almost the only organic remains are the occasional bones of reptiles, the tracks of reptiles preserved in dried-out mud and a few arthropods.

The New Red Sandstone was deposited through two geological periods—the *Permian period* at the end of the Upper Palaeozoic and the *Triassic period* at the beginning of the Mesozoic era. Throughout the whole of this time the sea invaded Britain only once, when an arm of the Permian sea spread for a time across northern England, depositing east of the Pennines a series of shallow-water magnesian limestones and marls containing a restricted fauna of brachiopods, lamellibranchs and gastropods. *Volcanic activity* was considerably less violent than that which accompanied the accumulation of the Old Red Sandstone; but during the Permian lavas and small intrusions of various kinds were formed in a number of regions.

4. Upper Palaeozoic Life

The most important event in the history of life during the Upper Palaeozoic era was the *colonisation of the lands* by both plants and animals. The first highly organised land plants had come into existence shortly before the beginning of the era. By the Carboniferous period, they included trees and fern-like plants of many different groups. Among them were the first *seed-bearing plants* or *spermaphytes* to which the dominant plants of modern times are related. In the Carboniferous, these types were still subordinate to *spore-bearing plants* or *pteridophytes* related to the modern club-mosses, horsetails and ferns. The Coal Measure forests included, for example, the huge *Lepidodendron* of the club-moss group and *Calamites* of the horsetails (Fig. 17.4).

The colonisation of the land by plants opened the way for a similar migration of animals and the

Fig. 17.4 *A Coal Measure forest*

Reconstruction showing tree-lycopods such as *Lepidodendron*, Articulatales such as *Calamites* and early ferns and seed-bearing plants

effects of this advance were shown by the appearance during the era of the first groups of terrestrial vertebrates—the *amphibia* and *reptiles* —and the first *insects*.

Among the *invertebrate marine faunas* great changes took place. The trilobites and graptolites were now rare, and both groups became extinct before the end of the Upper Palaeozoic. *Brachiopods*, on the other hand, had become more varied, the more complex articulate types increased in importance and most inarticulate types died out. *Corals* were prominent, especially the rugose corals which built numerous reef-knolls in the shallow seas in which the Carboniferous Limestone was deposited. *Echinoderms*, *gastropods* and *lamellibranchs* continued to develop without spectacular changes. Freshwater lamellibranchs rather like the fresh-water mussels of today are used as zone-fossils in the Coal Measures. Molluscs of the sub-division *Cephalopoda* (p. 135), evolved rapidly and became, by the end of the era, among the most important marine animals. The common Upper Palaeozoic forms were tightly coiled *goniatites* with simple zig-zag sutures (Fig. 11.14); these forms provide zone-fossils for marine successions, but as they lived mostly in the open seas they are not widely distributed in Britain. A group of large foraminifera, the *fusilinids*, provide zone-fossils for Upper Carboniferous and Permian marine successions in some parts of the world (Fig. 11.5).

5. The Mesozoic Era: Shallow Seas and Islands

Once the Hercynian orogeny had come to an end, the British area never again formed part of a mobile belt. The later history of this area is one of limited deposition, usually in shallow seas and under comparatively stable conditions, and of the gradual emergence and shaping of the modern land forms. Orogenic activity, of course, had not ceased throughout the world; during the Mesozoic a great new *Alpine geosyncline* was developing

Chapter seventeen

in southern Europe in the region where the Alps, the Apennines and other mountain tracts of the Mediterranean region were later to be formed.

The Mesozoic era is divided into three periods —the *Triassic*, the *Jurassic* and the *Cretaceous*. The Triassic deposits in Britain are part of the *New Red Sandstone* and have already been dealt with. At the end of the Triassic or beginning of the Jurassic, a great change was marked by the *return of the sea*. The long erosion of New Red Sandstone times had worn down the British area to a low-lying plain over which the advancing Jurassic sea later spread widely. The higher parts of the old plain still stood above sea-level as islands (Fig. 17.5); among them were the old resistant blocks of St. George's Land and northern Scotland and new high areas in Ireland and south-west England. The first marine deposits, the *Rhaetic* (classified as Jurassic by some geologists) were laid down in a dirty, badly aerated sea and contain a restricted fauna of stunted lamellibranchs.

The Jurassic Period

In Britain, the Jurassic period began with the *transgression* just noted and ended with a *regression* and the emergence of low swampy lands. The succession deposited during the period has a maximum thickness of about 2,500 feet and represents a single *cycle of marine deposition*. Many of the Jurassic strata are very fossiliferous and have been studied in very great detail—they were in fact the strata from which William Smith first deduced the two fundamental laws of stratigraphy. We therefore know a good deal about the small variations in lithology and fauna which record variations in the environment of deposition. It would be impossible to give all these details here, and we must confine ourselves to a few examples.

The Jurassic succession in Britain, with the names of the main divisions, is shown diagrammatically in Fig. 17.6. It will be seen that this succession is made very largely of *clays and limestones* and we can therefore deduce that the islands which stood above sea-level were low-

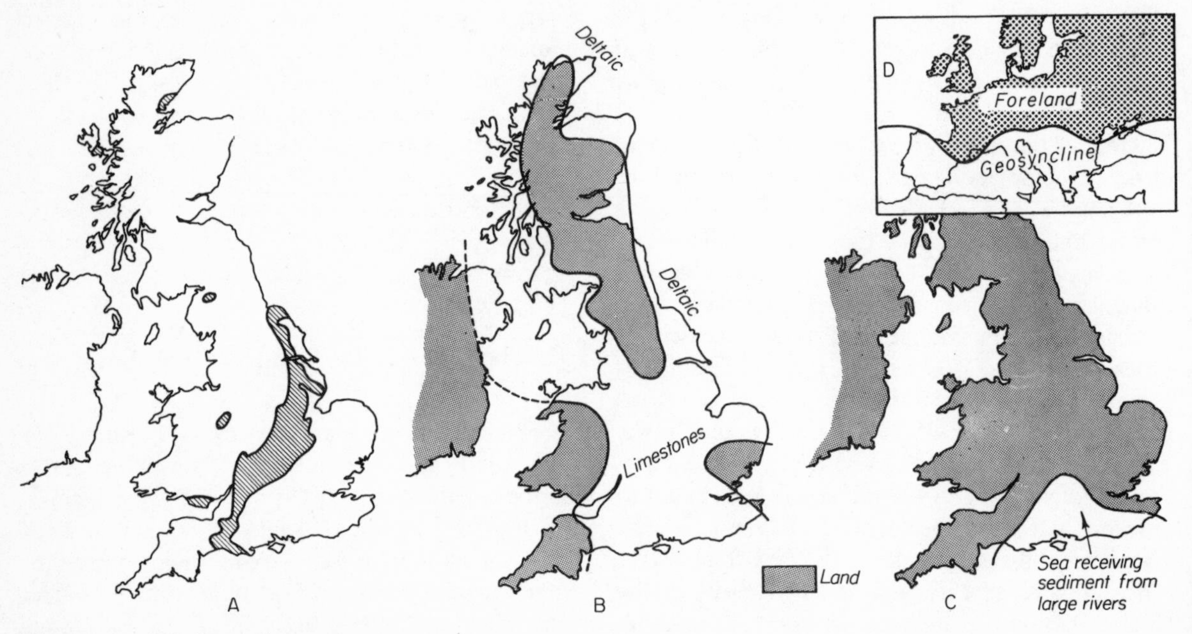

Fig. 17.5 *The Jurassic system*

A. The main outcrops of Jurassic
B. Palaeogeographical map for mid-Jurassic times
C. Palaeogeographical map for late Jurassic times
D. (*inset*) The position of the Alpine geosyncline during the Mesozoic era (after Wills)

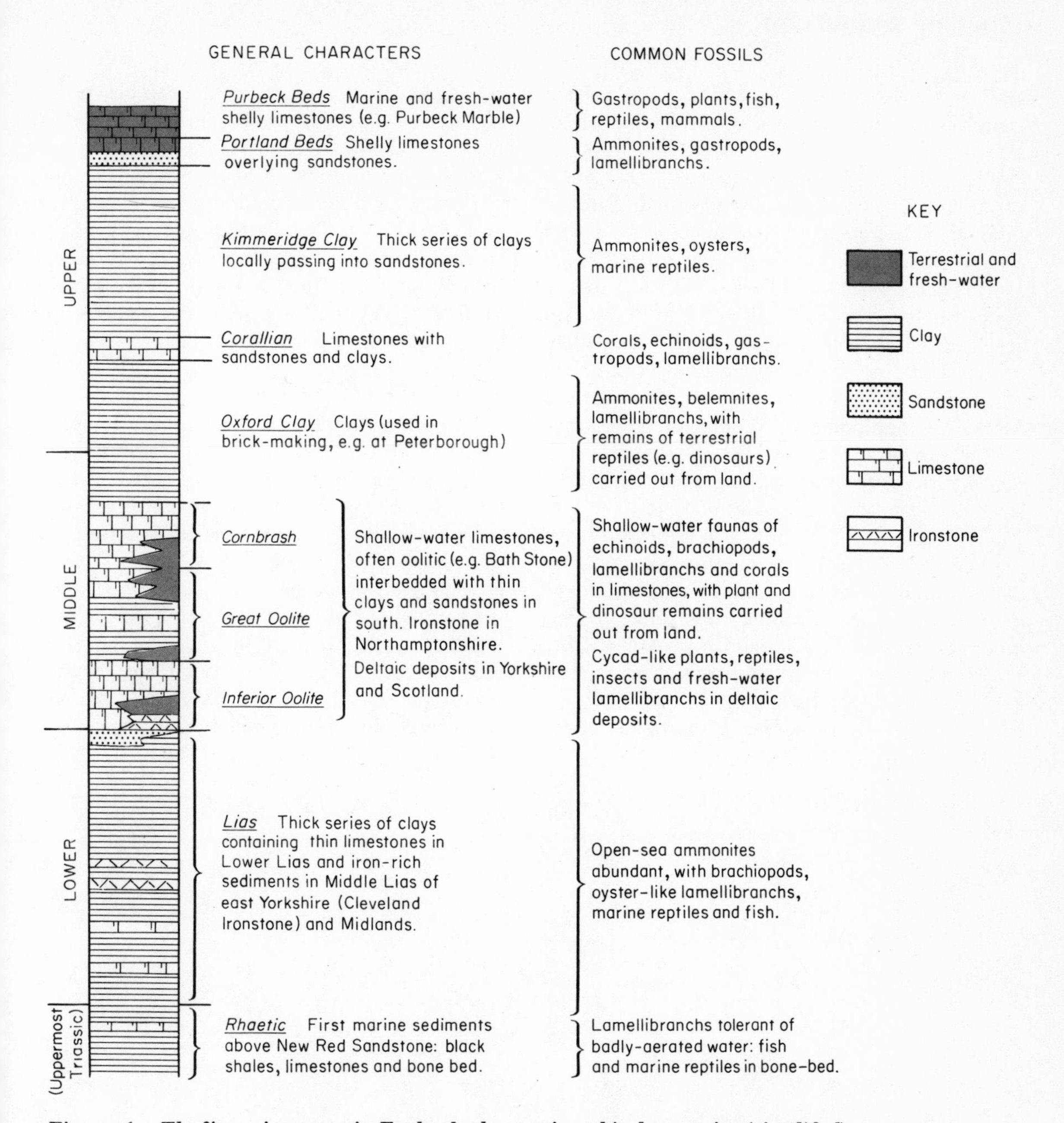

Fig. 17.6 *The Jurassic system in England: the stratigraphical succession (simplified)*

lying and that their rivers supplied little coarse sediment to the sea.

The *Lower Jurassic* clays which make the *Lias* were deposited in moderately deep water and contain among their fossils abundant *ammonites*, used as zone-fossils throughout the Jurassic. They are the first members of this very important Mesozoic group—descendants of the goniatites—which are found in Britain, although ammonites had been common in the geosynclinal seas of southern Europe during the Triassic period. The Lias also contains beds of *ironstone*, rich in siderite ($FeCO_3$) and the complex iron silicate chamosite, which are worked in Yorkshire as iron-ores.

The *Middle Jurassic* sediments are more variable and include in Southern England a number of limestones which are especially well seen in the beautiful ridges of the Cotswold Hills. During the deposition of these sediments the sea appears to have become shallower. The limestones are often oolitic (e.g. the Bath Stone, once widely used for building) and contain fossils of organisms which lived in clean shallow waters. Towards the north, the predominant Middle Jurassic limestones give way to iron-rich sediments (the *Northamptonshire ironstone*) and finally in Yorkshire to deltaic and fresh-water deposits.

The beginning of the *Upper Jurassic* was marked by a deepening of the sea and the deposition of thick clays containing abundant ammonites. The *Corallian* limestone which separates the *Oxford Clay* and *Kimmeridge Clay* is of shallow-water origin and consists in part of coral-reefs with their communities of molluscs and echinoids. After the deposition of the Kimmeridge Clay, regression of the sea began. The succeeding *Portland Beds* consist of shallow-water sandstones and limestones (*Portland Stone*, an important building stone, makes the peninsula of Portland Bill) and are confined to the southern part of England. The *Purbeck Beds* which follow are fresh-water limestones deposited in lagoons behind the retreating shore-line. The so-called *Purbeck Marble* used as a decorative stone is made largely of fresh-water snail-shells.

The Cretaceous Period

By the beginning of the Cretaceous, the sea had retreated from most of Britain leaving in the south and east a broad coastal plain. The first Cretaceous deposits were deltaic and fresh-water sediments laid down on this plain. After some time, however, the sea advanced again over much of the region which had been submerged during the Jurassic leaving, as before, large islands in the west and north. At about the middle of the period, the sea spread much more widely, covering almost the whole country. This mid-Cretaceous advance of the sea coincided with a very wide-

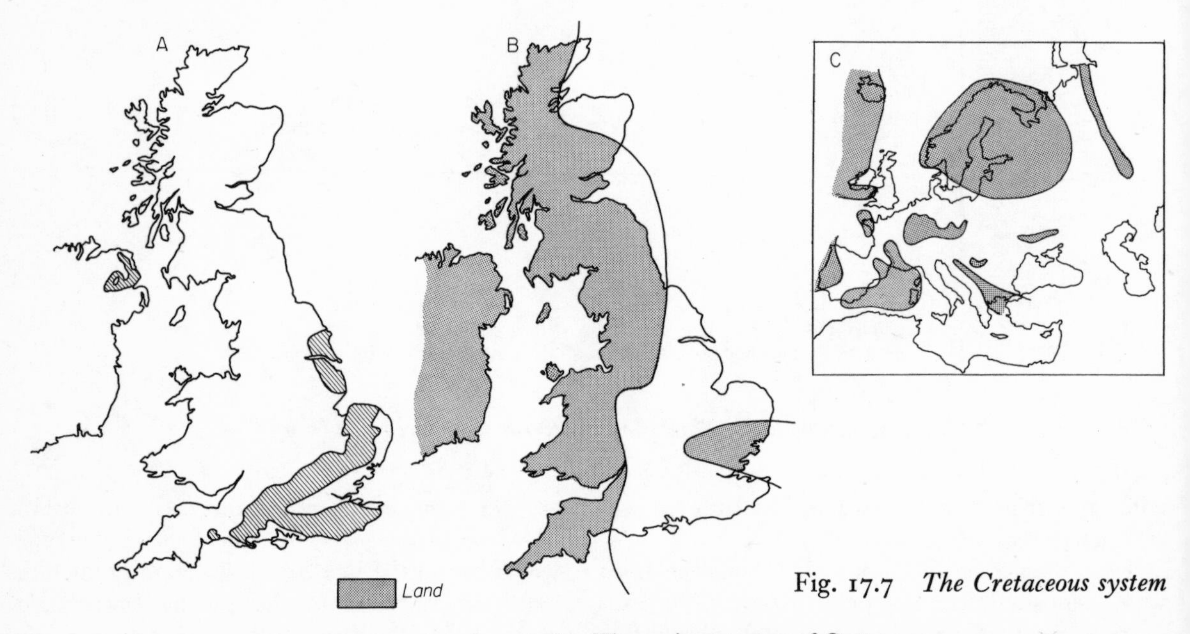

Fig. 17.7 *The Cretaceous system*

A. The main outcrops of Cretaceous
B. Palaeogeographical map for mid-Cretaceous times
C. Europe after the Cenomanian transgression (after Wills)

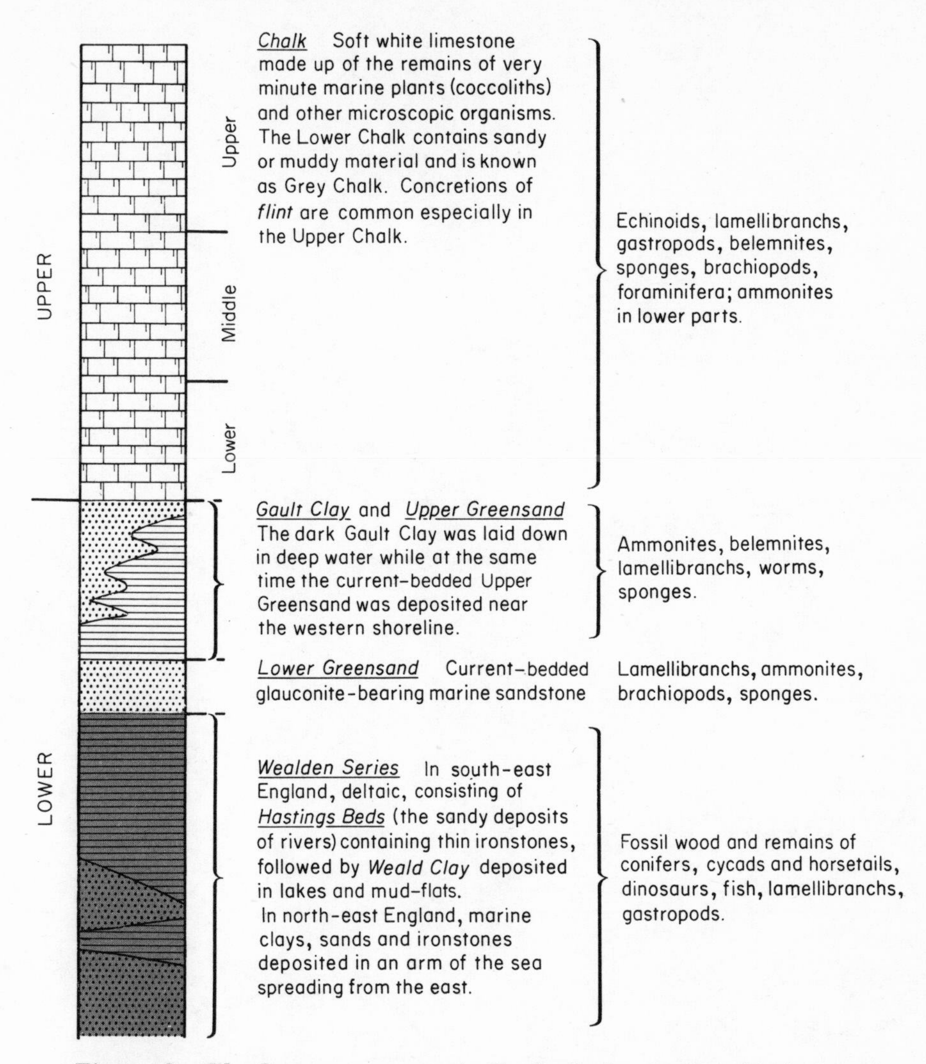

Fig. 17.8 *The Cretaceous system in England: the stratigraphical succession (for key see* Fig. 17.6)

spread *Cenomanian transgression* during which vast areas of Europe were flooded (Fig. 17.7).

The Cretaceous succession in Britain is represented in a simplified form in Fig. 17.8. The deltaic *Wealden Series* at the base is best seen in the Weald of south-east England, and consists of sandstones and clays with thin limestones and ironstones—these latter were mined as iron-ore from prehistoric times until the Industrial Revolution. The Wealden sediments contain plant-remains, reptile bones and fresh-water gastropods and lamellibranchs. The overlying *Lower Greensand* marks the return of the sea and consists of shallow-water marine sandstones and clays containing shelly fossils, especially brachiopods and lamellibranchs. The green coloration is due largely to the iron mineral *glauconite* formed on the sea-floor during sedimentation. After deposition of the Lower Greensand, the sea deepened and spread towards the west and north during the first stages of the *Cenomanian transgression*. The sediments laid down in its deeper

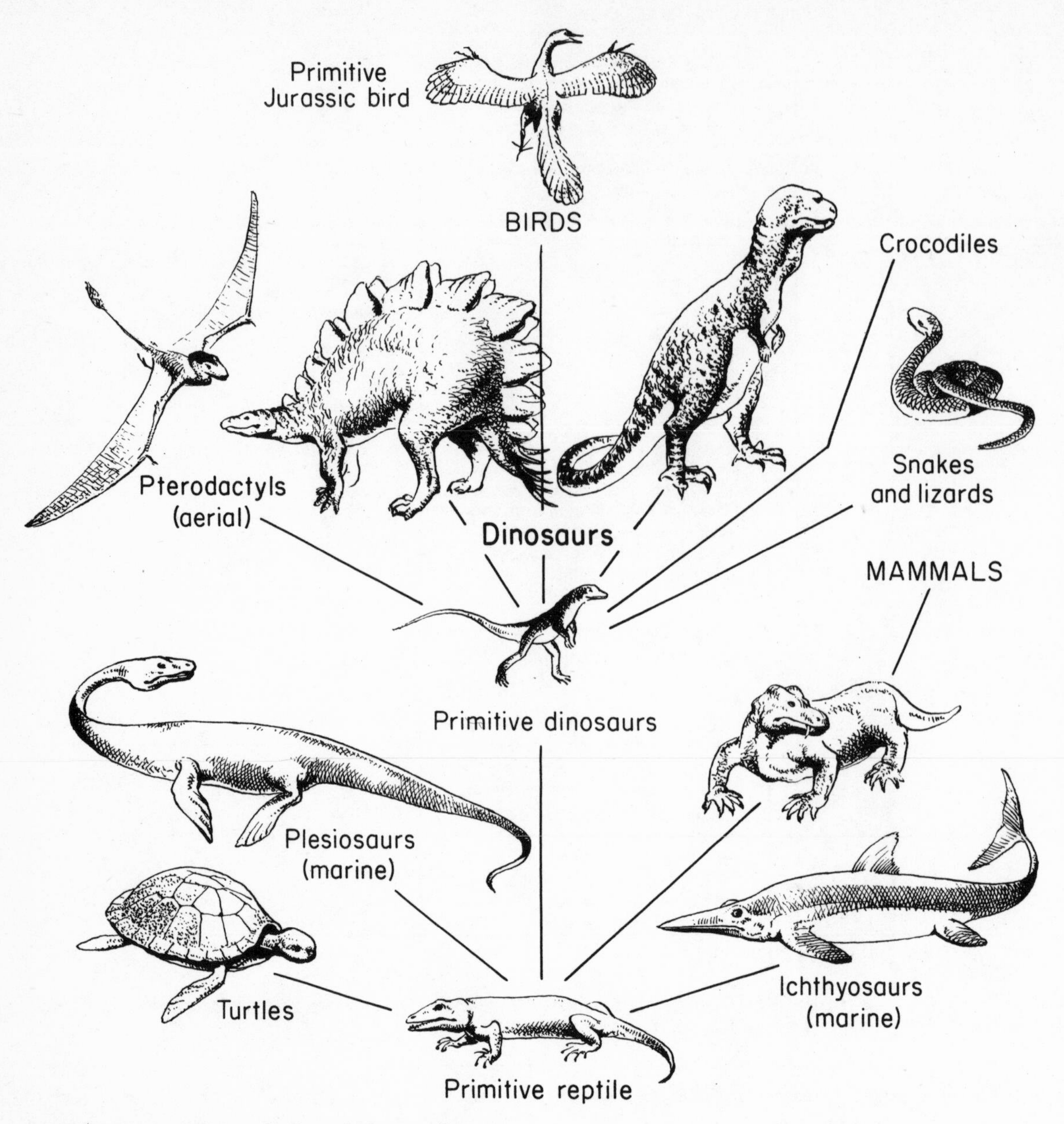

Fig. 17.9 *The evolution of the reptiles*

parts—that is, in Kent and Sussex—are mostly clayey, with ammonites and belemnites as well as shelly fossils, while those deposited at the same time near the margins of the sea—especially in Dorset—are largely greensands: thus, two formations of roughly the same age *but different sedimentary facies* were laid down, the deeper-water *Gault Clay* and the near-shore *Upper Greensand.* Because the sea of Gault times was more extensive than that of the Lower Cretaceous, its deposits overlap in many places on to pre-Cretaceous rocks and rest unconformably on them.

The formation of the Gault and Upper Greensand was followed by the deposition of soft white limestones which make up the remainder of the Upper Cretaceous. These limestones constitute the *Chalk*, whose characteristic scenery of rounded hills and dry valleys is seen in the great ridges of the North and South Downs, the plateau of Salisbury Plain and the belt of hills extending from the Chilterns to the Wolds of Yorkshire. Isolated outcrops are preserved as far north as Mull and west as Antrim. The great extent of the Upper Cretaceous sea (Fig. 17.7) is reflected in the uniformity of its deposits and in the fact that these deposits are almost entirely organic. The Chalk is a pure limestone, of very fine texture, made largely of the remains of submicroscopic plants known as *coccoliths*. Among larger fossils, echinoids are common enough to be used as zone-fossils, while brachiopods, lamellibranchs and belemnites are also widespread. The accumulation of the fine calcareous mud of the Chalk in quiet seas, into which no coarse sediment was being carried, brought the Mesozoic era in Britain to a peaceful end.

6. Mesozoic Life

In the course of the Mesozoic, the faunas and floras of the land and sea acquired a much more modern character. On land, the seed-bearing plants or *Spermaphyta* virtually ousted the pteridophytes. They were represented, however, mainly by the rather primitive conifers, cycads and allied groups: the flowering plants did not become important until late in the era. Among land-animals, the *reptiles* enjoyed pre-eminence and became specialised for life in a wide variety of environments (Figs. 17.9, 17.10). Both the *mammals* and the *birds* came into existence during the Mesozoic, but the great expansion of these groups did not take place until the Tertiary.

In the sea, the distinctive Palaeozoic groups of graptolites and trilobites and the tabulate and rugose corals had all died out. The *Mollusca* were now perhaps the most important phylum. *Ammonites* (Fig. 11.14) descended from the Upper Palaeozoic goniatites and *belemnites* (Fig. 11.15), both belonging to the *Cephalopoda*, were characteristic Mesozoic groups both of which became extinct at the end of the Cretaceous. *Lamellibranchs* (Fig. 11.12) and *gastropods* also increased in both importance and variety. The *brachiopods*, on the other hand, became restricted to a few divisions, notably complex types with brachial supports, related to *Rhynchonella* and *Terebratula* (Fig. 11.8). *Corals* were now represented by the *Scleractinia* (Fig. 11.6). Among the *Echinodermata* the *echinoids* increased greatly and now included both regular and irregular types (Fig. 11.17); but the fixed Pelmatozoa died out with the exception of the crinoids, and these were becoming increasingly uncommon.

Fig. 17.10
A Mesozoic reptile, Ornithosuchus, from northern Scotland: length about three feet

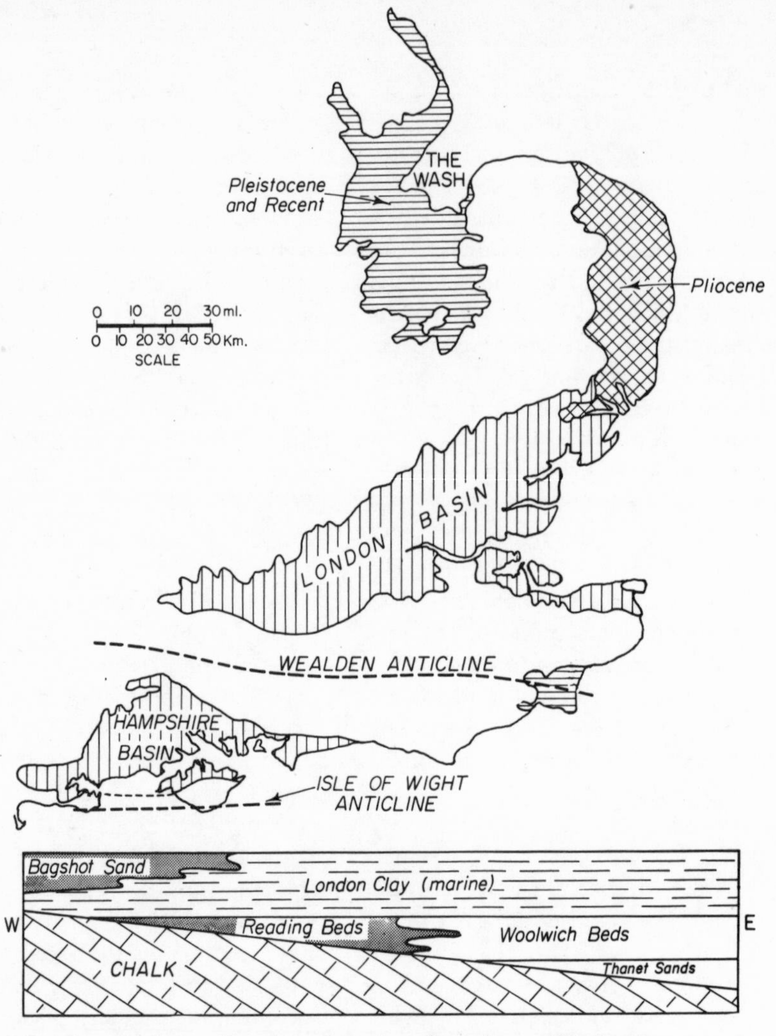

Fig. 17.11 *Tertiary and post-Tertiary deposits of south-east England*

Above The main outcrops and fold-structures

Below Diagrammatic section illustrating facies variations in the Palaeogene of the London Basin (based on Ager)

7. The Tertiary Era: the Making of the Lands

At about the end of the Mesozoic era, the long period of Chalk deposition was halted by gentle uplift and the sea retreated from much of the area which it had covered during the Upper Cretaceous. From this time onwards, deposition was restricted to relatively small parts of the British area, taking place principally in East Anglia and south-east England. Here, sequences of marine clays and sandstones were laid down, passing locally into estuarine or brackish-water deposits towards the west and north; the total thickness is little more than 1,000 feet. For most of the rest of Britain the Tertiary was, on balance, a time of erosion during which the sea made only limited advances—its greatest interest for us lies in the fact that it was the period in which the main features of the British land-surface were finally blocked out.

In Southern Europe the *Alpine mobile belt* which had developed as a geosyncline during the

Mesozoic and early Tertiary was subjected to violent orogenic folding, metamorphism and upheaval in mid-Tertiary times, with the production of the huge mountain ranges of the Alps, Apennines and Carpathians. The great *Alpine orogeny* was felt through much of the Alpine-Himalayan and circum-Pacific mobile belts (Fig. 2.8) and its effects have not yet entirely died away; as we have seen earlier (p. 14), these belts are still subject to unusually frequent earthquakes and volcanic eruptions. Britain, however, lay far beyond the limits of the Alpine mobile belt and the effects of orogenic pressures were shown only by the development of a few open folds in the new Mesozoic and Tertiary strata of Southern England. The chief of these structures are the *Wealden anticline* flanked by the *London Basin* and *Hampshire Basin*; and a sharp but localised monocline crossing the Isle of Wight which tilts the strata into a vertical position (Fig. 17.11).

The Tertiary era, however, was not a time of peace and quiet throughout the whole of the British Isles. In western Scotland and northern Ireland, igneous activity broke out on an impressive scale. This outburst, following on a long period of quiet, was associated with similar activity over most of the North Atlantic region and the igneous rocks produced by it are said to belong to a *North Atlantic Tertiary igneous province* (Fig. 17.12). This widespread activity,

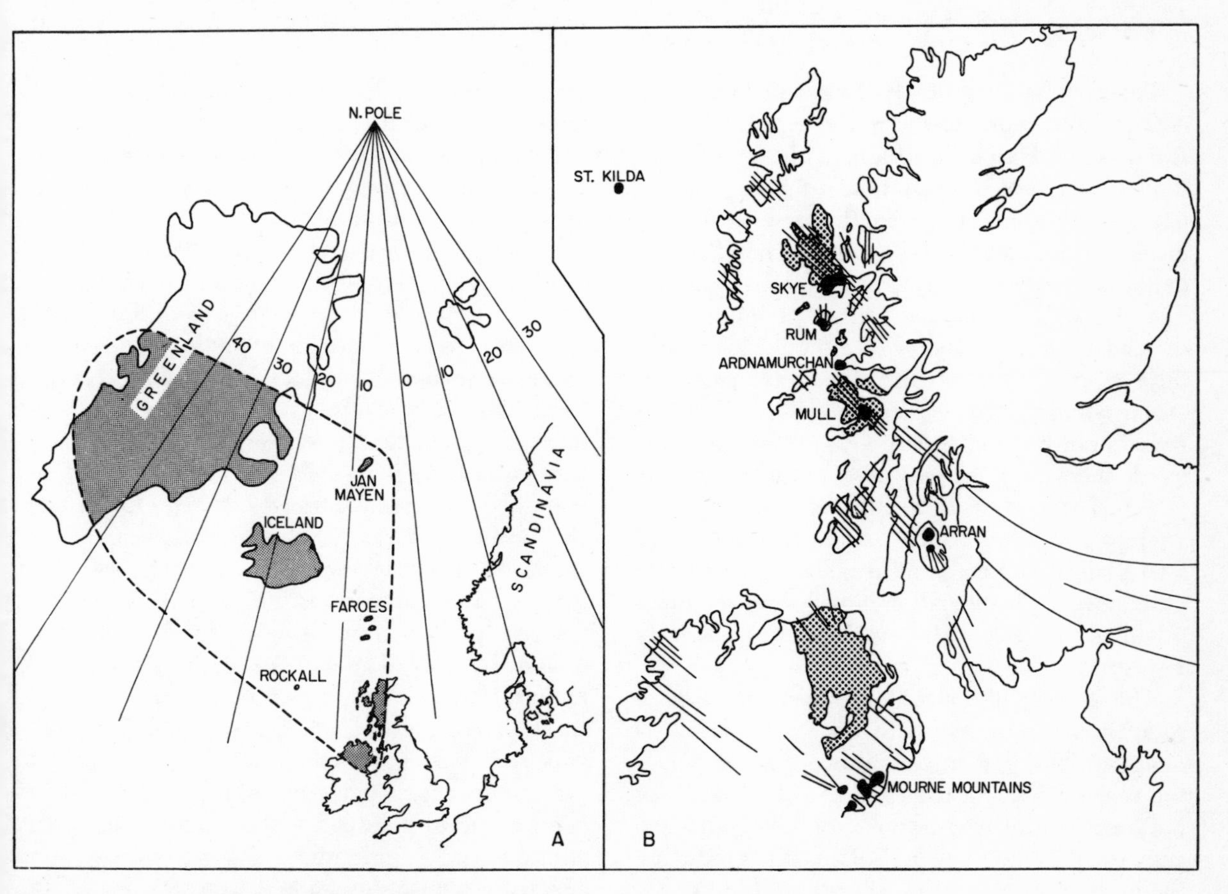

Fig. 17.12 *Tertiary igneous activity*

A. The North Atlantic igneous province

B. Tertiary igneous rocks in northern Britain showing plateau lavas (stippled), plutonic intrusions (black) and some of the dykes (lines)

which was predominantly basic in character, was probably connected with the enlargement of the Atlantic Ocean resulting from the westward drift of the continental masses of North America and Greenland. The disturbance of the deep parts of the earth connected with this continental drift may have been responsible for allowing magma to form and rise to the surface.

The Tertiary era is traditionally divided into four geological 'periods', the Eocene, Oligocene, Miocene and Pliocene. Each of these, however, was very short compared with the periods of the Palaeozoic and Mesozoic eras, and for our purposes it is better to adopt a twofold division of Tertiary times as follows:

Tertiary { *Neogene*=Miocene+Pliocene
Palaeogene=Eocene+Oligocene

Tertiary sediments in Britain. *Palaeogene deposits* were laid down in the regions of the London and Hampshire Basins (Fig. 17.11) in a shallow sea which from time to time became partially filled in by the south-eastward growth of deltas and lagoonal deposits. The Tertiary deposits rest with slight unconformity on the underlying Mesozoic and consist predominantly of sandstones and clays with a total thickness of little more than 1,000 feet. They are partly of shallow-water marine facies, with abundant lamellibranchs and gastropods. Especially towards the west, however, these marine sediments are interbedded with the deposits of lagoons, swamps and rivers which contain the remains of land-plants, mammals and fresh-water snails and other shells. This variation in facies can be illustrated by a schematic section as shown in Fig. 17.11.

The ending of Palaeogene deposition corresponded more or less with the great period of mountain-building in the Alpine mobile belt—the repercussions of the episode caused much of northern Europe to rise above sea-level, putting a stop to sedimentation in Britain and producing there the fold-structures shown in Fig. 17.11. This *mid-Tertiary upheaval* effectively restricted the sea to areas which were little more extensive than those of the present North Sea and English Channel. *Neogene deposits* are therefore found only in a few localities near the present sea-coasts. The most important are the *Crags* of Pliocene age in East Anglia: these rocks are shallow-water shelly sands packed with the remains of gastropods and lamellibranchs not unlike forms which still inhabit the adjacent parts of the North Sea.

Tertiary volcanic activity. The great outburst of basic vulcanicity in the North Atlantic Province began in early Tertiary times and has not yet died away—volcanic eruptions are still frequent in Iceland though they have long since ceased in Britain, where activity was largely of Palaeogene age. During this outburst, thick piles of *plateau basalt* were erupted from fissures and central volcanoes to make the Atlantic islands of the Faroes, Iceland and Jan Mayen, as well as large areas near the coasts of north-west Britain and southern Greenland. *Plutonic intrusions* were emplaced in the roots of the volcanic centres and at other sites within or below the lava plateaux.

In Scotland, Tertiary volcanic rocks are confined to the western regions flanking the Atlantic (Fig. 17.12). *Plateau basalts* cover much of Antrim in Northern Ireland and large parts of the Inner Hebrides including most of Mull and Skye; the lavas with their columnar jointing (p. 177) produce the remarkable scenery of the Giant's Causeway and Fingal's Cave. Plutonic intrusions are lined up along a roughly north–south axis that probably marks the course of a deep crustal fracture which served as a conduit for the magma. These intrusions include several *gabbros* (especially important are those forming the Cuillin Hills of Skye and the peninsula of Ardnamurchan) and a number of *granites* such as those of northern Arran, the Red Hills in Skye and the Mourne Mountains. *Dyke-swarms* of dolerite are associated with several of the plutonic centres and also extend widely across the north and west of Britain; these dykes generally run NW–SE, following systems of fractures opened up by the extension of the crust during the period of activity.

Tertiary life. The beginning of the Tertiary era was marked by the extinction of several important Mesozoic groups such as the reptilian dinosaurs, the ammonites and the belemnites, and the rise from relative obscurity of the mammals, the bony fish and the flowering plants. Both on land and in the sea, the faunas and floras thus acquired a very modern look so far as the balance of power between the major classes was concerned. Individual families, genera and species living at the beginning of the Palaeogene were not often the same as those of the present day, but by the late Neogene the bulk of the animals and plants had come very close to forms which we know today.

On land, the *flowering plants* established a dominance over all other seed-bearing plants, as well as over the dwindling groups of spore-bearing plants. The *mammals* established a similar pre-eminence over other vertebrates and became adapted to a great range of habitats (p. 145). The amphibia and reptiles, unable to compete with them, faded into insignificance.

In the seas the *Mollusca* were now represented mainly by abundant *lamellibranchs* and *gastropods* (Figs. 11.11, 17.13)—the cephalopods had had their day and were now rare. *Echinoids* remained the most important echinoderms and included numerous irregular types. *Scleractinia* continued to represent the corals. *Foraminifera* became larger and more abundant than in previous eras and provided useful zone-fossils—indeed, an alternative name for the whole Palaeogene period is the 'Nummulitic', after a very common disc-shaped foraminiferan *Nummulites* which flourished in Mediterranean regions (Fig. 11.5). Finally, *brachiopods* were now long past the peak of their development and had become scarce.

The shaping of the land

When we look back on the record of geological happenings discussed in this chapter and the preceding one, it is obvious that the shape and topography of the British Isles must have evolved gradually over a long period. We have seen that certain regions were established as land-masses early in geological history and have remained above sea-level, with only brief periods of submergence, since the mid-Palaeozoic. Examples of such regions are the Scottish Highlands and north and central Wales, both of which were

Fig. 17.13 *Tertiary gastropods of the genus* Turritella *in an Eocene sandstone from Sussex*

Fig. 17.14 *A late Tertiary erosion surface in Cornwall*

repeatedly subjected to erosion during Upper Palaeozoic and Mesozoic times. By contrast, the south-eastern part of Britain remained swamped by the sea, with only short periods of emergence, until mid-Tertiary times. Thus, there have been forerunners of what is now north and west Britain off and on for some 400 million years; but through most of that time the seas have covered what is now south-east England.

By mid-Tertiary times, however, the piling-up of marine sediments in Britain almost came to an end and from this time onward erosional processes continued the work of shaping the land. It is often possible to interpret the history of erosion by applying a knowledge of surface processes and of the land-forms which they produce.

In many parts of Britain, it appears that the oldest erosional processes still recorded in our scenery are ones which gave rise to extensive platforms or surfaces of very low relief. Some of these surfaces were *old peneplains* formed by sub-aerial erosion; others were *old platforms of marine erosion* cut by an advancing sea. The older platforms have themselves been *dissected* by further erosion due mainly to streams and rivers, so that the original surfaces were largely eaten away. The last-formed platforms are naturally the better preserved, since river-erosion has had less time to destroy them. Examples of relatively young platforms are very well displayed in western Cornwall, where a level table-land standing a few hundred feet above sea-level represents a late Tertiary plane of marine erosion (Fig. 17.14). Although the characteristic deep narrow valleys of the region have been chiselled out of this platform, much of the original surface is still unmodified and occasional patches of old beach-deposits are preserved on it. An example of an older and much more deeply dissected platform stands at about 800 feet in southern England. Almost the only portions of this surface which survive are the crests of the Chalk hills which rise to almost uniform heights in many places. These crests are capped by a layer of *clay-with-flints* which represents the insoluble residue left after long solution-weathering of the Chalk: from the presence of this residual deposit it is inferred that the 800-foot level of south-east England represents an old peneplain formed sub-aerially. Evidence of kinds similar to those outlined above suggests that erosional platforms were produced at several dates during the late Tertiary in Britain: a general tendency for the area to rise carried the earlier platforms upward, so that in general the older the platform the higher its elevation.

8. The Last Phase: the Great Ice Age

The last episode in the geological history of Britain was connected with a world-wide worsening of the climate. At the end of the Tertiary era, temperatures everywhere became lower and snow and ice began to accumulate widely in the polar regions and the high mountain ranges. Thus began the *Great Ice Age* during which *ice-sheets* spread over millions of square miles of northern Europe and the north of North America. So conspicuous are their effects that the episode of glaciation has been assigned to a geological period of its own—the *Pleistocene Period*—and even to a new 'Quaternary' era of geological time. The whole ice-age, however, lasted less than a million years and is therefore not really comparable in length with the other geological periods.

We have used the effects of the Great Ice Age to illustrate our study of the processes of glaciation (Chapter 8). It will be recalled that these effects included both erosion and deposition. *Glacial erosion* predominated over high ground, where the scenery bears witness to the scraping and moulding of the rock-surfaces by ice flowing over them (Fig. 8.4). Corries were excavated in the mountain sides (Fig. 8.3). The existing river-valleys were modified by the passage of ice to give the U-shaped cross-sections typical of glaciated valleys (Fig. 8.6). *Glacial deposition* took place especially in the valleys and on the lower ground. Many regions now have a blanket of boulder-clay or till, sometimes quite irregular, sometimes moulded into swarms of drumlins (Fig. 8.10). In addition, *fluvioglacial deposits* were laid down by melt-waters during phases when the ice-sheets were retreating. These deposits include the sandy and gravelly kames and eskers.

From the effects of glacial erosion and deposition, conclusions can be drawn concerning the history of the ice-sheets. The evidence shows that the Ice Age was not a period of uniformly cold climate. In fact, four or more cold periods were separated by three or more warmer *interglacial periods* during which the ice melted and temperatures rose to levels equal to or even above those of the present day. The last ice-sheet began to retreat only about 20,000 years ago and it is by no means certain that the earth has even now finally emerged from the Ice Age—our present mild climate could represent another interglacial period.

The fluctuations of climate during the Pleistocene are clearly recorded by the sequences of fossil plants and animals. When the ice-sheets were most extensive most of Britain must have been almost devoid of living things. But the deposits of the interglacial periods contain abundant organic remains, largely of pollen and gastropod shells but with occasional bones of mammals. By the identification of different types of pollen grains it has been found that at the start of each interglacial period an open heath vegetation adapted to arctic conditions first colonised the area from which the ice had retreated; this phase was followed by the establishment of a mixed oak forest and, finally, as the ice was beginning to advance again, by one of conifers and plants of distinctly arctic types. Some molluscan shells in interglacial deposits indicate climates warmer than that of Britain today—one such species now lives only in the rivers of North Africa. The mammalian remains include teeth and bones of early types of elephants such as the mammoth, and of musk-ox, reindeer, bison, rhinoceros and other forms, some of which were adapted to cold conditions. Early man himself also moved into Britain in the interglacial periods, as is shown by the occurrence of flint implements in certain deposits.

At its greatest extent, reached during the second and third most recent glaciations, the Pleistocene ice covered Britain as far south as the outskirts of London. Early in the Great Ice Age, an ice-sheet centred on Scandinavia impinged on the east coast of Britain and interfered with the free outward flow of the native British ice nourished in the mountain regions of Scotland, Wales and the Lake District. This struggle for space between competing ice-bodies is shown diagrammatically in Fig. 17.15. The directions of ice-flow recorded on the figure are derived from observations on striae, erratics, till-fabrics, drumlin-shapes and similar phenomena. During the last glacial stage ice covered a smaller part of the country and finally it shrank to small ice-caps and valley-glaciers in the northern hills.

During the maximum of the glaciation such an enormous bulk of water was locked up in ice-sheets that the level of the seas was lowered by 200–300 feet. The English Channel and southern North Sea—both shallow seas—were almost dried out and the southern part of Britain was

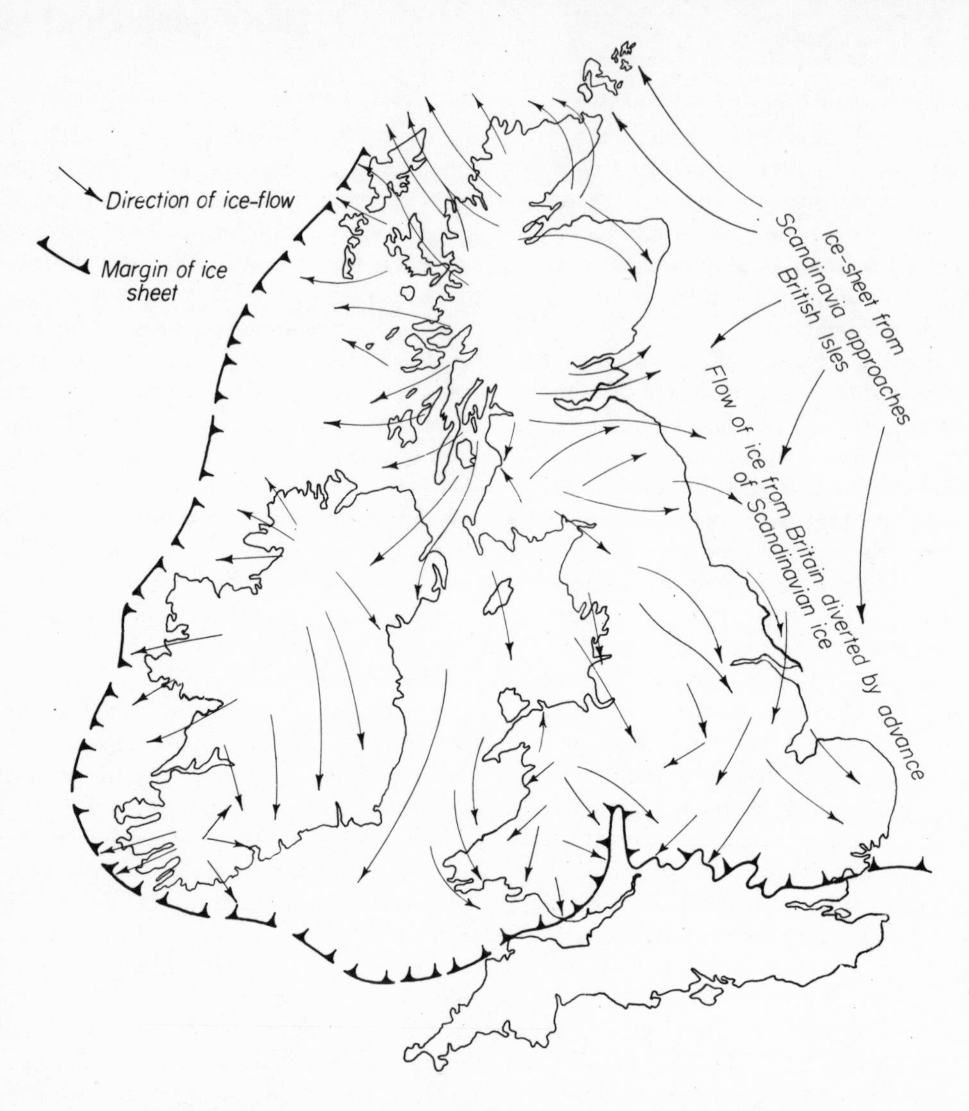

Fig. 17.15 *Maximum extent of the Pleistocene ice-sheet in Britain* (*based on Wills*)

therefore joined to adjacent parts of France and Belgium. A barren *periglacial region* in which the ground was deeply frozen fringed the ice-sheet. The effects of hill-creep and other phenomena produced by movements of partly frozen ground can be observed especially in southern England.

Changes of sea-level. One final effect of glaciation has to be mentioned. The mass of the ice-sheets which spread over northern Europe weighed down the crust to such an extent that subsidence took place. Towards the end of the glaciation, when the melting of ice was returning water to the sea, beaches were formed at appropriate levels around the shore-line of that time. As the land was relieved of its load of ice it began gradually to rise in order to restore isostatic equilibrium. The early-formed beaches were therefore carried up above sea-level to become *raised beaches* (Fig. 17.16). Pauses during uplift gave time for new beaches to be cut and these in turn were later raised like the steps of an escalator. This process continued until the sea stood well below its present level and forests were able to grow near the new shore-line. Subsequent advances of the sea flooded this shore and produced *drowned forests* now to be seen at some points off the shore. The sea then stood for some considerable time above its present level but now it has retreated once more. These ups and downs of land and sea are clearly recorded in the northern parts of Britain. In Scotland there are raised beaches at heights of about 100 feet, 50 feet and 25 feet; the submerged-forest episode is found to have taken place between the making of the 50-foot and 25-foot beaches. Associated with the various beaches are corresponding *alluvial terraces* formed on the sides of river-valleys when the base-level of erosion stood at different heights: each small phase of retreat of the sea caused the rivers to cut down their beds to lower levels. The flood-plains and beaches of the present day record the last stage in the process: but since geological history is still going on, we can be sure that these will not last for ever.

9. The Emergence of Man

During the Pleistocene period, for the first time in the course of organic evolution, a single species emerged as the dominant organism of the lands. That species was man, *Homo sapiens*. Human beings are unique because their agricultural and engineering activities, such as the ploughing of lands, the cutting or burning of trees, the damming of rivers and the building of cities, have entirely changed the appearance of huge portions of the continents. It seems safe to predict that

Fig. 17.16 *Raised beach with old sea-cliffs behind, Island Magee, Northern Ireland*

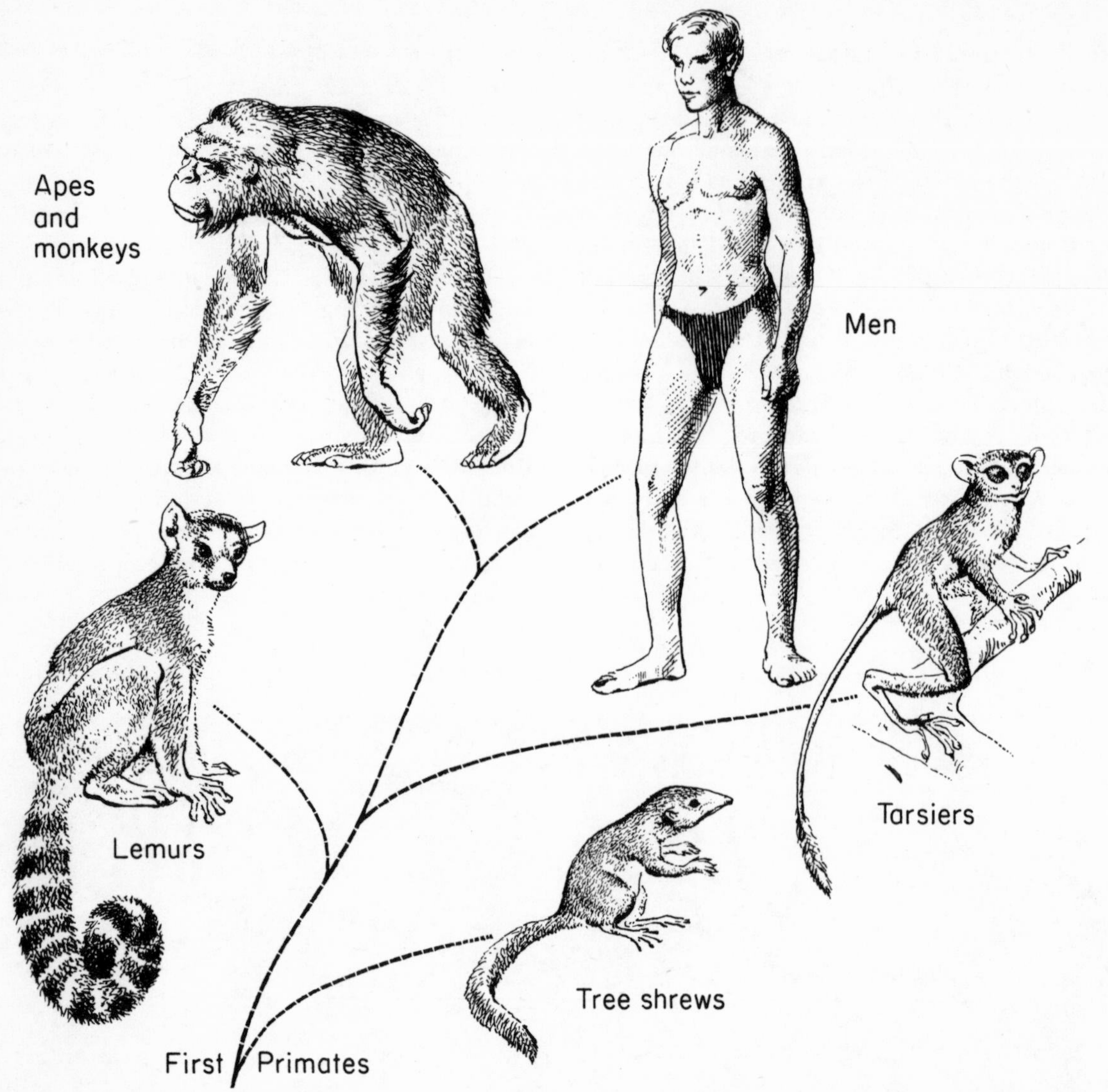

Fig. 17.17 *Evolution of the Primates*

while men remain they will influence the working of the surface-agents of erosion and deposition and so become geological agents themselves. We can end this chapter with a brief summary of their antecedents.

The *Primates*, the mammalian order to which man belongs, first appeared in the early Tertiary and remained through most of their evolution *adapted to life in trees*. This environment of life played an important part in developing characteristics which the human species inherits; two such adaptations are, firstly the modification of the fore-feet for handling objects, with the development of a hand with thumb opposed to the other fingers; and secondly the reliance on sight rather than smell, aided by reduction of the snout-like front of the head into a shorter face from which both eyes could be focused forward on to a single object. These modifications, of obvious value to creatures moving rapidly through trees, provided a basis for man's skill in constructing and using tools. A yet more important evolutionary change was a progressive *increase in the size of the brain* which culminated in the relatively enormous brain of man.

The earliest primates appearing at the beginning of the Tertiary were humble little creatures resembling the tree shrews of the present day, with a sharp snout and four clawed feet (Fig. 17.17). They were followed by the first *lemurs* and *tarsiers* which were already rather more advanced. In Oligocene times the group of *Anthropoidea* was developed from a line of early tarsiers: this group includes the *monkeys*, *apes and men* and is distinguished from the rest of the primates by the large size of the brain, the perfection of the hand as a grasping organ (with nails in place of claws) and the generally human appearance of the face.

In Miocene times, about 30 million years ago, the first truly *ape-like anthropoids* came into existence. The first and most primitive members of this line, from which the gibbon, orang-utang, gorilla, chimpanzee and man may all be derived, have been found in East Africa: but by Pliocene times their descendants appear to have spread to Europe and Asia. At about the beginning of the Pleistocene period, rather less than a million years ago, there appeared in east and southern Africa some very man-like forms which are known collectively as *Australopithecinae* ('the southern apes'). Their skeletons suggest that they walked nearly upright and the facial bones, jaws and teeth are very human in structure. Most significant of all, crude implements made of chipped stone have been found along with australopithecine fossil remains. It seems therefore that the Australopithecinae possessed the distinctively human attribute of making and using tools and they are regarded as the first members of the *Hominidae*, the zoological family of men. In the middle parts of the Pleistocene period a number of other members of this family came into existence, the best-known being 'Neanderthal man'. All are now extinct except for the one species *Homo sapiens* to which the races of modern man belong.

It will be seen that the evolution of the family of men was crowded into an insignificant part of geological time, less than a million years in length. The species of *Homo sapiens* has undergone little physical evolution since its first appearance, but its mode of living has been utterly changed by the development of more and more complex implements and tools. The further history of the human race as the geologist sees it is recorded by *artifacts* or manufactured objects rather than by fossil men. Such objects include many kinds of weapons and domestic and agricultural tools. They were at first made chiefly of stone, and so the early periods of human development are termed the *Palaeolithic* and *Neolithic*, or older and newer Stone Ages, followed later by the Bronze Age and Iron Age as men acquired the ability to work metals. Subdivisions of the record of primitive man are made according to the characteristic forms and styles of implements which can be used as zone-fossils. As the complexity of the human record increases, however, the geologist hands over to the archaeologist and the historian.

18

GEOLOGY AND MAN

Geological materials of one sort or another have been essential to man since he began to make tools. Modern industry is based on the exploitation of rocks and minerals in ever-increasing variety. We deal with the main classes of useful geological products more or less in the order in which man first encountered them. Thus we start with the geological factors affecting his supply of food and water, continue with fuels and materials used to provide shelter, and pass on to gems and other adornments, ending with the minerals and rocks on which our own industrial civilisation depends. We close with some comments on the political and social significance of mineral products.

1. The Uses of Geological Products

A working knowledge of geology first became necessary to man when he began to select stones suitable for the production of axes, spearheads and knives. We know that early man could recognise rocks which were good or bad for this purpose from the fact that implements made of certain rocks, notably flint, were imported into regions where the raw material was not available locally. Coming down to our own times, it is obvious that almost all the raw materials of modern industry are derived from the earth; coal, petroleum, iron-ore, the ores of other useful metals, salt and fertilisers, refractories, abrasives and a host of other products originate in the earth. Soil, building-stones and all materials of construction are mineral or rock products. The importance of the mineral industry today is illustrated by the figures for the output of industrial ores given in the following table:

ANNUAL OUTPUT OF GEOLOGICAL PRODUCTS

(in million tons)

Product	Output
Iron-ore	430
Manganese-ore	14
Aluminium	4½
Copper	4
Lead	2½
Salt	4 (each by U.S.A., China, U.S.S.R., Britain, France, W. Germany, India)
Potash salts	17
Gypsum	42
Sulphur	8–9
Coal (Britain only)	200
Petroleum	9,500 million barrels

There are many ways in which mineral and rock products can be classified. From a strictly geological point of view, they can be divided into (i) *exogenetic deposits* (*exo*, outside) formed at the earth's surface, mainly in connection with the processes of weathering and sedimentation; and (ii) *endogenetic deposits* (*endo*, within) formed within the crust, mainly in connection with igneous and metamorphic activity. From the point of view of the user of the products, they can be divided into (i) *non-metallic mineral deposits* worked chiefly because the materials have certain desirable physical properties, and (ii) *metallic mineral deposits* worked because they contain valuable chemical elements. For the purposes of our general survey, we shall adopt neither of these classifications; instead, we shall deal with the mineral deposits in a non-scientific order which reflects that of their utilisation by man.

Man's earliest preoccupations were with food, water, warmth and shelter. Not content with making utilitarian stone implements, he also adorned his person with native gold, silver and copper, or with precious and semi-precious stones, and painted his cave-dwellings with mineral pigments. From the beginnings of agriculture there grew over the centuries the need to protect the soil, to increase and control the water-supply and to seek for fertilisers.

After the age of stone implements man learnt, possibly by accident, how to smelt the associated ores of copper and tin and to use the resulting bronze for weapons and tools. Later came the discovery of iron-smelting and for a long period iron was the major mineral product. Early iron-smelting was carried out with charcoal as a fuel, but with the exhaustion of the forests *coal* and its derivative coke began to be exploited. The invention of the blast-furnace and the steam-engine laid the foundations of the Industrial Revolution; in this, countries such as Great Britain, France and Germany took the lead, quite largely because they held abundant home-produced supplies of both coal and iron-ore. As industrial production became more and more complex the range of useful geological materials increased. The great new development of our own century has been in the demand for *petroleum* as a fuel and a raw material for other chemicals. There is no doubt that further needs will go on arising in the future to make demands on the knowledge and ingenuity of the geologist.

2. The Soils and Fertilisers

Soil, the product of age-long weathering and organic processes (p. 61), is man's heritage and, because it cannot be replaced, its misuse through ignorance or greed is a crime. Huge areas of farm-land have been devastated by *soil-erosion*; indeed, some ancient civilisations have declined and fallen largely through the mismanagement of their irreplaceable soil. Soil-erosion is usually the result of over-cropping and over-grazing, which leads to the destruction of the protective plant cover and breakdown of the soil fabric. Sheet erosion and gullying by running water and 'blowing' or removal of fine particles by wind then strip off the loose surface material, the damage being especially rapid in dry climates. To combat these processes, some form of *soil-conservation* has to be put into operation with the object of decreasing the run-off and increasing the percolation of rain-water. This is achieved by terracing the surface, by contour-ploughing in which the furrows run around the contours, or by strip-cultivation in which arable and growing crops alternate up the slopes.

Fertilisers contain certain elements such as phosphorus, calcium, potassium, nitrogen and sulphur which are essential for proper growth of plants; and soils need to be replenished in these components if they are to keep their fertility through years of intensive cultivation. The earliest kinds of fertilisers were organic products —farmyard manure and seaweed. *Mineral fertilisers* are becoming of very great importance in a world where the greatest problem is to achieve freedom from hunger for the poorest countries. The chief raw materials for such fertilisers are phosphates, limestone, potash salts, alkali-nitrates and sulphur. The common fertiliser '*superphosphate*' is manufactured from sulphur and the naturally occurring phosphates, apatite and phosphorite. *Lime* is provided from powdered limestone and has the virtue that it corrects the acidity of the soil and improves its texture. *Potash-fertilisers* are produced from the 'bittern' of evaporate deposits; an important source is the great Stassfurt evaporate deposits mentioned on p. 78. *Nitrates* of sodium and potassium are worked in Chile from evaporates containing salts leached from nearby volcanic rocks. *Sulphur*, employed as a fertiliser and insecticide, is mostly obtained from the 'cap-rock' which covers some salt-domes where it is interleaved with and derived from beds of gypsum.

Chapter eighteen

3. Minerals of the Chemical Industry

We may mention some of the main materials used in the present-day chemical industry here because many of them are derived from the same sources as the mineral fertilisers. Evaporate deposits provide *rock-salt*, which is the main product of many dried-out salt lakes. Because salt is a light and very mobile substance when in bulk, it is often squeezed up through the cover of later rocks deposited on top of it to make intrusive plugs known as *salt-domes*. Other *sodium salts*, the sulphates and carbonates, are precipitated as evaporate deposits from lakes receiving alkaline waters leached from volcanic regions. '*Borax*', employed in many industrial processes, is obtained from natural sodium borate deposited in lakes receiving appropriate salts from their drainage-basin. *Anhydrite* ($CaSO_4$) and *gypsum* ($CaSO_4.2H_2O$) are other important substances derived from saline residues. Native *sulphur* and *pyrites* (FeS_2) are sources of sulphuric acid and other important chemicals of commerce. More important than all these however in the chemical industry is the great range of carbon compounds obtained from *coal* and *oil* which are dealt with later: these fuels are the source of plastics, man-made fibres, paints, dyes, drugs and a host of other articles.

4. Water-supply

Supplies of water come from three sources: from *groundwater* tapped by springs and wells; from *rivers and lakes*; and from *reservoirs*, artificial lakes. In Britain, the rainfall is fairly high and there is therefore no overall shortage of water: but even here it is difficult to provide enough for large towns, especially those such as Manchester, which lie at some distance from suitable source-areas. In more arid climates, shortage of water is an important factor limiting the possible extension of grazing, agriculture and industry, and one of the primary functions of the geologist is to advise on the use of water-supplies. It is obvious that planning on a national or even an international scale is needed if water resources are to be used most efficiently.

In Britain, minor supplies of water are obtained from joint- and fault-springs in the otherwise compact Pre-Cambrian and Lower Palaeozoic rocks and from massive igneous intrusions. Upper Palaeozoic sandstones of the Old Red Sandstone and the Carboniferous provide fairly copious supplies, as does fissured Carboniferous Limestone. But the most important aquifers are *Triassic sandstones* which supply water especially in the Midlands, and the *Lower Greensand* and *Chalk* of south-east England which are the aquifers of the *London artesian basin* (p. 64). The sands of the Tertiaries give supplies to small communities and still smaller supplies come from sand and gravel lenses in superficial deposits. Water from the Triassic sandstones and the Chalk is hard; that from the Greensand is soft.

London draws water not only from wells but also from the River Thames. Most other cities draw supplies from lakes and reservoirs in the upland parts of the country. It is obvious that the siting of a large reservoir, and especially of the dam holding up the water, demands detailed geological knowledge. The existence of porous beds, their attitude and the arrangement of planes of discontinuity such as bedding, joints and faults, all have to be taken into account if the reservoir is going to hold water. As an example of the proper siting of a dam we mention the Vyrnwy Reservoir of the Corporation of Liverpool. Here, the dam is carried down to the bedrock at a rise in the bedrock surface, thereby saving money on construction; it is placed where the strata dip up-stream, an arrangement which adds to its stability and water-tightness. Finally the reservoir is situated in a position where the rainfall is high and where the *catchment area* is large.

5. Fuels

The chief fuels used by man as sources of heat, light and industrial power have until recently been *of organic origin*—in modern times, the most important have been *coal*, *oil* and *natural gas*. It is possible that *nuclear energy* produced by the breakdown of atoms and dependent on 'fuels' such as uranium minerals or heavy water may ultimately replace the energy we now get from coal and oil. A few nuclear power stations and ships already use this source of energy but it is at present no cheaper than the traditional sources. *Water-power* is used to generate electricity in hilly countries and for this purpose many rivers have been dammed up, creating the same problems as those dealt with in the last section.

Coal. The most important source of coal in Europe and America is the Carboniferous system. Older coals are uncommon, thin and of poor quality, because terrestrial plants were scarce in pre-Carboniferous times. Younger coals are often very thick but because they have suffered less modification than the Carboniferous coals their rank is lower. Tertiary lignites and brown coals (p. 113) are extensively used in Germany and recent peat in Germany, Denmark and Ireland. The leading coal-producing countries are, in order, the United States, U.S.S.R., Britain and West Germany. The rest of the world produces altogether about as much as the United States.

We have already considered the way in which coal is made (pp. 113–14) and the history of the Carboniferous coal-swamps in Europe (p. 208). We saw that enormous forested plains lay along the borders of the Hercynian mobile belt and that during the late phases of the Hercynian orogeny the Coal Measures deposited in these plains were gently folded and subjected to erosion (p. 209). We may now finally look at some of the geological factors which affect the *mining* of the coal seams.

The Coal Measures are preserved in *coal basins*, the old synclines formed during orogenic movements (Fig. 17.3); they were removed from the crests of the associated anticlines by erosion soon after folding. Coal can be worked from the outcrop of the seam at the surface by *opencast* mining: but most coal is worked in mines going down as much as 4,000 feet below the surface. The accessibility of the coal depends on the dip and on the effects of folding and faulting.

In South Wales, the coals go down to unworkable depths in the centre of the wide coal basin and the mines are therefore mostly marginal. Coal seams are however locally brought up to a workable level by smaller folds and on the upthrow sides of certain faults. In the western part of the coalfield, faults with westerly downthrows have lowered the coals to depths at which they cannot profitably be worked. The intensity of the disturbances affecting the Coal Measures is important to the miner—too violent disturbance may convert coal into graphite, as in the western continuation of the coalfield in Pembrokeshire but, on the other hand, moderate disturbances may have something to do with raising the rank of the coal and may lead to the formation of valuable anthracite.

The remnants of the Coal Measures which escaped erosion after the Hercynian orogeny were naturally covered up unconformably by Permian and younger strata during subsequent geological periods. The *exposed coalfields* of the present day are seen where the cover has been stripped off again by more recent erosion. *Concealed coalfields* still lie hidden beneath the younger cover and can be reached by sinking shafts through this cover or by lateral penetration from an exposed coalfield. A valuable part of the great Nottinghamshire-Yorkshire Coalfield lies underneath a Permian cover: the unconformable base of this cover dips eastward and the Coal Measures therefore emerge at the surface towards the west.

Oil and gas. The *source-rocks* of oil and gas are, as we have seen, probably shales rich in organic matter—the *sapropelites*—deposited in stagnant basins deficient in oxygen. These rocks, however, are not those from which oil is extracted commercially. After formation, oil migrates into more permeable types of rock which constitute *reservoir-rocks*. It gathers finally in workable quantities in places where some structural arrangement of the rocks impedes its progress. All geological structures which cause *oil-pools* to accumulate are known as *oil-traps*. Locating these is one of the main preoccupations of the oil geologist who must therefore gain a thorough knowledge of the geological structure in depth, using information from

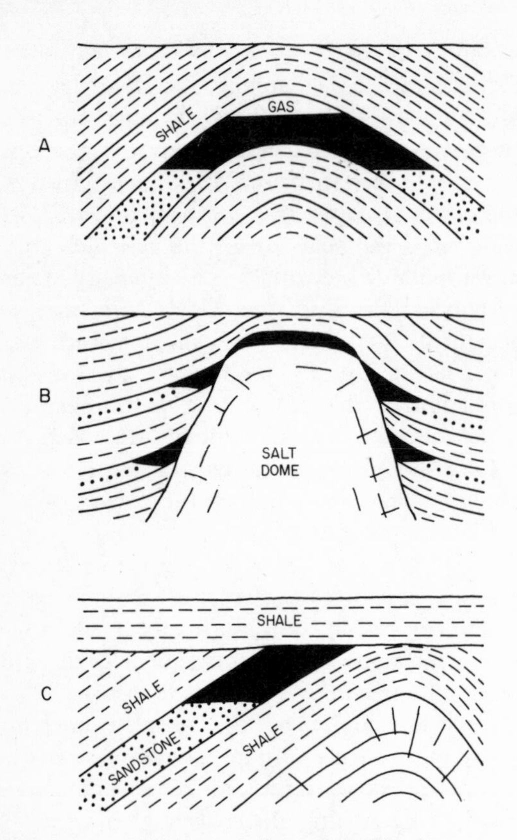

Fig. 18.1 *Oil-traps (oil-pools in black)*
A. Anticline B. Salt-dome
C. Unconformity

surface mapping, from bore-holes and from geophysical investigations.

Reservoir-rocks obviously must be permeable. They are usually rather coarse sandstones or porous or fissured limestones and dolomites. *Oil-traps* are usually produced by the sealing-off of part of a reservoir-horizon by an overlying impervious layer; since oil and gas are lighter than water they tend to accumulate, with the gas at the top, in the highest parts of the reservoir. A selection of common oil-traps is illustrated in Fig. 18.1. The most favourable structural trap is an *anticline* capped by impervious rocks. *Salt-domes* produce traps in the beds tilted up around the intrusive salt-plugs. Stratigraphical traps are produced where an impervious cover rests unconformably on inclined reservoir-rocks.

The chief producers of oil are the United States, the numerous states around the Persian Gulf, Venezuela and U.S.S.R. Both the U.S.A. and U.S.S.R. have large reserves but it seems probable that in future times the Persian Gulf area and Venezuela will become even more important as producers of oil. So colossal are the possibilities in the former region that it is to be hoped politicians understand their implications.

6. Building Materials

The construction of roads, bridges and buildings involves the use of enormous amounts of rock-materials which must all be selected as suitable for the function they are to perform. So vast are the tonnages of gravel, sand, limestone and other rocks excavated for construction works that man must be reckoned among the more potent agents of erosion. The traveller coming in to land at London Airport sees great clusters of water-filled hollows which are entirely man-made, resulting from the excavation of gravel. The following notes on geological materials are arranged for convenience according to rock-types.

Fine-grained *igneous rocks*, especially dolerite and andesite, are used as *crushed stone* for aggregate in concrete, for ballast and for road-making with tar-macadam. The coarser-grained igneous rocks, especially *granite* and *syenite*, make strong and decorative building-stones; they will take a high polish and, though difficult to work, have been used effectively for carving.

Among *metamorphic rocks*, *slate* makes excellent roofing and facing material. The type most widely used in Britain is the blue Welsh slate, derived from cleaved Cambrian and Ordovician muds; but green slates from Ordovician pyroclastics of the Lake District are even more attractive. *Marbles* can be easily cut and polished. Very pure varieties are used for sculpture (the white Carrara marble of Italy clutters up many cemeteries in Britain), but for building purposes streaky impure marbles are decorative; one of the best is the greenish serpentine marble of Connemara. Another metamorphic product that may be mentioned incidentally is *asbestos* used in insulating boards, roofing-sheets and so on; it is derived from fibrous minerals of the serpentine and amphibole families.

The most important raw materials of the modern building industry are *sedimentary rocks*. *Clays* are burnt for bricks and tiles, the largest concentration of brick-making in Britain being located on the Jurassic clays of the Peterborough district. *Sands* of the unconsolidated Tertiaries are used in mortar and cement. *Gravel* is dug from alluvial terraces along the main rivers as aggregate for concrete. The *sandstones* (especially Triassic and, in lesser degree, Carboniferous sandstones) form attractive building-stones—the Triassic stone used in Liverpool Cathedral is a good example. The *limestones* are perhaps the most important natural products for the building industry. They are widely used as 'dimension stone', two Jurassic limestones extensively used in the pre-concrete buildings of London being Bath Stone and Portland Stone (p. 214). Polished ornamental stones are made from crush-breccias of limestone and from *travertine*, the deposit of calcareous springs. Furthermore, limestones are used as crushed stone for aggregate and immense quantities are dug for *lime-burning* and especially for *cement*. In cement-making, one part of clay is ground up and burnt with three parts of limestone or chalk. The product is used with sand and aggregate to produce *concrete*—inspection of any large town will show that we are living in an age of concrete. Lastly, *gypsum* is used in the production of plasters, sheets and boards and for stucco work; four million tons of gypsum are mined annually in Britain.

7. Gold and Gems

The true value of this group of products lies in their beauty, but artificial values are given by scarcity, fashion, commerce and politics. We deal here only with gold and diamonds.

Gold is distinguished by its yellow colour, malleability and high specific gravity (pyrites, FeS_2, 'fool's gold', is brittle and cannot be cut with a knife). Gold occurs as the native element and also as tellurides and in gold-bearing sulphides from which it is extracted by complex smelting processes. The primary source of native gold is in veins, the *'reefs'* of the miner, where it forms grains in a matrix of quartz, pyrites and other minerals. Weathering and erosion of gold-bearing veins provide granules which are transported and redeposited in sedimentary accumulations or *placers*. Because of its high specific gravity, the gold is separated from quartz and other light components and concentrated with other heavy minerals, just as the prospector washes the gold out from the 'dirt' in his pan. Most placers are naturally *alluvial*, but *beach-placers* and even *eolian placers* are also known. Possibly the greatest gold deposits are those of the *Banket*, the gold-bearing conglomerates of the Rand, South Africa, which supply nearly half of the world's output. Some geologists hold that these conglomerates are ancient deltaic placers; others believe that the gold was introduced by migrating fluids long after the formation of the conglomerates. Next after South Africa, the chief gold producers are U.S.S.R., Canada, U.S.A., Australia and Ghana.

Diamonds owe their value as gemstones to their high refractive index and power of dispersing light. Their great hardness (10 on Mohs' scale) also gives them an industrial value as abrasives and for arming drilling-tools and diamond-saws. The primary source of diamonds is in small ultrabasic intrusions of a rock called *kimberlite*. These are strange rocks. They form vertical *pipes* apparently rising from great depths and containing lumps of very dense *eclogite* made of varieties of garnet and pyroxene. It is thought that the eclogitic inclusions may be derived from the mantle beneath the crust: the diamonds, the denser form of carbon (p. 19), may themselves be produced under great pressures in the mantle. Alluvial deposits of diamonds concentrated by stream-action are derived from erosion of kimberlites. The most important diamond-fields are those of Kimberley in South Africa, with alluvial deposits derived from primary sources at the mouth of the Orange River. Smaller deposits occur in other parts of Africa and in U.S.S.R.

Chapter eighteen

8. Metals for Industry

Metalliferous mineral deposits are often called *ore-deposits* and a whole branch of geology is concerned with the occurrence of such deposits. An *ore* is material which it pays to work for a metal. It is a mixture of the desired *ore-mineral* with unwanted minerals called the *gangue*. In the gold-bearing reefs just mentioned, gold is the ore-mineral, quartz the gangue. All ore-deposits are the result of some geologial process which has *concentrated* the desirable ore-mineral, either by the action of surface-agents or by magmatic or metamorphic processes in depth. The importance of the concentrating processes is obvious from the fact that most valuable metals are very scarce in the crust as a whole: copper forms only 0·007%, tin 0·004% and lead 0·0016%. The two principal means of concentration, by igneous and metamorphic processes on the one hand and by surface processes on the other, give the two distinct classes of ore-deposits, the *endogenetic* and *exogenetic* deposits mentioned at the beginning of this chapter.

The close association of *endogenetic deposits* with igneous activity is shown by the fact that certain ore-deposits are arranged in *zones* around igneous intrusions. In the region of the *Hercynian granites of Cornwall*, for example, tin deposits are found in or against the granites, copper farther out, zinc still farther away and lead farthest of all (Fig. 18.2). The crystallisation of the granitic magma led to the concentration of the metals in the residual fluids. These ore-bearing solutions then streamed out into the country-rocks and deposited ore-minerals in succession as each species in turn began to crystallise. The ore-deposits were formed as *veins* in fissures, as scattered *impregnations* or as *replacements* of soluble country-rocks.

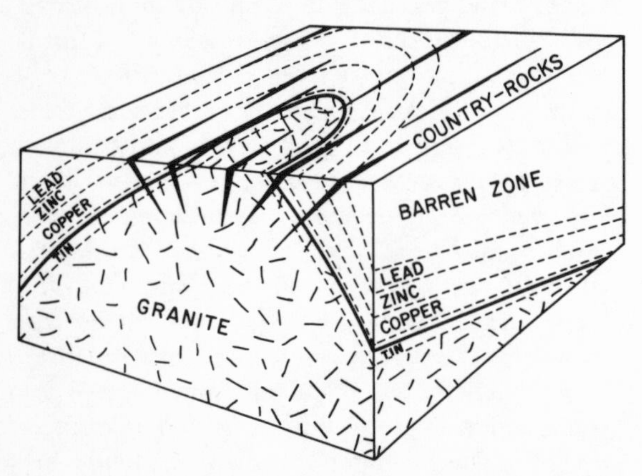

Fig. 18.2 *Zoning of metalliferous ore-deposits around a Hercynian granite: pegmatite veins shown, diagrammatically, in black*

Other endogenetic deposits are formed by the *segregation* of ore-minerals which crystallise early in a magmatic intrusion. Such deposits are usually associated with basic or ultrabasic rocks and are sources of chromium and nickel: the early-formed mineral chromite ($FeCr_2O_4$), and minerals containing traces of nickel, tend to sink in the magma and pile up as ore-deposits near the floor of the intrusive body.

The *exogenetic mineral deposits* formed at the earth's surface are, in effect, simply special kinds of sedimentary rocks; we have already encountered examples in the gold placers. The exogenetic deposits fall into the same categories as the sedimentary rocks themselves, namely *residual deposits* illustrated by bauxite, the ore of aluminium; *detrital deposits* illustrated by placers; and *chemical–organic deposits* illustrated by bedded iron-ores.

In the Table opposite there is given a summary of the chief characters of the common ore-minerals of the six important metals, iron, tin, copper, zinc, lead and aluminium. These details will not mean much unless the reader is able to examine specimens of the minerals as he reads them; but the minerals are all common ones, readily obtainable at low cost, and are on display in most museums.

Iron forms the foundation of modern industry, being used as cast-iron, wrought-iron and steel: the numerous grades of *steel* owe their properties to the addition of small amounts of *alloy-metals* such as manganese, chromium, nickel, tungsten and titanium. These alloy-metals are now so

THE COMMON ORE-MINERALS

Metal	*Mineral*	*Composition*	*Colour*	*Crystal system*	*Cleavage*	*Hardness*	*Specific gravity*	*Common form, etc.*
IRON	Magnetite	Fe_3O_4 iron oxide	iron-black streak black	cubic	poor	6	5·18	octahedra, massive
	Hematite	Fe_2O_3 iron oxide	steel-grey streak red	hexagonal	poor	6	4·9–5·3	tabular, kidney-shaped, earthy
	Limonite	$2Fe_2O_3.3H_2O$ hydrated iron oxide	brown–yellow	none	none	5	3·6–4	earthy, concretionary
	Siderite	$FeCO_3$ iron carbonate	brownish	hexagonal	rhombohedral	4	3·8	rhombohedral, massive
TIN	Cassiterite	SnO_2 tin oxide	black	tetragonal	—	6–7	6·8–7·1	massive, granular, water-worn grains
COPPER	Chalcopyrite (Copper pyrites)	$CuFeS_2$ copper iron sulphide	brass-yellow	tetragonal	—	3·5–4	4·1–4·3	massive, softer than pyrites, crumbling
	Malachite	$CuCO_3.Cu(OH)_2$	bright green	monoclinic	—	3·5–4	3·9–4	massive, stalactitic
ZINC	Blende	ZnS zinc sulphide	black, brown	cubic	rhombdodecahedral, good	3·5–4	3·9–4·2	tetrahedra, massive
	Smithsonite	$ZnCO_3$ zinc carbonate	white, grey	hexagonal	rhombohedral, perfect	5·5	4–4·5	massive, stalactitic, granular, earthy
LEAD	Galena	PbS lead sulphide	lead-grey	cubic	cubic, perfect	2·5	7·4–7·6	cubes, massive or granular
ALUMINIUM	Bauxite	$Al_2O_3.2H_2O$ hydrated aluminium oxide	white–brown	Amorphous, in earthy granular or pisolitic masses				

important that the distribution of their ores is a matter for national and international concern.

Iron-ores are of several different kinds. Some are segregations from magmas, like the great magnetite-deposits of Sweden. Others are formed at the contacts of igneous intrusions. Still others are sedimentary in origin. In Britain, the Jurassic of the Midlands includes workable ironstones in which the ore-minerals are mainly hydrous iron silicates (p. 214). In early Pre-Cambrian times sedimentary iron-ores were formed on an enormous scale: examples are provided by the Lake Superior deposits of North America which have suffered later regional metamorphism. Still other iron-ores are residual deposits usually made chiefly of limonite. The output of iron-ore in normal years is over 400 million tons, the main producers being, in order, U.S.A., U.S.S.R., France, China, Canada, Sweden, West Germany and Britain. Great new sources are being opened up in north-west Africa.

Tin has one ore-mineral, cassiterite, which is deposited in primary veins and impregnations from residual fluids associated with granites and porphyries (Fig. 18.2). Alluvial placer deposits are derived from the erosion of tin-bearing veins. The chief uses of tin are in the manufacture of tin-plate for cans and in certain alloys. Supplies, derived about equally from primary and alluvial deposits, come principally from Malaysia, China, Bolivia, Indonesia and U.S.S.R. Tin-ore is one of the few important ores not produced in significant amounts in the United States.

Copper is an essential metal in the electrical industry, used both as a conductor and for electrical machinery, and is also of importance in the construction of machinery generally, in the motor-car industry and in chemical engineering. It is a component of important alloys such as bronze and brass. The sources of copper are the native metal and the various sulphides, especially chalcopyrite and its weathering product malachite. These ore-minerals occur in magmatic segregations, in ore-bearing veins, in deposits at igneous contacts and occasionally in sedimentary beds. Malachite is important in the great Katanga 'Copper Belt' of Zambia and the Congo and it is probable that the deposits of this belt represent the weathered upper portions of enormous disseminations of primary copper sulphides. The Copper Belt is the main producing region, followed by the United States, Chile, U.S.S.R. and Canada.

Zinc is used chiefly for coating (*galvanising*) iron and for the production of various alloys, the chief of which is brass. The common ore-mineral, blende, occurs intimately mixed with the lead-ore galena, and mechanical separation or 'dressing' is necessary before the ores can be used. Blende and galena occur as replacements of limestone, in veins and in contact-metamorphic deposits. Smithsonite, the other zinc ore-mineral, occurs in the same situations and is probably formed by alteration of the primary sulphide. The chief producer of zinc-ore is the United States.

Lead is used in making accumulators, for sheeting, pipes, cable-covers and ammunition and for certain alloys such as pewter and solder. Lead compounds are extensively used in paints. The principal ore-mineral, galena, occurs along with blende as already described. From this primary sulphide other minerals are formed in zones of oxidation including the carbonate, cerussite, and the sulphate, anglesite. The dominant producers of lead are the United States and U.S.S.R., followed by Australia, Mexico, West Germany and Canada.

Aluminium is produced in electric furnaces by the reduction of the oxide, alumina, obtained from bauxite which is the only ore. Aluminium is used where lightness and strength are important, especially in the manufacture of aircraft: the chief light alloys are made with zinc, copper and magnesium. Bauxite is produced by the subaerial decay and weathering of aluminium-rich rocks, often of igneous origin, and is formed as a residual or transported deposit under tropical conditions (p. 61). The chief sources are the Guianas, Ghana, Jamaica, the East Indies and France. The production of aluminium metal from bauxite needs cheap electrical power and is therefore mostly carried out in countries where hydroelectric power is available, notably the United States, U.S.S.R., Canada, France, West Germany, Norway, Japan and Scotland.

9. Geology and Mankind

Our survey of the applications of geology to the affairs of mankind could be expanded indefinitely, but even from the little that has been said some general conclusions can be drawn. It is abundantly clear that from their very character the natural resources of geological materials must be unevenly distributed between the nations. We have noticed that even the United States, which has immensely rich and varied mineral resources, lacks tin—we may add that it is also short of diamonds, nickel and manganese. No nation is completely self-sufficient in mineral products nor can it be made so; and many wars and invasions of the past can be traced to the attempts of some power to secure supplies of a particular product.

Whether any nation is capable of becoming a great industrial power depends as much on its geology as on its government. But mineral deposits, even the largest, are wasting assets and their value may be reduced (or enhanced) by technological changes. Britain first rose to industrial greatness because she was plentifully endowed with coal and iron-ore. These resources she can still use, but she is almost lacking in the more modern fuel, oil, and is more and more dependent on foreign supplies of other industrial mineral products. The future of Britain, like that of most European powers, depends now on the application of her technical skill and industrial know-how: the primary mineral products must come from overseas. Clearly, the problems of discovering and using mineral resources are problems which concern not only this nation but the whole world.

Suggestions for further reading and reference

Blyth, F. G. H., *Geological Maps and their Interpretation*, Arnold, 1965

Davies, A. M., *An Introduction to Palaeontology*, 3rd edn., revised by C. J. Stubblefield, Murby, 1962

Dury, G. H., *The face of the earth*, Penguin Books, 1959

Harker, A., *Petrography for Students*, 8th edn., Cambridge University Press, 1954

Hatch, F. H., Rastall, R. H. and Black, M., *The Petrology of the Sedimentary Rocks*, 3rd edn., Allen & Unwin, 1938

Hatch, F. H., Wells, A. K. and Wells, M. K., *The Petrology of the Igneous Rocks*, 12th edn., Allen & Unwin, 1961

Holmes, A., *Principles of Physical Geology*, 2nd edn., Nelson, 1965

Jones, W. R. and Williams, D., *Minerals and Mineral Deposits*, Home University Library, 1948

Pettijohn, F. J., *Sedimentary Rocks*, 2nd edn., Harper, 1956

Read, H. H., *Rutley's Elements of Mineralogy*, 25th edn., Murby, 1962

Read, H. H. and Watson, J., *Introduction to Geology*, Vol. I: *Principles*, Macmillan, 1962

Romer, A. S., *Man and the Vertebrates*, 2 volumes, Penguin Books, 1954

Shepard, F. P., *Submarine Geology*, 2nd edn., Harper, 1963

In addition to these general works, the following series are of interest to British students:

British Regional Geology: handbooks published by the Geological Survey and Museum (for details see footnote, p. 188)

British Palaeozoic Fossils
British Mesozoic Fossils
British Cainozoic Fossils
} Handbooks published by the British Museum (Natural History) London (see footnote, p. 123)

Geologists' Association Guides: a series of excursion guides to selected British localities (see footnote, p. 188)

INDEX

Page numbers shown in italics denote illustrations

Index

Index

Index

Index

Index

Index

Index